The Hurricane and Its Impact

The Hurricane and Its Impact

ROBERT H. SIMPSON and
HERBERT RIEHL

LOUISIANA STATE UNIVERSITY PRESS
BATON ROUGE AND LONDON

Manufactured in the United States of America

Designer: Albert Crochet
Typeface: VIP Caledonia
Typesetter: G&S Typesetters, Inc.
Printer and Binder: Thomson-Shore, Inc.

LIBRARY OF CONGRESS CATALOGING IN PUBLICATION DATA
Simpson, Robert H.
The hurricane and its impact.

Includes index.
1. Hurricanes. 2. Hurricane protection.
I. Riehl, Herbert, joint author. 2. Title.
QC944.S55 363.3′492 80-13911
ISBN 0-8071-0688-7

Contents

Figures

Tables

Preface and Acknowledgments

Every year the tropical hurricane—a major source of natural disaster—becomes a seasonal topic of fascination, awe, and sometimes fear, as it reminds residents of the coastal zone of the need for awareness of hurricane hazards. However, because a hurricane occurrence at any one location is a rare event, public response to it is more that of fascination than of fear, and uncertainty about the need for individual actions.

Moreover, with the growth of population in the United States and of affluence and mobility in all levels of society, a larger percentage of the population is discovering the climatic and recreational advantages of living near the seacoast. The result has been that the increases in population- and property-at-risk at the seashores are disproportionately higher than elsewhere.

Statistics compiled by the National Hurricane Center at Miami in 1977 show that in 20 years the population had risen in the Houston area (Harris County) from 750,000 to more than 1,750,000 and in the Miami area (Dade County) from 500,000 to more than 1,300,000; and in many coastal counties of the middle Atlantic states, it had doubled. This escalation of hurricane risks presents a hierarchy of problems, responsibilities, and challenges that—if the potential for hurricane disasters is to be kept at acceptable levels—requires an effective response from government at all levels, from the professional engineer and architect, the meteorologist and oceanographer, the industrial and residential developer, and the users of coastal properties.

This book attempts to define and describe—numerically, where practicable—the physical nature of the hurricane and its impact at sea and in the coastal zone. We do not propose unique solutions or responses to the hurricane problem. Rather, our objective has been to provide fundamental knowledge and some methods for estimating and analyzing the threat as a basis for deciding the best course of action to protect life and property and, in the public interest, the natural resources of the coastal zone.

Since knowledge of the structure and dynamics of hurricanes and the nature of their impact has advanced substantially since the 1950s, a new book to present this knowledge in a form suitable for a wide circle of users is timely. Directed in part toward undergraduate instruction at the junior and senior level for students with at least a limited background in physics, this volume is also designed to serve as a reference for technical users, including architects, professional engineers and building contractors, those concerned with planning and management problems in the coastal zone, and those involved in disaster preparedness.

The physical processes that transform a tropical rain disturbance into a dangerous hurricane and steer it to a landfall are discussed with just enough mathematical reasoning to permit and encourage the serious reader to compute for himself the probable range of hurricane extremes that may occur at any given locality, and to understand the distributions of winds, tides, and waves—in deep water and at landfall—with reference to the center.

Parts I and II are primarily concerned with the hurricane as a meteorological event, and Parts III and IV with its impact at the coast and at sea. Part V deals with the challenge and the problems of planning effective coexistence with the hurricane hazard. The hurricane-prediction and -warning capabilities are discussed together with the problems of social response in the face of warnings and the need for preparedness actions.

Addressing an audience with such broad and varied concerns, we recognized from the outset that no one aspect of the subject could be treated in depth from first principles, and that such an audience would not find all chapters equally useful and satisfying. However, our experience during the last 3 decades in analyzing the effectiveness of preparedness and warning measures for hurricanes has led to the conclusion that adequate protection cannot be achieved by proper applications of modern science and technology alone, by more effective programs of hurricane awareness, or by sound coastal-zone management per se. Effective coexistence with this hazard depends upon a comprehensive awareness of the potential impact of the hurricane, of the need and justification for land-use planning and regulation, of the limitations of science in providing early warnings, and of the role that behavioral science must play in improving human responses during a hurricane emergency. Our goal here is primarily to provide this awareness and to suggest the direction in which the search for solutions to a wide range of hurricane problems may be fruitful.

H. Riehl prepared Chapters 1, 3, 5, 6, 7, and 10; R. H. Simpson, Chapters 2, 8, 9, 11, 12, 13, and 15. The remainder of the book was a joint effort.

We wish first to acknowledge Louisiana State University Press for the access given us to materials from Gordon E. Dunn and Banner I. Miller's *Atlantic Hurricanes*, published in 1960. Further, we wish to express appreciation for the encouragement and assistance of our colleagues at the National Hurricane Center and the Hurricane Research Laboratory, both in Miami, and elsewhere for informational materials and for discussions and suggestions for the content of the manuscript. We are particularly grateful to Gordon Bell, Neil Frank, William Frank, Chester Jelesnianski, Roger Pielke, Herbert Saffir, Robert Sheets, and Joanne Simpson—all of whom reviewed various parts of the manuscript; and to Mary-Scott Marston for editorial and typing assistance in the production and collation of the final draft of chapters prepared by R. H. Simpson.

Note on Measurements

This book has used the Système International d'Unités (the so-called S.I. system of units, currently the accepted form of the metric system), except in connection with those few topics of which the preponderance of current literature, research, and basic understanding, mainly in engineering, still depends upon earlier forms of metric or English units for effective communication.

As an aid in relating the obsolescent units to the new international standard units, two tables are provided in Appendix B.

PART I

The Nature and Importance of the Hurricane: An Overview

CHAPTER 1

The Tropical Cyclone—An Extreme Event

Throughout the middle and high latitudes of the world, summer and winter succeed each other with undoubted certainty, sometimes milder and sometimes harsher, but with inexorable continuity. Summer is the time for respite, because each winter brings long, cold nights and some very heavy storms, especially at sea and along coastlines. The storms attain and exceed wind velocities, officially termed *hurricane-force*, of 74 mph (64 kt, 33 mps), which may persist for many hours. The landscape is shaped by and conditioned to these enormous events. Tree growth is low, often only bush, and the sparse branches extend only toward east, that is, away from the direction from which the hurricane winds come. Coast dwellers know these winds and endure them quietly. Because their homes are built to withstand such onslaught, their only other concern is for the seashore. Great dike installations protect many coasts from the inundations threatened by great high tides, but certain questions remain: Will these coastal bastions survive the storm? What can be done to shore up weak links?

At the opposite margin of world climates, people have other concerns. Vegetation grows profusely in the tropics, where rains are frequent and heavy enough to sustain the plant cover: other areas, however, are arid. And in the monsoon countries especially, the main questions are whether the rains will come and whether there will be enough water for the crops. But into the tropical life—pleasant, monotonous, and seemingly timeless—occasional stark interruptions of extreme wind do occur, even stronger than those experienced in polar latitudes. Great floods and huge ocean waves and surges overrun the beach areas and flat inland areas 100 km and more. These visitations are rare. Due to foresight, ports occasionally have harbor protection designed for this kind of storm; most buildings, however, are not erected to withstand it, and vegetation, with little memory, spreads in all directions, ready prey to the overpowering wind.

This most unwelcome visitor is the tropical revolving storm, news of which spread quickly through Europe with the expeditions of Columbus. Countless sailing vessels became victims of the great violence stirring up huge waves over the deep sea as well as the surrounding shores. Even today, a crew on a fast seagoing ship or aircraft takes most respectful notice when the captain hears of a tropical revolving storm on or near his projected route. The revolving storms visit preferred areas in all tropical oceans from time to time—all except the South Atlantic. Among the picturesque local names for them, the term *hurricane* (Spanish *huracan*), from the Atlantic area, is thought to have been fashioned from names in use by local Indian tribes. In the western Pacific they are called *typhoons*, and in the other oceans *tropical cyclones*. The Australian term *willy-willy* is in rare use. At all latitudes the word *cyclone* denotes any large rotating bad-weather area with low pressure in the middle. The special meanings of *severe local weather system* and *tornado*, common in the Middle West of the United States, are not employed in this text. However, the term *hurricane* has acquired a generic meaning; it denotes a revolving storm with hurricane-force velocities, located anywhere in the global tropics.

Why are we interested?

One part of the answer to this question is obvious. Because hurricane incidence at any one location is rare—a century may pass without a direct hit, in sharp contrast to the winter storms of higher latitudes—safeguards against it are minimal or nonexistent in most tropical countries. They are also inadequate along the equatorward margins of middle latitude continents, which experience hurricane landfalls as the storms come out of the tropics. Because of this defenselessness, enormous loss of life and an untold variety and amount of damage can result from a hurricane passage. Some precautions can always be taken if people are advised and are willing to follow advice; at least they can leave the most threatened areas. Technological development of weather satellites allows detection of all hurricanes around the globe except those that sometimes appear on short notice near a coast; and warnings can always be issued as to existence, position, and current movement. Beyond that, it is important to know just where a hurricane will strike, how intense it will be, and how long it will last. The fully grown hurricane may well have an outer envelope with a diameter of 1,000 km, or the distance between Chicago and New York. At an average travel rate—quite variable, to be sure—the hurricane will be a 2-day event. But, the bad time normally lasts only a small fraction of the 2 days. In fact, at the beginning of pressure fall, cloudiness may be suppressed and the weather much finer than normal. This is the deceptive "quiet before the storm," which has made it difficult for the hurricane forecasters of past centuries, often highly trained and observant missionaries, to persuade local populations that a violent storm was in the offing. The period of most severe weather, including the eye if it passes directly, will seldom exceed 6 hours. But the aftermath, including possible flooding, road clearing, and restoration of all public services and commerce, may drag on for a long time indeed.

Besides these highly practical matters of concern, the meteorologist is interested to learn why the number of hurricanes in any year is very low compared to the number of cyclones outside the tropics. Ten full hurricanes is well above the mean for the tropical Atlantic. But the number of cyclones crossing from America toward Europe in the north will average at least ten times as many, or two per week. The question of low frequency will be a recurrent theme in the book. In one sense, the people of the tropics should consider themselves very lucky that there are so few of these storms. If there were two per week, tropical life would be far different from what it is. On the other hand, the rareness of the event encourages the development of many activities that can be knocked out in a single devastating 6-hour interval.

Formation

Although hurricanes are small in number, the frequency of opportunity for their development per season is always high. Formation is rarely a surprise occurrence; the cloud and rain areas that are always wandering across the tropical oceans are potential candidates for storm development. On the present-day satellite photos all these rain areas can be seen around the whole globe—a marvelous technological achievement of the twentieth century. On the average, the cloud systems, which are relied on for bringing most of the needed rain, cover at least an area of about 500 × 500 km. Thus the initial disturbances to be watched are of readily observable size.

In the presatellite era one relied mainly on ship observations to deduce strengthening of a rain area. Forecasters were keen to search for indications of unusually low pressure; of pressure falling more than normal along the path of ships; of winds exceeding 25 mph (11 mps); of winds from "unusual" quadrants, especially in the normally very steady trades; and of unusually heavy rain reports. Nowadays, the satellite shows particular cloud masses becoming surrounded by large clear areas, while the cloud takes on an oval or circular pattern with a contracting intense core. In the Atlantic and West Pacific oceans reconnaissance aircraft are sent to suspicious areas as soon as possible for exploration; the use of aircraft for such dangerous tasks, with very few losses over the years, is another instance of highly beneficial application of modern technology.

The story of formation begins by noting whether the suspicious rain system is situated over a part of the ocean where surface temperature exceeds a recognized threshold of 26°C–27°C and whether the atmosphere of the general environment is without temperature inversions (temperature increasing upward) or very dry layers that evaporate and debilitate cloud towers trying to grow to great heights. Then the search passes on to "adjoining" weather systems over a radius of at least 1,000 km. As in the case of manmade engines, a starting mechanism and energy source must be available. Such a mechanism, internal to the rain areas, is difficult to identify if it exists. Many investigators have become convinced that external forcing is uniquely the source of all development over the world's oceans. The requirement of these external systems, whatever their shape, is that they provide energy through sinking of relatively cold air in their own cores and accelerate surface flow in the direction of the potential development area.

Increasing mass flow through the center of development furnishes a strong indication of at least some intensification. Concentration of ascending cloud masses in tight curving bands can lead to some warming of the rain area, with fall of surface pressure to 1,000 mb and gale-force winds. The ocean becomes agitated with strong waves and spray. Presence of clockwise revolving flow in a layer centered on the 12-km (200-mb) level is helpful in accelerating the mass of air risen in deep convective cloud outward, thus providing balance and even an excess of outflow against the externally forced increased inflow.

For the final stage the "burner" must be turned on underneath the "kettle." Through enhanced energy transfer from the agitated ocean, rising air masses can follow successively warmer ascent paths and establish a temperature difference between the core, which now takes on circular shape, and the wider surroundings. This difference assumes the function of the normal temperature contrast between high and low latitudes, utilized by developing cyclones of middle latitudes for their growth. Depending on the amount of temperature difference, the resulting hurricane may be minimal, with pressure of 990–980 mb at the center; or pressure may fall to 950 mb or less, which corresponds to winds ranging up to 150, even 200, kt (103 mps). The famous hurricane eye, well known before the days of satellite photography, forms; downdrafts develop, maintaining the eye and providing another mechanism for extremely low surface pressures of 900 mb to become established.

The main reason for the high frequency of extratropical, compared to tropical, cyclones is thought to lie in the fact that the energy source of the former is drawn from *preexisting* temperature differences across the latitude circles, whereas the main part of the temperature difference must be *created* in tropical weather systems for a hurricane circulation to become possible. The oceanic heat source, though only a few percent of the total energy transaction in hurricanes, herewith is seen to take on a highly important function. "Efficiency" of the hurricane machine should be computed in terms of this critical source rather than, for instance, in terms of total rainfall.

Structure of the Mature Hurricane

The most outstanding feature of the mature hurricane is that it resembles a narrow funnel. Although its periphery may grow to a radius of 1,000 km, almost everything of importance occurs within the 100-km radius. Here surface winds, which only rarely exceed hurricane force at 100 km, rise with increasing steepness of slope to their maximum. Highest speeds may be confined to a narrow ring of 1- or 2-km width around the edge of the eye and are, on the average, about 30 km distant from the very center, though with great variability from 10 km to around 80 km. Most of the inward temperature rise associated with rising cloud masses occurs in this interval, as well as the rapid drop of the barometer. One speaks of an "eye wall cloud," spectacular on many radar photographs, as the most solid inner feature of a hurricane. However, not all hurricanes have a complete eye wall, and even in well-formed storms one observes continuous fluctuations of eye wall and wind speed on the order of 10% during a period of 1 to 2 hours.

Nearly all inflowing air escapes upward in the funnel, which may be vertical up to 5–10 km height and from there curves outward. But some mass does break through into the eye and forms very pretty low-level cloud patterns, such as one or two inferior spirals that may end in a taller central cloud, whose top may be as high as 2 km. This height, then, is the limit of the penetration of descent of air in the eye at upper levels. Sometimes there is even a high overcast. But when the eye becomes visible on satellite photos, large masses of high-level cloud, all cirrus (ice cloud), may be pictured, streaming rapidly toward the center and descending as great ice cascades.

At the surface, winds die down and clouds lighten. In earlier days people thought that the end of the storm had come, only to be surprised by the second, often stronger, half. The ocean waves, arranged in neat, narrow, parallel rows along the wind outside the eye, break down into a confused pattern inside. Temperature, in spite of the very low pressure, is about the same as outside the hurricane.

The great revolving vortex—counterclockwise in the Northern Hemisphere and clockwise in the Southern—extends at nearly undiminished strength up to at least 7–8 km height (400 mb). In contrast, the inflow, which produces the spectacular radar-seen spirals, is concentrated in the lowest levels. In the middle atmosphere the air rotates around the hurricane axis like a carousel. Above 7–8 km height the inward pressure drop lessens in connection with the temperature difference between inside and outside; the air begins withdrawing from the center and so the rotational speed lessens. At a radius of 150–200 km the winds blow straight out in the height range of 10–15 km, the top of the outflow. Rotational speed here is zero, even though at low levels revolving winds of 30–50 kt may prevail at this radius. Frictional retardation of the inflow by the ocean within these radii accounts for the difference that is centrally important for coherence and maintenance of the whole storm in the mature stage for days and sometimes for weeks. Beyond the 150–200-km radius high-level flow rotates in the opposite direction of the inflow, or clockwise in the Northern Hemisphere. Some of the long spiral bands of the hurricane seen in satellite pictures are a part of the outflow. Since the satellite looks at the event from above, the dense clouds surrounding a center—a part of the outflow—sometimes obscure the rain bands of the inflow.

A final word may be reserved for hurricane precipitation, once thought to be the most disastrous of all extreme precipitation. It has been amply demonstrated that tropical storms of lesser intensity can produce as much or even more rainfall than the hurricane.

Further, numerous rain situations occur, without any revolving storm or even any distinguishable moving synoptic-weather system altogether, that lay down 50 to 100 cm of rain, compared to an average of 35 cm for a moderate hurricane making a direct hit. Hurricanes thus make little contribution to total rainfall at tropical stations, and even less contribution to beneficial precipitation, for most of the water runs off in floods, unless there is a large reservoir to catch the water, as in Hong Kong. But, the flood-producing capability of the hurricane should never be underestimated. Inside the 200-km radius, computed rainfall from a moderate storm in one day has been shown equal to average *annual* discharge of the Colorado River at its point of largest flow. The problems of flood protection may be imagined!

Hurricane Motion

As noted well in the nineteenth century, hurricanes follow a basic parabolic path (Fig. 1). They are "steered" by the great anticyclones overlying the tropical oceans. In the deep tropics they move

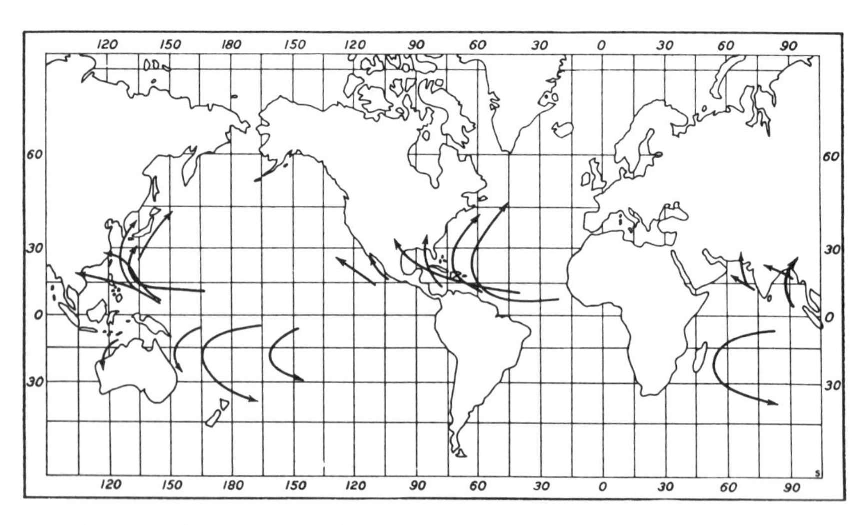

Fig. 1. Areas where tropical cyclones form, showing principal directions of paths. Gordon E. Dunn and Banner I. Miller, *Atlantic Hurricanes* (Baton Rouge: Louisiana State University Press, 1960), 33.

westward in the easterlies, coupled with a tendency to drift poleward. This drift sooner or later brings them to the limit of the tropics where the east winds change to west winds. Accordingly, the hurricanes "recurve," even if they have made prior landfall, and retrace the same ocean over which they had first traveled westward.

Though basically correct, this scheme is grossly simplistic. In the seasonal picture the trade wind belts expand poleward from spring until midsummer, then retrograde equatorward. Correspondingly, the long westward tracks are found mainly in midseason. Early and late season motion may be directly poleward from the start but soon acquires an eastward component of motion.

Upon attaining the subtropics, hurricane tracks are forcefully influenced by surrounding extratropical circulations of large size. When strong upper troughs extend deep into low latitudes, hurricanes are drawn out of the tropics early on their eastern side. With flat east-west flow of the westerlies, westward hurricane tracks may be greatly extended; they may never recurve. Given only weak flow between easterlies and westerlies in the subtropics, a hurricane, without any forcing mechanism acting on it, may become almost stationary for days, in a state of indeterminacy.

Major forecasting problems arise mainly because the flow patterns of higher latitudes themselves are not steady. The successful hurricane forecast depends in large measure on a correct estimate of what happens at distant latitudes poleward. A strong trough may start to draw a hurricane poleward on one day. But on the next day the trough may have disappeared, to be replaced by a high-level ridge with easterlies on its equatorward side. The hurricane, first moving poleward on a "normal" recurvature path, stops dead in its track; it may execute a cusp or even a loop, and then it picks up speed with a sharp turn back toward west. Such an event is most enervating, yet quite frequent. Other sequences of even greater complexity could be named. A look at a chart of individual tracks for a given part of the hurricane season over a number of years looks like a veritable jumble of crisscrossing tracks, though the basic undertone of parabola remains.

Even when a storm follows a fairly reasonable track, it is likely to develop wavy wobbles for perhaps 6 hours or longer. Near a coast these can make differences in the place of landfall by 100 km or more. Because most of the hurricane's destructive power is packed inside the 100-km radius, however, the accuracy in landfall prediction needs to be better than 100 km. Again, this is in contrast to middle latitude cyclones, whose weather pattern is usually broad, and acceptable "area of uncertainty" in prediction large. Hurricanes also have a marked affinity for remaining on water if they can, to preserve the oceanic heat source underneath. Thus, there are cases in which a hurricane has directly approached a coast line, halted, and then sharply turned away. In contrast, there are other situations in which the storm has marched right inland when it reasonably should have skirted the coast.

In summary, while climatic hurricane tracks ordinarily give the pleasing picture of parabola, nothing of the sort can be assumed for any particular hurricane, which only too often turns out to be an individualist and must be treated as such.

Terminal Stages

Hurricanes that follow a parabolic track over the oceans will, for the most part, become vigorous extratropical storms outside the tropics and may reach the polar oceans. Upon landfall, the central hurricane core always weakens, whether over flat coastal areas or mountainous terrain. The underlying ocean heat source—the means for the storm to retain its warm core—suddenly is extinguished. Previously, the weakening was ascribed to surface friction over land, a hypothesis that has not been found valid except along large mountain ranges. The friction of the wildly agitated ocean with large masses of spray in the air may equal or exceed that over land. However, the effect on the wind profile of the lowest layer will differ, since the water is mobile and the ground solid (except for flying debris).

The intensity of a hurricane crossing a short stretch of land such as an island normally is regained quickly when the storm moves once more over water; but there are notable exceptions that as yet defy understanding. When a hurricane reaches a continent, much depends on the weather patterns present there. During indifferent flow the hurricane may die out in perhaps 2 days on a slow path inland. Sometimes, only the lower portion dies while a strong upper vortex remains, tends to drift westward slowly, and retains the capacity of delivering heavy precipitation for days. If the hurricane arrives when the flow constellation is such that an extratropical cyclone is about to form anyway, strong regeneration will follow. Maximum wind and weather catastrophes on land have occurred in consequence of such new energy infusion. Even under considerably more subtle circumstances old hurricanes have rallied enormously and delivered 50 to 75 cm precipitation, *i.e.*, up to one-half or more of mean annual precipitation in the affected area during a center passage lasting 12 to 18 hours.

Seasons of Occurrence and Variability of Frequency

Seasons. As Figure 1 indicates, hurricanes are known in all tropical oceans except the South Atlantic. In general they tend to be most frequent in late summer and early fall. Ocean temperatures then are at their peak; temperature and humidity of the air near the surface are still at their highest. With the start of winter cooling of the air in high latitudes, troughs in the westerlies that serve as external starting mechanisms for forming hurricanes intrude more frequently and vigorously into the tropics than during the midsummer lull. The seasonal lag with respect to the summer solstice is larger in the Northern Hemisphere than in the Southern, which has much smaller seasonal variations in general on account of the predominance of ocean surface.

Exceptions to the predominant pattern occur in the western Caribbean as well as in the Bay of Bengal and the Arabian Sea. There, advances and retreats of the equatorial trough zone are correlated with hurricane occurrence. The trough seasonally intrudes twice into the Caribbean: in May and June, and in late September and October. In midsummer it is displaced by a strong Atlantic subtropical anticyclone with fast trade winds on its equatorward side. In the Arabian Sea and the Bay of Bengal the equatorial trough, or monsoon advance, also occurs in May and June; its retreat in September and October is often protracted in the southern Bay of Bengal well into November.

Average Frequencies. Due to gross inadequacy of observations over the wide ocean expanses prior to the satellite era, average frequencies are most uncertain and will remain so until an adequate period of satellite observation has passed. It is well known that frequencies in several oceans have been greatly underestimated. Long period increases have been noted in most oceans and are thought to be due to the increased number of observations from long-distance sea and air travel, even well before the days of the satellite. In Australia the fictitious trend is particularly obvious prior to 1960. Mean annual world frequency of tropical storms, formerly estimated as fifty, may well be one hundred, still only a fraction of cyclones outside the tropics.

In spite of all shortcomings of the data base, the western North Pacific in the past has been rated as the most productive area of the Northern Hemisphere, with about twenty-six tropical cyclones per year (App. E). Of course, its area is very large; and no land mass intrudes, as do Central America and the northern part of South America, which break the main hurricane area of the Western Hemisphere into two parts—the Atlantic and East Pacific Oceans. The North Atlantic average of eight cyclones per year for the period 1885–1975 is thought to be most reliable. The East Pacific frequency has been doubled from earlier estimates to fifteen per year, based on the period 1966–1977. One also tends to regard Atlantic plus East Pacific generically as a single hurricane region, especially since it has been shown that the East Pacific storms are all (or nearly all) derived from Atlantic impulses crossing Central America. The combined region then averages twenty-three tropical storms per year and rivals the western Pacific. Frequencies in South Pacific and South Indian oceans are too uncertain to warrant an estimate at the present time.

Seasonal and Secular Variability. In the Atlantic Ocean, a 10-year survey from satellite observations for 1968–1977 has demonstrated that the number of rain systems with potential for hurricane development is very close to one hundred per season, with little change from year to year. The number of storms—not all of them full hurricanes—averaged only eight per season, or 8%, with considerable variability of 50%.

In large measure this variability is connected with persistence of sharply contrasting flow patterns in the upper parts of the troposphere, most clearly seen near 12-km height or 200-mb pressure. In "good" hurricane years warm air with clockwise-curving (anticyclonic) flow overlies the western Atlantic; in "poor" years cold air is aloft with upper west winds that curve cyclonically.[1] The hurricane-poor years tend to be coupled with drought; the rain yield of the one hundred migrating weather systems is held below average by the unfavorable high-level winds and temperatures.

Perhaps the only ocean with a reasonably reliable hurricane history for the last century (there is none earlier) is the Atlantic. There the records suggest that tropical storm frequency was above the 100-year annual average of eight before 1900; dropped to six from 1900 to 1930; rose dramatically to ten from 1931 to 1960, and then dropped back again to average or below. The reason for the long-term variation has not been ascertained; attempts of correlation with ocean temperature and other potential controls have not proved convincing; and there is at this time no clue that can be offered for the future.

Naming Hurricanes

This chapter should not close without brief comment on the custom developed since World War II of naming hurricanes. Its main purpose, of course, is convenience of communication, especially with the public, and accuracy in identification when more than one hurricane is present at the same time. The particular custom of choosing girls' names is partly in reaction to the novel *Storm* by George R. Stewart, published in 1941.[2] Stewart invented the idea, practiced in the novel by the junior meteorologist at the San Francisco office of the United States Weather Bureau.

> The first sweeping glance (of the map) assured him that nothing exceptional or unforeseen had happened in the twenty-four hours since

> he had prepared the last similar map. Antonia had moved above as he had expected. Cornelia and the others were developing normally. Not at any price would the Junior Meteorologist have revealed to the Chief that he was bestowing names—and girls' names—upon those great moving low-pressure areas. But he justified the sentimental vagary by explaining mentally that each storm was really an individual and that he could more easily say (to himself, of course) "Antonia" than "the low-pressure center which was yesterday in latitude one-seventy-five east, longitude forty-two North."

When, in the end, the Chief found out, it turned out that he had the same vice, though he chose famous historical generals and world conquerors.

The first actual use of girls' names for storm systems was by the United States armed forces in the Pacific during World War II. The commander of the reconnaissance aircraft that first discovered a typhoon was allowed to assign a girl's name to the system (usually the name of his wife or close girl friend). In 1953, the practice of assigning girls' names to Atlantic hurricanes in alphabetical order was initiated by the United States Weather Bureau. With the passage of time, different countries adopted different regional sets of names. In 1979 a new system was introduced for the United States, alternating girls' and boys' names for successive hurricanes. The use of an alphabetical sequence, beginning each new year with an *A* name is convenient in keeping tabs on the storm number for the year. One problem, however, especially in the Atlantic, where only the first part of the alphabet is run through in the normal year, is that the last half of the alphabet is rarely reached. Moreover, in order to preserve individuality, the names of the more memorable hurricanes need to be retired, so that suitable names ultimately become in short supply. Nevertheless, Stewart's reason for naming severe storm systems is as valid as ever, and names provide an identity that is useful in communicating the public need for awareness and preparedness actions as the hurricane advances toward landfall.

REFERENCES

1. H. Riehl, *Climate and Weather of the Tropics* (London: Academic Press, 1979), 583–87.
2. George R. Stewart, *Storm* (New York: Random House, 1941), 12.

CHAPTER 2

Impact of the Hurricane

The hurricane, when compared with severe extratropical storms, is a rare event. Nevertheless, its impact is sufficiently important to the United States to require a continual reassessment of the threat to coastal areas and to many inland areas subject to flash and river flooding. From the Gulf of Mexico to New England, the threat escalates each year as the property- and population-at-risk increase. Yet the impact of the hurricane includes a number of benefits.[1] Rain accumulations from a hurricane in a broad coastal zone may in a single day exceed the total volume of water transported by some major rivers in an entire year. In many coastal areas, water resources, agriculture, ranching, and some aspects of the bioecological chain are substantially influenced by, if not dependent upon, the recurrence of hurricanes, which therefore must be regarded as double-edged swords, holding the potential for catastrophic disaster locally while playing a supporting role in the regional ecology of the coastal zone.[2]

Globally, a hurricane, or tropical cyclone, is one means of maintaining the orderliness of circulations in lower latitudes, serving as a kind of escape valve for the transport of accumulated heat and momentum from the warm tropics to the colder middle and higher latitudes. Figure 2 is a weather satellite picture of cloudiness, with temperatures 15°C–20°C above normal streaming from the top of the hurricane into higher latitudes.

However, by comparison to the size and the transports that occur in an extratropical cyclone, Figure 3 being an example, the hurricane is indeed a small, though intense, circulation, not as

Fig. 2. Satellite photo of hurricane Agnes (June 1972) moving inland across the Florida panhandle. Here the cloudiness in the outflow layer is being conducted northeastward over the eastern seaboard of the United States, carrying a warm effluent to higher latitudes.

likely to cause significant changes or adjustments in large weather-breeding planetary waves in the westerlies as are the cyclones of higher latitudes.

The Threat at Ocean and Bay Shores

The major damage and loss of life from most hurricanes occur within 100 to 150 km of the landfall position, the point where the low-pressure center crosses the coastline. An exception is when torrential rains continue after a hurricane moves far inland, causing extensive river flooding.

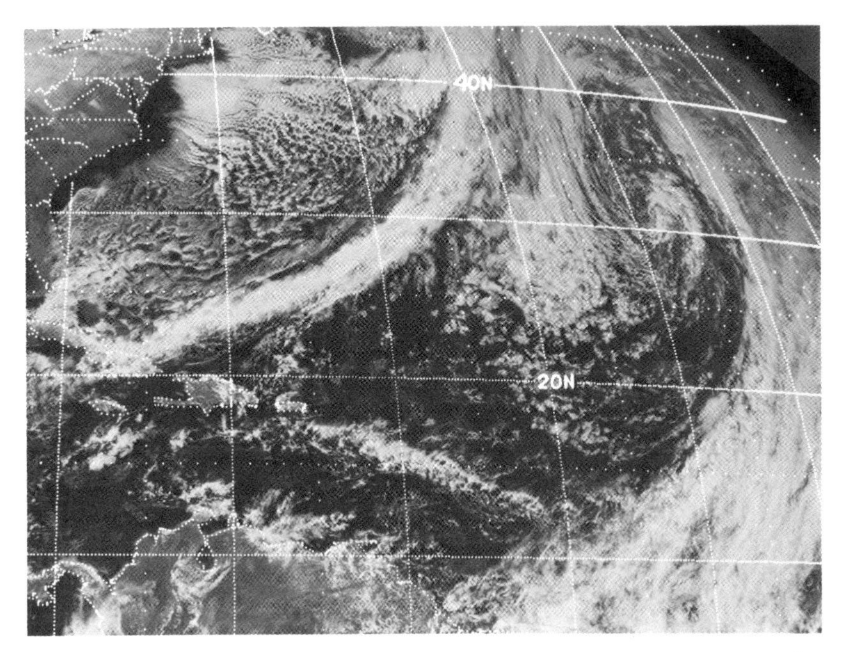

Fig. 3. A large extratropical cyclone in the middle Atlantic on 10 January 1976. The circulation in this cyclone (center near 31° N, 35° W), now past its prime of life, is enormous compared to the largest of hurricanes; it encompasses more than 35° latitude. Worst weather and strongest winds occur at a much larger distance from the pressure center than in hurricanes. National Environmental Satellite Service photo.

To the person who has never experienced one, the hurricane is generally perceived as a spectacle of wind fury. However, nine out of ten lives lost and the preponderance of damage at the coast result from inundation by the hurricane tide, which commonly rises more than 4 m above mean sea level. An uncommon example is that of Camille (1969), when tides rose to a peak height of 7.5 m. At open exposed shorelines, damage and destruction begin with the erosive scour and battering from large breaking waves, which may remove as much as 35 m^3 of beach and dune sand per meter of water frontage. As tides rise, these waves may then extend their influence inland to attack foundations and structural members near grade level. Heavy debris, tree trunks, telephone poles, and pilings carried by these waves and by the currents that accompany tidal surges become battering rams that may cause serious structural damage to the lower floors of buildings. Yachts, working boats, and some smaller ships torn from their moorings may be carried hundreds of meters inland.

Wind damage to structures is approximately proportional to the square of the wind speed. Wind pressures, or the dynamic forces caused by strong winds moving over or around an irregular portion of a building like a roof or a cornice, produce external forces similar to the aerodynamic lift on an airplane wing. This may cause roof failure and, at times, the outward collapse of curtain walls and glass openings. However, the losses due to wind alone, in the absence of an earlier rupture or weakening of a structure by rising water, can be virtually eliminated (with small increases in building costs) by using appropriate building standards like those designed by the state of Texas in 1975.[3]

Peak sustained surface winds at the open coast rapidly diminish inland and, thereby, the damage potential for low-lying structures decreases. Only in unusually strong hurricanes does widespread wind damage occur to structures located inland more than 100 km from the water's edge. A local exception is when small hurricane-generated tornadoes touch down. These have short path lengths and usually occur in advance of landfall, before the arrival of sustained gale- or hurricane-force winds. Other exceptions are the agricultural losses, especially of citrus and tropical fruits, that may occur hundreds of kilometers from landfall. An enormous amount of citrus was lost in central and north central Florida when hurricane Donna swept northward across the peninsula in 1960. Near total losses of truck and other crops occurred in south Texas due to flooding when hurricane Beulah (1967) moved inland and stalled, depositing more than 600 mm of rain over 15,000 km^2 in a 72-hour period.

Inland Flooding

Although the threat generally diminishes rapidly inland as the hurricane loses its oceanic heat source, sometimes the hurricane encounters an environment with significant horizontal temperature variations that supply auxiliary sources of kinetic energy to maintain hurricane-force winds great distances inland. Such circumstances may also cause local flash flooding or protracted periods of widespread river flooding. A prime example is that of hurricane Agnes (1972), which caused property losses of almost $3.5 billion from river flooding after it had traveled more than 1,600 km overland from its landfall near Apalachicola, Florida, where total losses at the coast were less than $10 million. Table 1 lists the losses of property and of lives from the twelve most disastrous hurricanes to reach the United States coastline. One of the more costly storms, Agnes (1972), was important mainly because of river flooding.

Table 1. Losses of Life and Property in the United States from the Twelve Most Disastrous Hurricanes of the Twentieth Century

Year	*Hurricane*	*Coastal area*	*Lives lost*	*Damage ($10⁶)*
1900		Galveston, TX	6,000	30.0
1919		Florida Keys; Corpus Christi, TX	> 600	20.0
1928		Palm Beach, Okeechobee, FL	1,836	25.0
1935		Florida Keys	408	> 6.0
1938		New England	600	306.0
1957	Audrey	Southwest Louisiana	390	150.0
1960	Donna	Florida; New York; New England	50	387.0
1961	Carla	Texas	46	408.0
1965	Betsy	Florida; Louisiana	75	1,420.5
1969	Camille	Louisiana; Mississippi; Virginia	256	1,420.7
1972	Agnes	Florida; Virginia; Maryland; Pennsylvania	122	2,100.0
1979	Frederic	Alabama; Mississippi; Florida	5	2,300.0

The Economic Impact

Figure 4 shows the trends in monetary losses due to hurricanes. The sharp rise in recent decades reflects the increase in the property-at-risk that has accompanied the migration of population

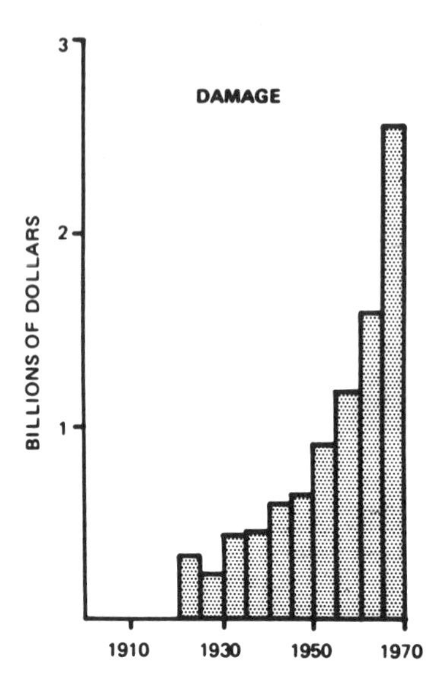

Fig. 4. Trends in hurricane-related damage in the United States (5-year averages). R. H. Simpson, "Hurricane Prediction," in *Geophysical Predictions* (Washington, D.C.: National Academy of Science, 1978), 143.

to seashores. The lack of effective land-use regulations and building practices in coastal zones significantly increases the hurricane damage potential. In many areas, protective dune lands at the oceanfront have either been removed to facilitate land development or have become the platforms upon which small structures have been erected without the foundational support of penetrative pilings. Either action reduces the natural protection afforded by the dunes and increases the potential for property losses.

Threat to Life and Life-Style

A survey by the National Hurricane Center in Miami in the mid-1970s revealed that population migrations and the increasing residential use of coastal lands have led to exceptionally low hurricane-experience levels among coastal residents.[4] In some of the more populous coastal areas of Florida, less than one person in ten had experienced a hurricane threat or understood what such a threat comprised. This raises many questions about human response to warning advices during a hurricane emergency and what this might portend with regard to the potential for disaster. At least this survey emphasized the importance of programs for public awareness and preparedness to face a hurricane crisis.

Minimizing Hurricane Losses

For effective reduction of the impact of a hurricane and the risks of living in a hurricane-prone area, coordinated action at several levels of government is needed, beginning with some form of land-use planning and regulation. In mid-1970, the state of Texas led the way in developing model hurricane-resistant building standards to protect against extreme winds of all but the most severe hurricanes.[5] The increased cost of construction using these standards is small, estimated to range from 2% to 10% above the cost of using common-practice building codes. Protection against erosion and the battering of waves is more costly but economically justified for most residential structures in low-lying areas subject to flooding.

Equally important in defending against a hurricane disaster are public-awareness programs to promote an understanding of the hurricane and its potential impact and of the measures that coastal residents must take to face the threat of a hurricane with minimal risk to their lives and property. Perhaps the most important single element in the defense against hurricane disaster is the hurricane prediction and warning service and its associated systems for communicating and interpreting warning advices to those who must take emergency actions on short notice to protect life and property.

Finally, there exists the possibility, as yet unproven, for directly reducing the hurricane damage potential by strategic cloud-seeding methods. If these measures can be validated and maximum winds of the hurricane reduced by as much as 10% to 20%, the damage from extreme winds can be reduced by 20% to 40%, with savings often exceeding the cost of the seeding operation by several orders of magnitude.

The escalating use of coastal lands for residential construction, and, thereby, the exposure of millions of new coastal residents to future hurricanes, poses a unique challenge to the government at all levels. To roll back or reduce the residential use of coastal lands is not politically viable. And prospects for increasing the accuracy or extending the timeliness of hurricane warnings are not presently within reach, scientifically or technologically.

In the chapters that follow, we shall discuss the hurricane as a meteorological event and as a natural hazard whose probable impact at the coast and on vessels at sea portends disaster potentials well beyond what the United States has yet experienced. As we shall see, however, the risks of these disasters can be greatly reduced by intelligent planning.

REFERENCES

1. P. J. Hebert and G. Taylor, *Hurricane Experience Levels of Coastal County Populations—Texas to Maine*, National Weather Service Southern Regional Technical Report (Silver Spring, Md.: Department of Commerce-NOAA, 1975); A. L. Sugg, "Beneficial Aspects of the Tropical Cyclone," *Journal of Applied Meteorology*, VII (1968), 39–45.
2. G. L. Clark, *Elements of Ecology* (New York: John Wiley and Sons, 1954); G. W. Cry, "Effects of Tropical Cyclone Rainfall on the Distribution over the Eastern and Southern United States," Professional Paper 1 (Department of Commerce, ESSA, 1967).
3. Texas Coastal and Marine Council, *Model Minimum Resistant Building Standards for the Texas Gulf Coast* (Austin, Tex.: General Land Office, 1976).
4. Hebert and Taylor, *Hurricane Experience Levels—Texas to Maine.*
5. Texas Coastal and Marine Council, *Model Minimum Resistant Building Standards.*

PART II

The Worldwide Setting

CHAPTER 3

The Summer Tropical Atmosphere

The Energy Cycle

Radiation and Surface-Air Exchange. The combined system, earth plus atmosphere, is a heat source at low latitudes within about 30° latitude from the equator, while the higher latitudes are a cold source. This creates a continual interplay of the winds across latitudes 30° N and 30° S, strongest in winter and carrying warm air from the tropics poleward and cold air from higher latitudes equatorward.

Between the Tropics of Cancer and Capricorn, the sun crosses the zenith above any latitude circle twice. Especially in the summer hemisphere the solar beam remains so nearly vertical at noon that the radiation incident at the outer limit of the atmosphere is nearly constant. Of the incident radiation, a small amount is absorbed in the stratosphere; over 20% is absorbed by contaminants such as dust, salt water drops, and foam, and by water vapor. Reflection from the troposphere, notably through water drops and ice particles, is estimated at 17% for the tropics on the basis of satellite observations. Reflection from the ground is as low as 5%, on account of the large expanse of oceans, which reflect only 2%–3% of sunlight. Therewith, 55%–57% of the sunlight that reaches the earth's atmosphere is absorbed at the ground.

The atmosphere itself is a known cold source, and it would get colder all the time from radiation fluxes alone. In the tropics the radiation heat loss is larger than that outside the tropics and cools the air by about 1°C per day or even a little more. Additional energy sources are needed to restore temperature in the tropics to the observed constant level. These sources must come from the sun's radiation stored within the earth.

Two known methods of energy exchange effect virtually all of the transport from ground to air. One of these is contact cooling by air adjacent to warmer soil or water (sensible heat transfer). The other is evaporation from water surfaces plus transpiration from plants (latent heat transfer). Whenever the pressure of water vapor at the surface of water bodies exceeds the pressure of water vapor in the air, the water body does work using stored heat from the sun to evaporate water into the atmosphere. The addition to the water vapor in the air heats the air, not directly, but when it condenses and then falls out as precipitation; hence we have the distinction in calling this part of the energy exchange *latent*.

It is of greatest importance that the tropical belts of the earth as a whole are about 75% to 80% covered by water and only 20% or a little more by land. Thus, the land-water ratio is close to 0.2 and the ratio of sensible to latent heat transfer into the atmosphere is also near 20%. The continents provide most of the sensible heat transfer. Such heating directly warms the atmosphere above the ground; very high desert temperatures occur when there is no water to be evaporated. Over a flat water surface, saturation vapor pressure and mixing ratio are strongly temperature dependent (Tab. 2). The higher the temperature, the higher the vapor pressure and, thereby, the capability of the water to evaporate. In fact, the increase becomes so large above 25°C that the 28°C isotherm is the highest that can generally be drawn over the tropical oceans, whereas 30°C is observed only in a few places as a seasonal mean.

A decisive factor for hurricanes is that most of the energy gained by the atmosphere from the earth in the tropics is latent heat from the oceans, which is mobile and therefore available to be sucked into a great vortex. The formation of cloud droplets, which does raise the air temperature, may occur sooner or later and rarely happens at the place where the evaporation has occurred. The water may be carried as gaseous vapor for thousands of kilometers.

Table 2. Saturation Vapor Pressure and Saturation Mixing Ratio as a Function of Temperature over a Flat Water Surface

T (°C)	*Saturation vapor pressure* (*mb*)	*Saturation mixing ratio* pressure 1,000 mb* (*g/kg*)
40	73.8	49.8
35	56.2	37.3
30	42.4	27.7
25	31.7	20.4
20	23.4	15.0
15	17.0	10.8
10	12.3	7.8
5	8.7	5.5
0	6.1	3.8
− 5	4.2	2.6
−10	2.9	1.8
−15	1.9	1.1

* Mixing ratio is defined as the mass of water vapor in a specified mass of dry air, commonly grams per kilogram.

Condensation and Precipitation. When dry air moves upward in the atmosphere toward lower pressure, it will cool at a rate of nearly 1°C per 100-m height increase due to expansion; a corresponding temperature rise will occur when the air sinks toward higher pressure. However, when the air contains water vapor, relative humidity increases during ascent as saturation vapor pressure decreases with the temperature. Sooner or later, a level is reached where the vapor carried upward becomes equal to the saturation value, or the limit of the amount of moisture that the air can contain in vapor form. Without going into the complexities of cloud physics, when atmospheric humidity and saturation humidity become equal—*i.e.*, when relative humidity attains 100%—small water drops form. During further ascent all water vapor in excess of the saturation value continues to be condensed.

What is the consequence of cloud-forming ascent on temperature? Consider typical values of temperature and moisture as widely observed over the tropical oceans: temperature 27°C at a surface pressure of 1,010 mb, relative humidity about 78%, mixing ratio of 17 g/kg. The air, rising, will reach the 940-mb level with a temperature of 21°C, which, from Table 2, corresponds to saturation there. We may follow the ascent further from 940 mb, making the assumptions that (a) the air is dry and (b) it is saturated and condenses water vapor on its way up, producing a cloud. The temperature of the cloud ascent increases continually upward compared to the dry ascent, and by large values (Fig. 5). At 500 mb the difference is no less than 24°C, a very large value that brings out the powerful warming effect of condensation on air temperature.

How does the tropical atmosphere react to these two very different modes of ascent? Primarily we need to consider the western

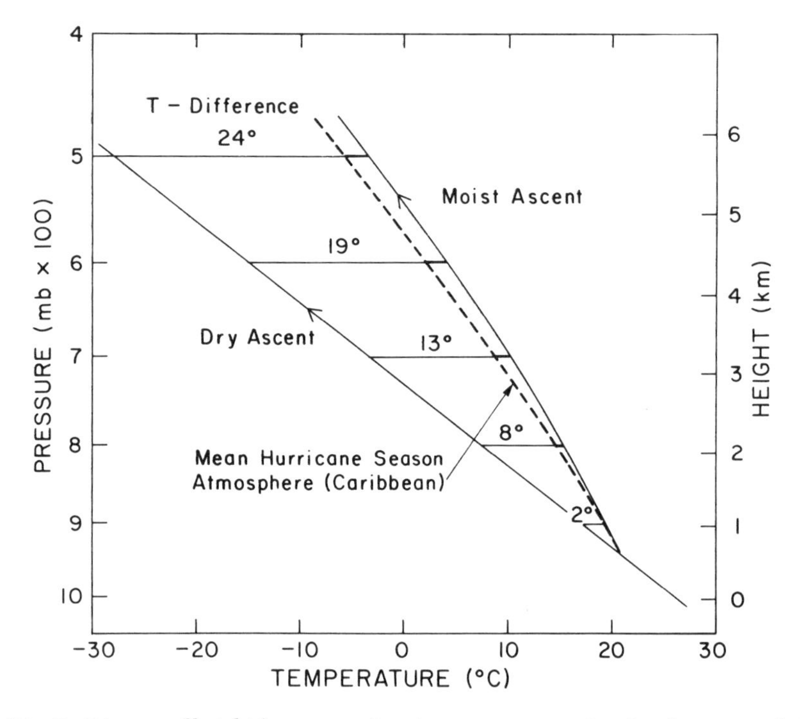

Fig. 5. Diagram of height (or pressure) against temperature showing dry ascent of air in atmosphere (*solid straight line*); ascent with condensation (*curved solid line*); mean hurricane-season Caribbean atmosphere (*broken line*); and temperature difference between dry and moist ascents arising from condensation of water vapor for several levels.

parts of the oceans in the warm or rainy season, and parts of the equatorial zone. Over the eastern parts of the oceans a temperature inversion—temperature increasing upward—usually prevails in the low levels together with a drop of humidity to very low values. Such an atmosphere provides formidable resistance against clouds growing to great heights. Hurricanes do not form there as long as the inversion is present, which is true nearly all of the time.

In contrast, over the western parts of the oceans in summer one encounters a temperature structure as given by the broken curve of Figure 5, with relative humidity near 80% at the surface and 50% at 500 mb. The mean hurricane season atmosphere, drawn for the Caribbean, is close to that of the Pacific and Indian Ocean hurricane areas, in fact, of all primary generating areas for tropical storms on the globe. It is very evident that the temperature structure determined from many balloon ascents (radiosondes) is almost identical with the cloud-forming ascent of Figure 5! It follows that the action of many thousands of clouds rising to many different levels all the time fashions the mean structure of the atmosphere.

Fig. 6. Trade-wind cumulus-cloud street over the ocean.

Fig. 7. *a*, Incipient cumulonimbus over the ocean. Note high top on right starting to turn to ice. Suppressed cloudiness around big towers. *b*, Cumulonimbus, late stage, fully developed, with large overhanging anvil produced by shearing of wind with height.

Of course, the actual atmosphere is not identical to the mean atmosphere from day to day. Actual structure varies slightly in temperature and considerably in moisture. On many days the air above 1 to 3 km becomes very dry; on some others it is moist to high levels. Accordingly, one distinguishes between *shallow* and *deep* convection. Low cumulus clouds, often arranged in neat rows, are present on nearly all days (Fig. 6). Their main function is to remove the heat and moisture gained at the ocean surface to higher levels and mix them through the *moist layer* occupied by the cumulus clouds. At other, far more rare, times cumulonimbus clouds (Figs. 7a, 7b) extend to 300, 200, and even 100 mb; *i.e.*, their vertical thickness ranges from 10 to 16 km. Over land, they often occur in preferred locations, especially along mountain ranges and along the forward edge of sea breeze penetrations inland. Over the ocean, they are found in small or large groupings. These will occupy us a great deal in this book, since most hurricanes develop in association with some of these groups of very large towers. Satellite measurements locate all these groups around the globe. The higher a cloud, the lower its temperature. Centers with a temperature below −50°C attain at least 200-mb tops and represent the deepest convection seen; those with −40°C are intermediate, and centers with −30°C attain only 300-mb or 10-km height and are relatively weak, though they may still yield rain as heavy as any of the others.

In Figure 8 we see satellite radiation measurements for a particular day in the Atlantic Ocean. The shading is inverse to that of satellite photographs presented on television screens. There the intense clouds appear very white. In Figure 8 the areas with deep clouds are marked dark, whereas the broad white warm areas indicate very little cloud. A very clearly marked dark band with several cold centers stretches across the Atlantic. Over South America and the Caribbean the band widens out and has many more dark centers, some induced by the Andes and the mountains of Central America. The narrow channel formed over the eastern Atlantic broadens and becomes diffuse, as happens also in the western Pacific and the South Indian oceans.

The action of different cumulus cloud types in transporting energy upward in the atmosphere may be represented in model form

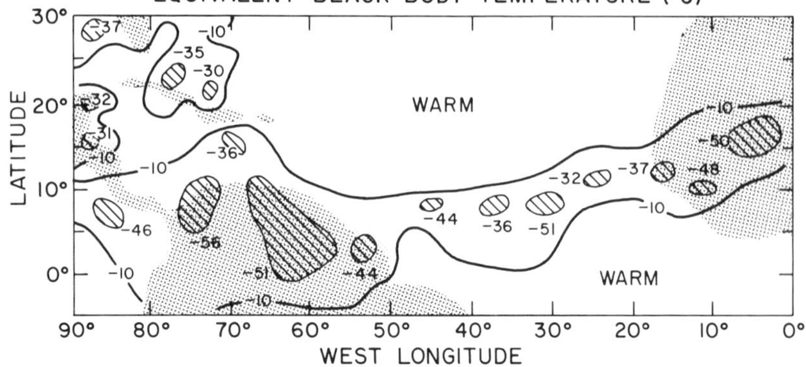

Fig. 8. Satellite-measured temperatures equivalent to black-body radiation. Cold temperatures denote high cloud tops. Approximate heights: 9 km for −30° C, 10.5 km for −40° C, and 12 km for −50° C. Temperatures averaged over square of 110 km × 110 km.

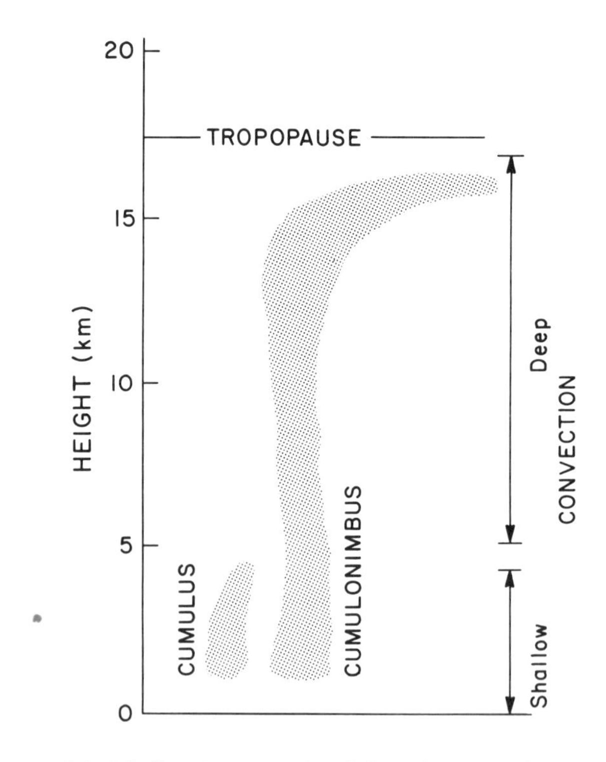

Fig. 9. Model of shallow (see Fig. 6) and deep (see Fig. 7) convection.

(Fig. 9). Shallow cumulus clouds or rows of cumuli moisten the layer up to 3 to 4 km and maintain temperature constant there against radiation of the air to space. Above these levels deep convection provides the balance. Net export of energy to higher latitudes may come from both these layers. Sensible heat (warmer air) as well as latent heat (high water-vapor content) may be exported across latitudes 30° into the high latitudes.

Figure 5 contains information about the *efficiency* of latent heat, or its availability, in generating wind systems, in starting and maintaining strong wind systems such as hurricanes. At the extreme, the whole temperature difference between dry and moist ascent would be available, and this would be a very large amount. However, since the observed atmosphere lies close to the moist and far from the dry ascent, this whole big temperature difference is not available; the deep dry ascent never occurs. Instead, the efficiency will be measured by the 1°C–2°C difference between moist ascent and actual atmospheric structure. Hence, at most 10%, and usually less, of the total latent heat release is "free" for conversion to such energies as energy of motion. This tiny fraction of the whole condensation will provide our focus in later chapters. Even the very small efficiency is not easy to realize, and very often one encounters areas with heavy precipitation but no generation of wind at all. It is, therefore, erroneous to quote the whole latent heat release by the atmosphere in comparison with man-created energy as sources to drive machines or generate motions. As the next step we shall inquire into the wind systems in which the convection occurs, and how wind systems and cloud masses interact.

Surface Wind, Pressure and Temperature

Pressure and Wind. Our example will be the Atlantic Ocean; however, the same features in principle are found in North and South Pacific oceans and in the South Indian Ocean. A surface high-pressure center—the subtropical anticyclone—extends east-west across the ocean (Fig. 10). The streamlines indicating mean surface wind direction point from high toward lower pressure everywhere. South of the center of highest pressure, winds blow from northeast to east-northeast and are very steady; that is, they will be encountered there most of the time. They were named *trade winds* in sailing ship days because vessels could make good progress in contrast to the region inside the high pressure center itself, where winds may be light for long periods and ships becalmed. In the Southern Hemisphere the corresponding winds, known as "southeast trades," blow from southeast and are just visible at the southern border of the map.

In or near the equatorial zone and extending around the globe is a belt of low atmospheric pressure toward which the wind systems of the trades in all oceans converge. Flow from high to low pressure will accelerate air; however, the contact with the ocean offers frictional resistance. In general, acceleration by the pressure force and retardation by ground friction are about equal, so that winds blow with fairly uniform speed, averaging 12–14 kt (6–7 mps).

At the southeastern border of the chart a turning of winds from southeast toward southwest and the continent of Africa is evident. Such deflected trades are generally called *monsoon*, a term originally used in India to denote seasonally reversing winds—south-

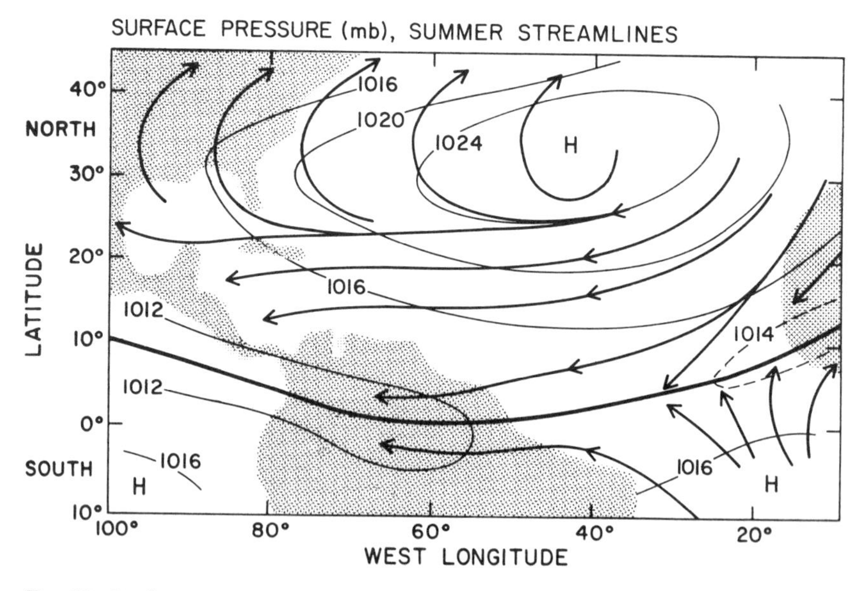

Fig. 10. Surface pressures (mb) and streamlines over the Atlantic Ocean in July. *Heavy solid line* marks equatorial low-pressure trough.

west in summer, northeast in winter. The original definition has gradually been widened to denote any large wind system (*i.e.*, larger than sea breezes on beaches) that blows from cold to warm regions. One even speaks of a "summer monsoon" from east and southeast in the southern United States. In winter the outflow from the cold land toward warmer oceans similarly deserves the name monsoon.

Temperature and Wind. The air imported into the tropics from higher latitudes is relatively cold at all seasons, including summer. Thus, over the eastern parts of the oceans the trade winds carry cold air into the tropics that gradually warms up. In Figure 11 temperature in the east is about 22°C and then gradually rises to 26°C–28°C in the course of a week needed by the surface air to traverse the ocean. The warming is accomplished by heating from the ocean, which shows very nearly the same temperature pattern as the air at the surface but at slightly higher values. Sea-air temperature difference is generally within 1°C, so rapidly does air ad-

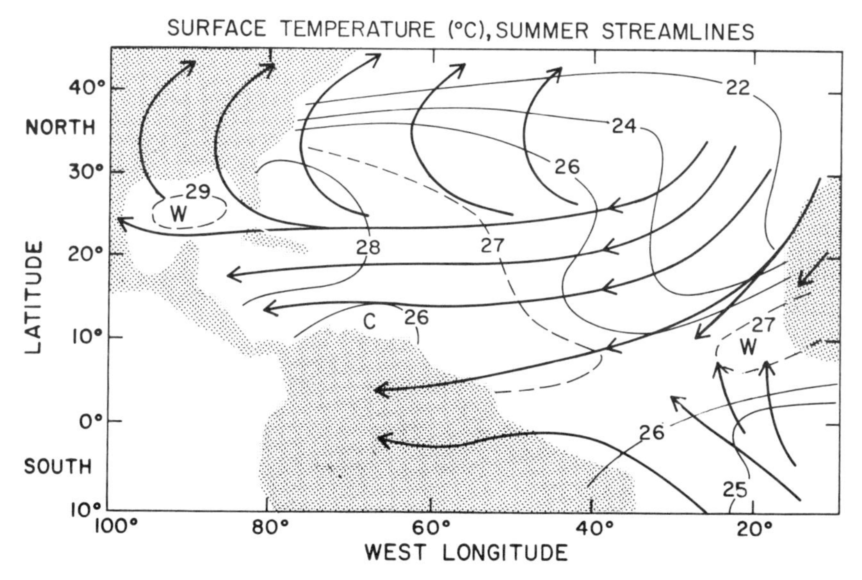

Fig. 11 Surface air tenperatures (°C) and streamlines for the Atlantic Ocean in July. Warmest temperatures coincide with the equatorial low-pressure trough in Figure 10, east of longitude 70°.

just to ocean temperature. A separate chart of ocean temperatures is not needed for this reason. Over the western parts of the oceans the reverse wind flow carries warmer air northward and in the Southern Hemisphere southward.

It is of particular interest that the region of warmest water in the East Atlantic, greater than 27°C, coincides with the low-pressure trough of Figure 10. Such coincidence is generally found to hold true. The trough aligns itself so as to coincide with the location of warmest air and ocean water. Further, about 90% of hurricanes form at temperatures of 27°C or higher. At lower temperatures air is unable to rise to the high troposphere; the ascent would lie on the cold side of the broken curve in Figure 5.

Humidity and Wind. The air imported from high latitudes into the tropics is not only cold but also relatively dry. From Table 2 ever-increasing evaporation should take place from the ocean surface as ocean temperatures rise along the air flow. The air tries to adjust to the ocean's vapor pressure but does not succeed nearly so well as in the case of temperature. This is due partly to the rise of saturation vapor pressure with temperature and partly to overturning of the atmosphere, which always imports drier air downward from upper levels (Tab. 2). Thus, evaporation remains large along the trade-wind flow. The moisture cycle becomes the most important factor for keeping tropical temperatures high against radiation cooling.

Convergence and Divergence. As noted from Figure 10, the mass of surface air diverges (goes apart) from the subtropical high-pressure center and converges (comes together) in the vicinity of the equatorial low-pressure trough. The importance of these motions lies in the fact that, since the mass of the atmosphere is conserved, air diverging near the ground must sink (so that a hole does not develop) and converging air must ascend. Symbolically, one may write

Eq. 3.1
$$w_h = -\overbrace{\mathrm{div}(V)}\,h$$

where w denotes vertical air velocity at the height h, and div (V) the horizontal divergence over the layer between the surface and the height h, denoted by the caret. By convention, divergence is defined positive and convergence negative. Equation 3.1, a simplified form of the general equation of mass continuity, may be regarded as valid for the layer of air between the surface and about h = 3 km height.

For visualizing the meaning of Equation 3.1, consider an area with convergence (more mass comes in than goes out). Since the amount of mass itself remains constant, the convergent mass becomes squeezed and must escape upward or downward. If the convergence takes place near the ground, the air cannot move downward through the solid surface, so that its entire motion is upward. Conversely, in an area with divergence (more air goes out than comes in), also located near the surface, the air must sink, or else a hole would develop in the atmosphere. Rising air expands and cools and clouds form when the saturation vapor pressure (Tab.2) is reached. The opposite holds for sinking air. It is compressed and warmed, and its relative humidity decreases. We therefore associate cloudy skies with converging motion, and clear skies with diverging motion, in the lowest levels above the ground.

In Figure 12 the principal areas of convergence and divergence at the surface are depicted for the summer season. The numbers should be understood in the sense just discussed. For instance, the label −2 indicates that the distance of two masses of air moving in the same direction shortens by 20% in a day, and the same would hold for two masses coming from opposite directions. For the + numbers the distances will get larger. Consider h = 1,000 m for a sample evaluation of Equation 3.1. Given surface convergence = -2×10^{-6} sec^{-1}, or roughly 0.2 per day, then $w_h = 0.2 \times 1000 =$ 200 m/day upward, if the surface convergence is constant through the layer from the surface to 1,000 m. This is a rather small value as will be seen later. Nevertheless, patterns of convergence and divergence in Figure 12—a climatic chart for the whole summer or rainy season—are well ordered. Most of the divergence is found in the trade-wind zone and in the subtropical high-pressure center, while most convergence lies along the equatorial low-pressure trough. From Equation 3.1 we can say that on the average, air is

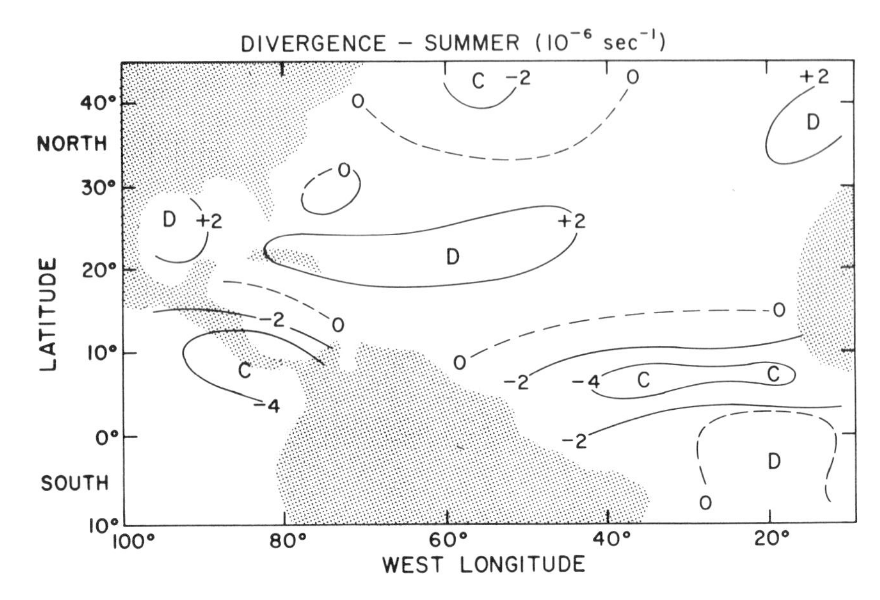

Fig. 12. Horizontal convergence and divergence of surface winds over the Atlantic Ocean in July. *D, positive,* divergence; *C, negative,* convergence. For units see text.

descending in the lowest part of the atmosphere in the trades and is rising in the equatorial wind convergence. From the discussion early in this chapter we should expect to find most rainfall in the low-pressure zone and fine weather in the trade winds. We are interested in the fields of convergence and divergence in Figure 12 because of their possible relation to the distribution of clouds and rain.

Mean Rainfall Chart. A glance at Figure 13 shows that the actual relation between the vertical motion fields just deduced and precipitation is indeed impressive. Unfortunately, the rainfall chart is only for the whole year, but most rain occurs in the summer season. Regions of convergence have most rain; regions of divergence are driest. Over a large part of the oceans precipitation is as little as one-tenth of that in the equatorial zone including equatorial South America. It follows that the presence of a water surface does not insure precipitation. On individual days (*cf.* Fig. 8) centers of cold temperatures and presumably heavy rain are also found in cli-

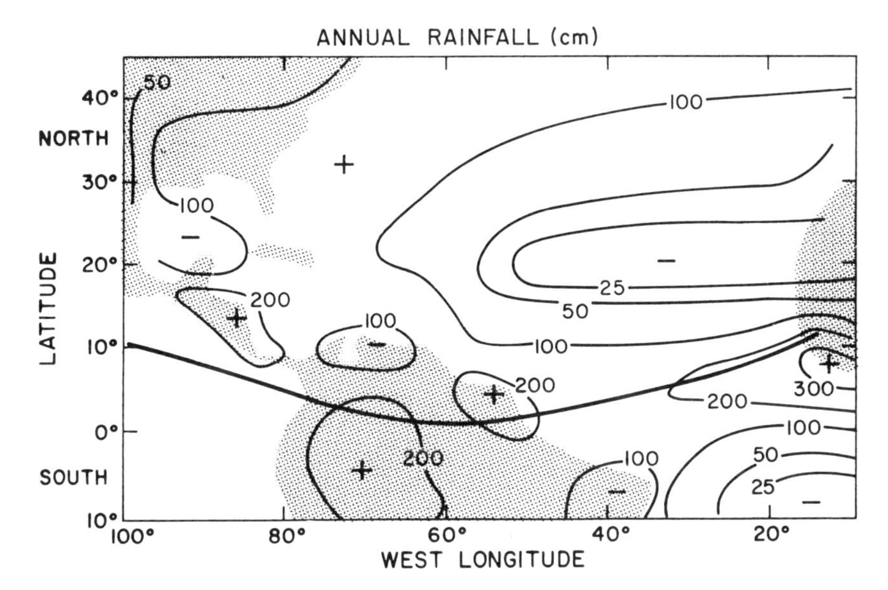

Fig. 13. Mean annual rainfall (cm) for the Atlantic Ocean (estimated) and adjacent continents.

matically unfavorable spots in addition to the main band along the low-pressure trough. But it must be remembered that climatology only furnishes a general background; one must still examine the particular state of affairs on any given day.

In summary, the trade winds diverge from the subtropical high-pressure centers toward the equatorial low-pressure trough; the motion from high to low pressure maintains their speed against friction over the ocean and land surfaces. Along the trajectories air warms up and acquires moisture, so that in the end the warmest and moistest air, most favorable for hurricanes, is situated over the western parts of the tropical oceans. Heaviest precipitation occurs where the surface winds converge, and the lightest where they diverge.

The High-Level Return Flow

The description of low-level conditions over the tropical oceans is quite valid throughout the hurricane season and, as emphasized, largely duplicated in other oceans. The trade-wind inflow toward the equatorial low-pressure trough is strongest in the lowest kilometer, then decreases quickly with height and vanishes above 2 to 3 km height, where the flow becomes largely easterly. Over the principal hurricane-generating regions the speed of the trade winds decreases slowly upward above 3 km (700 mb) and often gives way to west winds above 10 km (300 mb). However, the marked steadiness of the flow at low levels is not observed in the upper troposphere. Rather we find large troughs (see Fig. 14) extending into the tropics from higher latitudes and moving with time, and occasionally upper closed-cyclonic and anticyclonic centers that travel and may grow and decay. In short, we have a picture that by and large has much resemblance to *turbulent flow*, if by this term we mean very variable wind systems changing from day to day.

Therefore, the presentation of seasonally averaged upper charts, such as those for the surface layer, will not illuminate, and perhaps will obscure, the circumstances attending hurricane formation, motion, and decay. We shall therefore forego such charts, though they do have a place in some texts on tropical meteorology. Instead, Figure 14 illustrates the type of high-level flow encountered in hurricane situations over the Atlantic at 200 mb, the level of greatest development of the high-altitude flow patterns. High-level winds observed by balloon from upper-air measuring stations have been reinforced by the reported aircraft winds. Streamlines indicating the direction of flow have been drawn. More upper winds beyond the edges of the map assure the accuracy of the streamlines near the borders.

Compared to the simple flow of the surface layer, the high-tropospheric winds indeed give an appearance of turbulent flow on a very large scale. In the early days of meteorology such patterns, which are the rule rather than the exception, were not anticipated. The main feature is a very large trough, indicated by a heavy slanting line in the middle of the ocean, with northerly winds on the west side and south to southwest winds on the east side (a trough is so defined).

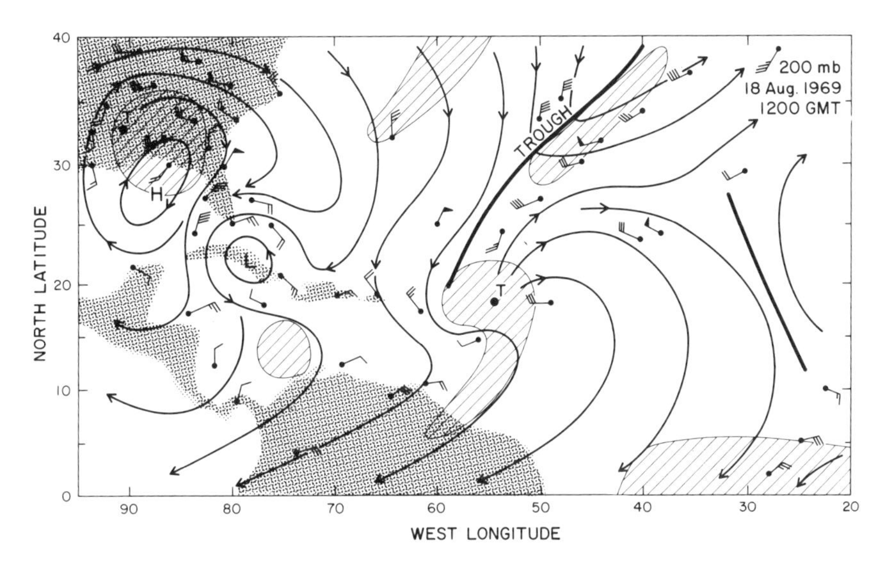

Fig. 14. Streamlines of the upper winds at 200 mb (12-km height) for the Atlantic Ocean area, 18 August 1969. *Heavy lines* denote troughs (cyclonic shear zone); *T*, tropical hurricane center; *H*, high-pressure center; *L*, low-pressure center; *light shading*, satellite-seen major cloud masses. Winds drawn for direction from which wind blows: *short barb* denotes 5-kt winds; *long barb*, 10 kt; *heavy triangular barb*, 50 kt.

The date has been chosen to coincide with the entry of hurricane Camille into the southern United States. At high levels it is ringed by clockwise or anticyclonic flow, and there is divergence of mass from the center. A second hurricane in the middle of the Atlantic has a similar outflow pattern of mass. An upper center with counterclockwise or cyclonic flow overlies the Greater Antilles. A secondary trough is indicated near the eastern end of the map, close to Africa. It is with such maps, and with the cloud patterns now furnished by satellite observations, that one must work to understand and predict hurricanes. In the later chapters we are concerned with the most relevant details instead of with such huge maps. It may be noted, however, that the second hurricane in mid-Atlantic lies by and large midway between winds from the west at latitude 40° and winds from the east from about latitude 10° southward. Such a central position in that broad field is very favorable for hurricane development, and, indeed, numerous storms formed in 1969. In many other years the very large-scale winds are arranged differently, for instance, with a west wind over all the southern part as far as the equator—an arrangement decidedly unfavorable for hurricanes. Because the very large flow often persists in a certain mode, once it is established, for such long periods as a whole hurricane season and even longer, there is some hope that estimates can be made for many years whether a season will or will not be active. Such background information is valuable; of course it gives no advice on timing or path of individual hurricanes. The markedly different "general circulation" in different years points to large rainfall variabilities.

Rainfall Variability

In the early days of meteorology the tropics were seen as a vast moisture reservoir in which precipitation, hurricanes apart, occurred by random process—that is, without any systematic evolution in time or space—and from which the higher latitudes could always draw an abundant supply of moisture. The passage of time has shown that this simple picture does not hold and that very large changes in precipitation, the most important climate factor of the tropics, do occur and on any time scale that one may choose.

Annual Rainfall. In parts of the tropics, notably monsoon countries, there is only one rainy season of perhaps 1 to 3 months duration that delivers the water supply for the entire year. In some other areas we find two rainy seasons, a long one and a short one. Two chances are better than one and, if the main season fails, a second one that may bring relief follows within a few months. Nevertheless, large variations in yearly precipitation are found everywhere. Consider New Orleans on the southern coast of the United States, where there are summer and winter rains, with a good chance for precipitation in every month. Mean annual precipitation has been 60 inches (152 cm) over the period of historical record. Regarding the 100-year time interval 1870–1969 (Tab. 3) we see at once that even under these favorable circumstances there is

Table 3. Frequency Distribution at New Orleans of 100-Year Precipitation, 1870–1969, in 10-Inch Class Intervals

Class *(inches)*	*(centimeters)*	*Percent*		
<35	<89	1	13	dry
35–40	89–112	12		
45–54	114–137	21	73	normal
55–64	139–162	31		
65–74	164–187	21		
75–84	190–213	12	14	wet
⩾85	⩾215	2		

NOTE: Mean = 60 inches (152 cm) per year.

a wide spread of annual rains when they are examined in 10-inch intervals. To be sure, the distribution is a normal one. Mean, median, and mode coincide at 60 inches; the modal class of 55–64 inches contains 31% of the observations, and 73% of the years fall within 15 inches of the mean, *i.e.*, in the range 45–64 inches. However, 27% remain outside this range, almost equally on the high and low sides. If the margin of 15 inches from the mean is regarded as too large from the viewpoint of normal water supply and, say, a 10-inch limit would be considered more realistic, the number of cases in the range 50–69 inches is only 58%. With this definition, only 60% of the years would have "normal" precipitation. The remaining 42% are divided in half between low and high cases. Thus 21% of the century analyzed would have some measure of drought. The extreme year had only 31 inches, or half of the average.

Long-Term Trend. To detect longer-period trends within the century, the deviations of individual annual values from the mean may be summed cumulatively. Over the century, the calculation must start and finish at zero, since the sum of all deviations from the mean must be zero. When the cumulative deviations are plotted against time, a slope toward higher values denotes above-average values, a slope toward lower values below-average values. From

Figure 15 the period 1870–1890 had precipitation averaging 13% above the mean. This period was followed by sharply lower values setting in abruptly after 1890 and lasting for 15 years, with an average annual deficiency of 20%—a very marked trend. Thereafter normal precipitation prevailed until 1940, with only minor swings; from then on precipitation averaged 11% above average until 1970.

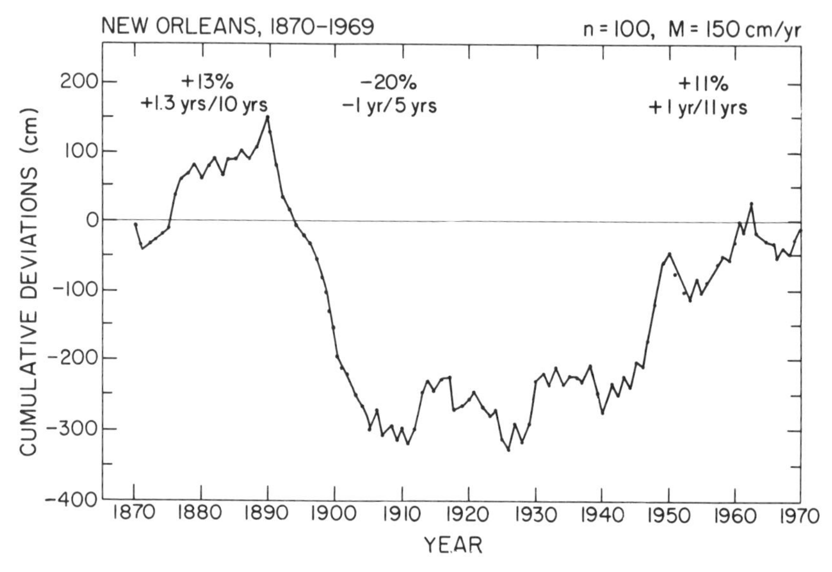

Fig. 15. Cumulative deviations of yearly New Orleans precipitation from the 100-year annual average of 150 cm, 1870–1969.

In spite of the normal frequency distribution, Figure 15 reveals significant short-period climate fluctuations of water supply; a change from +11% to −20% is not negligible when sustained over a number of years. Similar variations, often more extreme, are found in most rainfall analyses. Below-average periods of 20% and more may last on the order of 30 to 50 years and more. Besides, climate anomalies tend to cover considerable areas because they are related to variations of the general atmospheric circulation. Precise causes of general circulation changes over decades, centuries, and longer have not been ascertained, nor can they be predicted reliably. It has been pointed out that in areas with marginal rains, man's activities— his agricultural and grazing customs— have accentuated the spread of deserts. Cutting down of forests, notably equatorial rain forests, also leads to dramatic changes in the water budget; the drainage basins of the Amazon and Congo rivers are especially vulnerable to such cutting, which results in a large impact on climate.

Droughts and Floods within a Season. As noted above, New Orleans has a good chance of rain in every month of the year; hence, annual rainfall is held to narrower limits than those encountered when the lowest year in a century may be well below half the average and the highest year about double the average. On a shorter time basis, of course, a succession of several dry or wet months may be experienced anywhere. Dry spells will be noted, especially during the growing season and in summer, when evaporation is at its highest. Examination of the period 1960 to 1970 for May to September at New Orleans shows that in 6 out of 10 years below-average precipitation lasting 2 or more months in succession did occur. In 1 year, 1968, all months had below-average rainfall. Water storage systems to tide a city over such short-term deficiencies can be, and have been, constructed in many locations. A 15-year drought is quite another matter!

The question may be raised what monthly deficiencies or excesses of precipitation signify. At New Orleans, similar to most other cities, the day with largest rainfall in a summer month with cumulus precipitation contributes generally 25% to 50% of the monthly total, an average near 35% (Fig. 16). This is a very high percentage, and on a few occasions just about the whole monthly rain has come down on a single day, most likely in a single shower. Although hurricanes may easily produce a month's average rain or sometimes even the average yearly rainfall during passage, the high values of Figure 16 result in part from analysis of a single rain gauge that is not representative of, say, a whole city. Preferably one should analyze a small network of stations with rain gauges. In the future, area values of precipitation may be expected to be furnished by radar. Extreme values, thereby, will be reduced. But the general type of distribution seen in Figure 16 remains, indicating that tropical rainfall is not a random process but occurs in well-organized rain episodes sometimes labeled rainstorms. Monthly precipitation above or below average very much depends on whether that one heavy precipitation day will or will not occur!

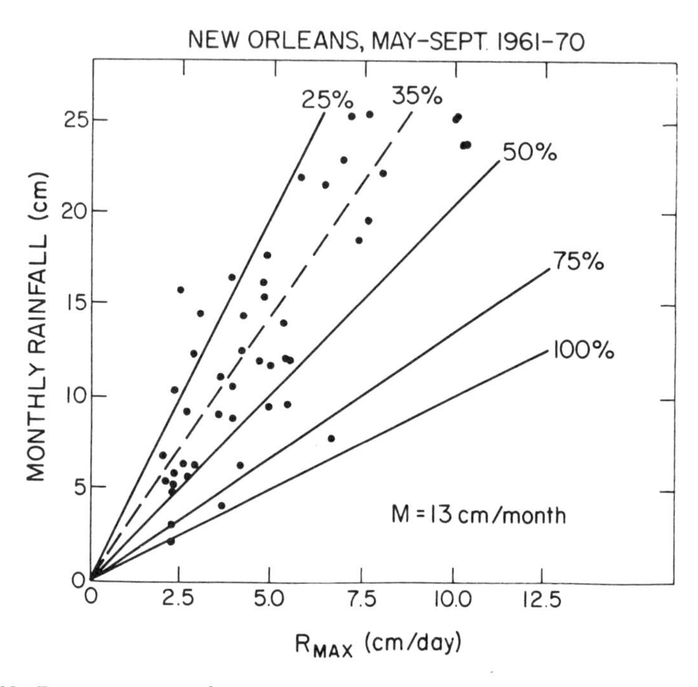

Fig. 16. Percentage contribution of the day with heaviest rain of the month to the total precipitation for the month at New Orleans. Data from 50 individual months, May to September, 1961–1970.

Rainstorms. A "rainstorm day" can be detected easily in a climatological tabulation of daily rainfall for a month, prepared, for example, for all stations in each state of the United States. In most months one immediately recognizes a few days when at least half, and sometimes all, stations report precipitation amounts well above the daily average. Enhanced precipitation at half the stations serves as statistical definition of a rainstorm event. One then finds from weather charts that these rainstorm days are embedded in weather systems that extend 1,000–2,000 km and have a lifetime of several days; these systems are the tropical counterpart of middle latitude cyclones. The precipitation area occupies about one-tenth of the area of such a synoptic system, which is so called because it is visible at one glance at wind and pressure fields over large distances. Nearly all hurricanes originate from synoptic systems.

Duration of a rainstorm at one station usually is 1 to 2 days, seldom 3 or more. In addition, many tropical stations show daily rainfall of a small amount during a month; sometimes there is rain on practically every day. But these small amounts contribute very little to monthly rainfall; if they occur in the middle of the day when evaporation is at its highest, they may be reevaporated at once and thereby not benefit vegetation or water supply at all.

One can see the relative contribution of rainstorms and small daily air-mass rains by ordering the rain events for a month or for several months in ascending or descending order of precipitation recorded. As stated above, rainstorms are best defined from a network of stations. For single stations the value of the definition becomes somewhat marginal. However, the procedure, still valuable in various analyses, was tried out for San Antonio in western Texas, which was compared to Beaumont in eastern Texas close to the Louisiana border. The interesting result was that small rains disappeared almost completely from east to west; almost all rain at San Antonio comes from synoptic systems, even in the middle of the summer season.

At Beaumont, the months of May to September 1976 and 1977 were analyzed together, yielding a sample of 300 days. Precipitation occurred on 93 days, or 31% of the sample, still rather small contrasted to some deep tropical climates. Half of these, or 15% of all days, were rainstorm days as defined above. The number of rainstorms was thirty, an average of three per month as found in many other tropical climates; some months of course had more and some had none. Most rainstorms lasted either 1 or 2 days; thus, 45 out of 93 rain days were rainstorm days; the others were days with small precipitation.

When the rainstorms themselves are ordered from the largest (4.48 in/11.6 cm) to the smallest (0.30 in/0.8 cm), and percent storms versus percent rain is plotted on a cumulative basis, the curve of Figure 17 results. Half of the precipitation attributable to rainstorms was delivered by 20% of the events, or six in number. All of these totaled 2 inches (5 cm) or more per storm. Events with 1 to 2 inches (2.5—5 cm) contributed another 25% of precipitation, while the fifteen lowest cases, or 50% of all storms, brought only the final 25% of the rain. Since the latter, plus small rains, serves mostly as "background noise" that occurs rather regularly, the variations in total rain that decide excess or deficit of available water, are entirely controlled by the fifteen largest storms in the 10 months analyzed. As found in a wide range of climates, the occurrence or absence of one or two events of great magnitude determines whether there is ample, sometimes excessive, or deficient water supply.

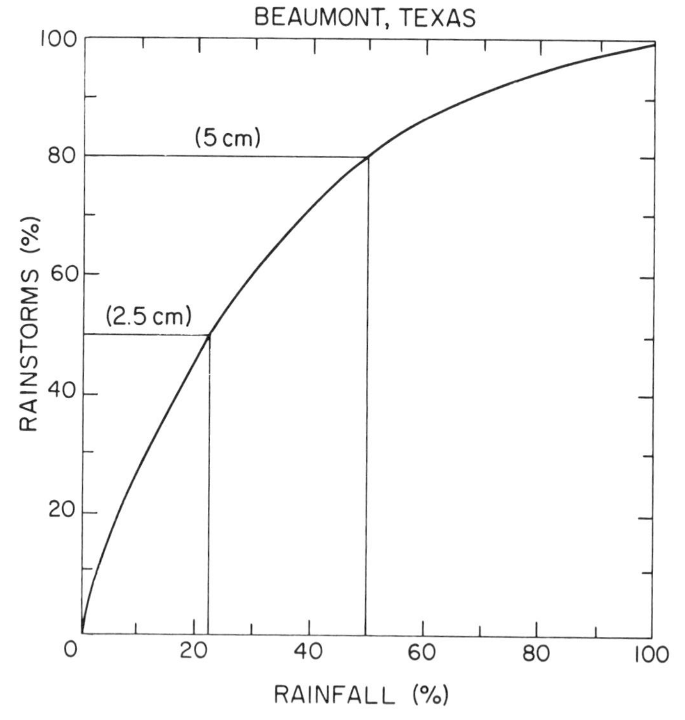

Fig. 17. Cumulative percent rainstorm versus rainstorm events for 10 summer months, May to September, 1976 and 1977, at Beaumont, Texas (30° N, 95° W).

CHAPTER 4

The Origin of Hurricanes

Tropical hurricanes develop from persistent but migratory rain disturbances usually in an oceanic circulation environment like that in Figure 10. These disturbances constitute an array of towering cumulus and associated altostratus clouds covering an area 200 to 600 km in diameter. Some develop in the equatorial zone, where trade winds of one hemisphere converge with the monsoon flow from the opposite hemisphere; often, however, their first appearance over the ocean is within the trade-wind circulation. In these disturbed areas, pressure tends to be slightly lower than in the trade-wind environment, and the air movement often acquires a wavelike cyclonic curvature. Only a small percentage of the migratory disturbances produce dangerous windstorms. However, when the processes at work in this benign setting cause surface pressures to fall, the wavelike character of circulation first grows in amplitude and then forms a nearly circular vortex, and winds tend to increase to destructive speeds.

This much has been known about hurricane development almost since the turn of the century. The processes that cause this metamorphosis, however, remain difficult to detail explicitly and completely. Nevertheless, the contributions of technology, especially from improved networks of upper-air sounding stations, surveillance by weather radar, direct probing by reconnaissance aircraft, comprehensive surveillance by weather satellites, and the use of high-speed computers to model and simulate the processes that generate severe storms, have allowed significant milestones to be reached, a few of which deserve mention here. Others will receive broader discussion later in the text.

Milestones of Knowledge about Hurricane Formation

In the 1930s, Gordon Dunn developed a procedure for identifying and tracking migratory tropical disturbances and the progress of their development by following the movement of 24-hour pressure-change patterns.[1] In the 1940s, Herbert Riehl described the dynamics of the wave system that generally embraces and supports the well-developed rain disturbance; he later identified a number of necessary conditions for the development of the wave circulation into a hurricane.[2] In the 1950s, Joanne Malkus and Riehl successfully answered a number of questions concerning energy sources and the transformations required to develop and maintain the hurricane.[3] In the 1960s, the weather satellite supplied for the first time a means of continuously monitoring the movement and development of tropical disturbances and for accumulating a meaningful climatology of them.[4] In the 1960s and 1970s, computer models of the development process provided some additional insight to the problem; but the results were difficult to interpret and of uncertain value because, first, these models almost invariably produced a full hurricane from every disturbance, and, second, problems were encountered in parameterizing the unique contribution of cumulus clouds, both in releasing and redistributing energy, which is vital to the support of the hurricane wind system.[5] In the late 1970s, W. M. Gray and his collaborators used many years of data from aircraft flights through hurricanes to derive a different physical basis for initiating the pressure falls that lead to hurricane development.[6] Despite these milestones and other achievements, there remains perhaps more scientific disagreement or uncertainty concerning the details of the dominant physical processes responsible for hurricane development than in any comparable problem faced by meteorologists.

We shall begin with a discussion of the large-scale setting and the climatology of migratory tropical disturbances in the Atlantic.

Then we shall consider the constraints to motion and the energy changes that are critical to the support of rain disturbances and their metamorphosis into severe windstorms.

Tropical Atlantic Weather Systems

In the so-called doldrums, a region we shall refer to as the equatorial trough, and in the immediately adjacent trade-wind region, the weather satellite nearly always shows an abundance of rain clouds, sometimes grouped in what appears to be randomly spaced clusters and other times in long lines. Most are short-lived and disappear as synoptic-scale events within 24 to 36 hours. Others, however, are migratory rain disturbances, 200 to 600 km in diameter. They move westward in the trade winds about 7° of longitude per day and may be tracked from satellite photographs for 5 to 10 days. Figure 18 is an example of a procession of disturbances extending across the Atlantic and the Caribbean Sea, some of which have developed into tropical cyclones. These migratory systems do not have a common source; a quite remarkable source region is the continent of Africa. Others, conceived at higher latitudes, may be the product of an upper-level trough or pressure center, spun off from the prevailing westerlies. Some form along the trailing edge of old cold fronts or shear lines.

The National Hurricane Center in Miami maintains a continuous watch of the persistent cloud systems. A migratory disturbance that maintains its identity in successive satellite pictures for a minimum of 48 hours, irrespective of its source, is given a number and is tracked as a *hurricane seedling*—a migratory rain disturbance in search of an environment that may transform it into a storm or hurricane.[7]

Atlantic Sources of Seedling Disturbances

In the eastern Atlantic, the seedling rain system is generally, but not always, connected with and supported by a wave disturbance in the trade winds. A disturbance that has its origin in the equa-

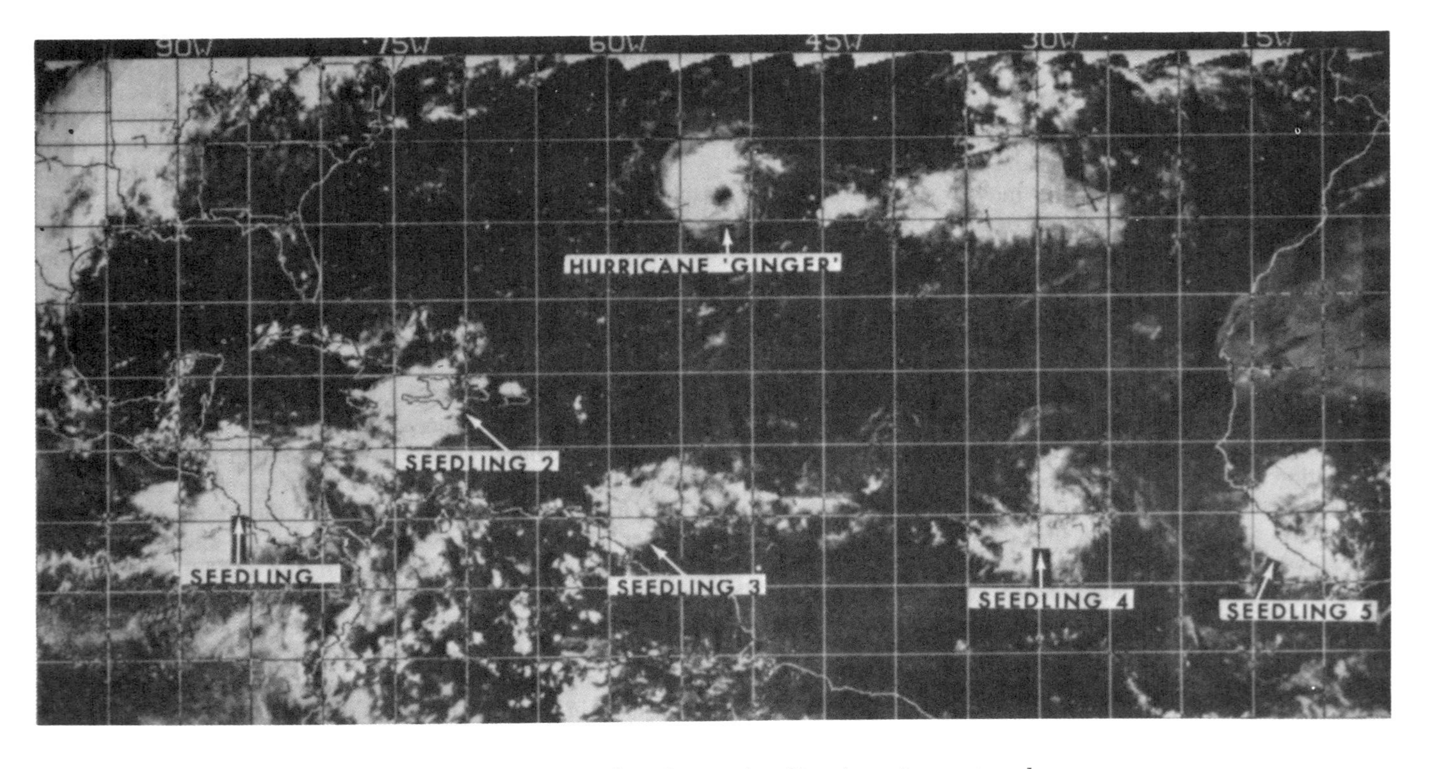

Fig. 18. A procession of migratory rain disturbances (seedlings) moving westward across the tropical Atlantic. One seedling has become hurricane Ginger, now at a higher latitude. R. H. Simpson, "Hurricane Prediction: Progress and Problem Areas," *Science*, CLXXXI (1973), 899–907.

torial trough, an upper-level trough or vortex from higher latitudes, or an old cold front may generate its own wave disturbance and then languish and meander until overtaken by a larger-scale wave disturbance emanating from Africa.

Figure 19 shows the origin of seedling disturbances in the Atlantic and Caribbean regions observed during 1977. These disturbances were tracked from satellite pictures and by analysis of wind soundings at Dakar, Barbados, and San Andres. It is interesting to note that the majority of hurricanes that form in the eastern Pacific can trace their origin to Atlantic seedlings. Figure 20 shows the census of such disturbances in 1977.

Climatology of Atlantic Disturbances

The climatological statistics in Table 4 are of particular interest. This shows an average of 104 tropical systems, ranging from 85 to 113, moving across the Atlantic or Caribbean each year, the standard deviation being only 5% of the mean. Of these, however, only 9 became dangerous windstorms, but with a standard deviation

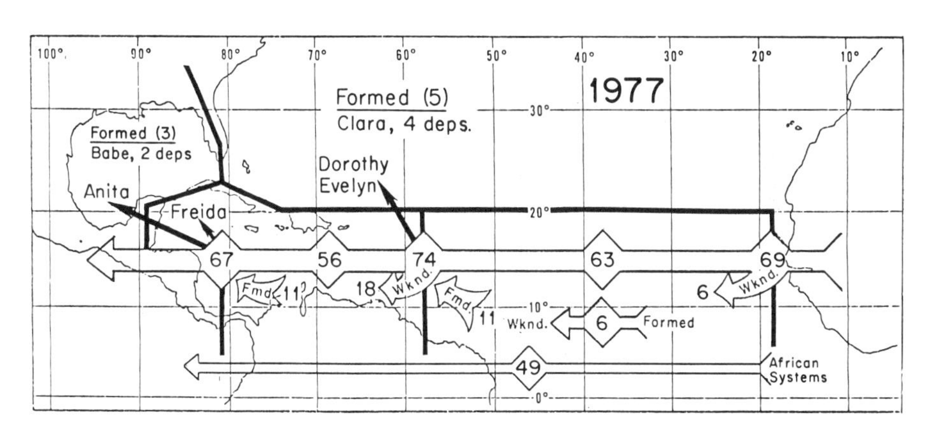

Fig. 19. Summary of Atlantic tropical migratory disturbances for 1977 that formed, disappeared, or transited one of three source regions; Africa, the tropical Atlantic, and the Caribbean. Those that developed into hurricanes are identified by name. After N. L. Frank and G. Clark, "Atlantic Tropical Systems of 1977," *Monthly Weather Review*, CVI (1978), 559–65.

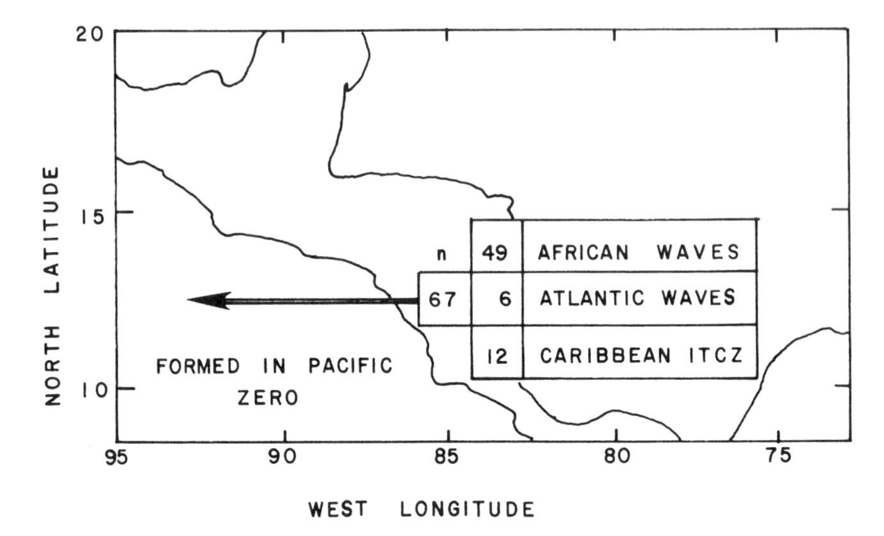

Fig. 20. Summary of tropical disturbances of 1977 from each of three Atlantic area sources that survived to become hurricane seedlings in the eastern Pacific Ocean. All 1977 hurricanes in this area developed from Atlantic area seedlings. Adapted from N. L. Frank and G. Clark, "Atlantic Tropical Systems of 1977," *Monthly Weather Review*, CVI (1978), 559–65.

Table 4. Sources of Atlantic Tropical Weather Systems, 1968–1977

	10-year average	*Range*
Source Region		
Western Africa	59	52–69
Western Atlantic	59	44–74
Western Caribbean	53	40–67
Total Systems	104	85–113
Development		
Depressions	24	22–34
Named Storms	8	4–13

NOTE: Numbers refer to systems that develop in or complete a transit of the region designated as a source (see Fig. 19).

SOURCE: After Neil L. Frank and Gilbert Clark, "Atlantic Tropical Systems," *Monthly Weather Review*, CVI (1978), 559–65.

50% of the mean! With such a regular procession of seedlings, why are there so few hurricanes? And why the great year-to-year variability? These are intriguing questions to which we must continually return in our discussions of source regions for hurricanes, of the environmental processes that nurture and develop the seedling, and of factors that influence hurricane recurrence.[8]

At this juncture, we need to step back out of the forest depths to gain a clearer perspective of the setting in which its trees have grown. We begin with the fundamentals of large-scale atmospheric motions.

Initial Development: Dynamic Constraints

The Earth's Rotation. As illustrated in Figure 21 for the Northern Hemisphere, the earth's rotation about its axis may be divided into two components: the component A, parallel to the earth's surface, and the component B, perpendicular to the surface. The component A, which acts in the vertical plane in the direction of gravity, does not concern us here. Component B, with rotation on the earth's surface, is indeed capable of developing rotating wind systems. Consider a rather large area on the earth's surface initially

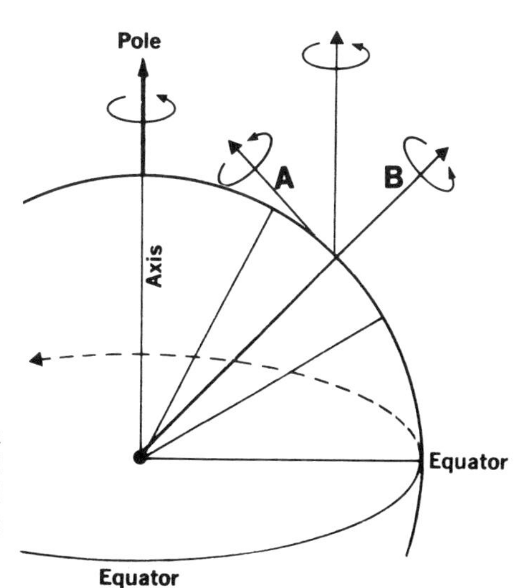

Fig. 21. Cross section through earth showing axis and earth's rotation, and components of rotation parallel and perpendicular to the earth's surface at an intermediate latitude.

without wind and well removed from the equator (Fig. 22). If the boundary is arbitrarily contracted (Fig. 22a), the mass above the area will start revolving cyclonically or counterclockwise; if its area is increased by outward motion, it will revolve anticyclonically or clockwise. The same effect can be produced by a ball rotated with a string: If the string is shortened, the ball will rotate faster in the initial direction; if it is lengthened, it will slow down.

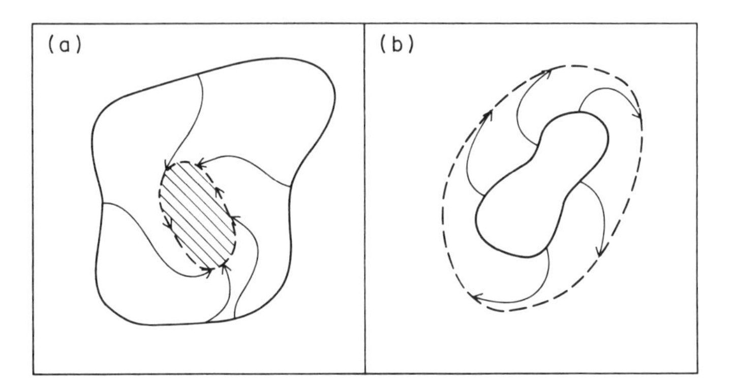

Fig. 22. Development of rotating wind systems during *a*, contraction, and *b*, expansion; *hatching* denotes residual after contraction from initial outer boundary.

The cyclonic or anticyclonic circulations denoted by the broken lines in Figure 22, when averaged over the entire areas which they enclose, indicate what is known as the *vorticity* of these areas. But, this vorticity of the winds must be relative to the earth. When the earth rotation itself and the *relative vorticity* are added, we have the *absolute vorticity*, so called because this is what an observer from space would see when looking upon the earth. From the elliptical shapes of the broken lines, or the envelopes around the rotating masses in Figure 22 one sees that two aspects of the wind field contribute to relative vorticity: shearing motion illustrated in Figure 23a and revolving motion in Figure 23b. With these conventions one can then also define vorticity of the wind field at a point.

Shear and curvature need not act in the same sense. A combination of cyclonic curvature and anticyclonic shear may result in zero relative vorticity. In Figure 22a, for instance, cyclonic flow curvature is largely canceled by anticyclonic shear—the rotational wind speed increases inward. Thus, relative vorticity, in contrast to low and high pressure, is not automatically discernible from weather charts. It is always of value to compute vorticity charts, an arduous task in former days but readily accomplished by modern computers.

It is evident now that the B component of the earth's rotation, shown in Figure 21, is a very necessary ingredient if the cyclonically revolving flow of Figure 22a is to be started. On the equator this component must be zero, because the earth's surface there is parallel to the earth's axis. As the inclination of the surface increases with latitude, the component increases. It is called the *Coriolis parameter*, denoted by

Eq. 4.1 $$f = 2\omega \sin \varphi$$

where ω is the rotation of the earth about its axis in angular measure, that is, 2π radians per day or, in common usage, 7.29×10^{-5} per second. Latitude is denoted by φ. Its sine ranges from zero at

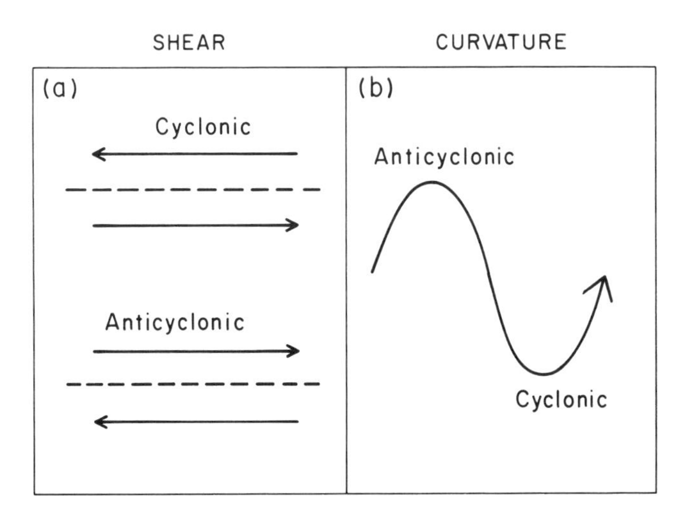

Fig. 23. Shear and curvature contributions of winds to relative vorticity in Northern Hemisphere.

the equator to 1 at the pole (90°). It is 0.5 at latitude 30°, roughly the poleward limit of the zone in which tropical hurricanes form. At latitudes 5° to 6° $\sin \varphi \cong 0.1$, and this is the lower limit for development of most, though not all, hurricanes.

Centers of Relative Vorticity. As knowledge of the atmosphere advanced, it became obvious that tropical cyclones, especially those of hurricane intensity, are very rare indeed, if the two criteria cited—a heavy cloud mass and distance from the equator—are sufficient to determine the birth of such storms. These criteria are met very frequently; thus one or more inhibitory factors must be acting to keep the number of hurricanes down. In the 1920s Norwegian meteorologists suggested, analogous to their ideas about the development of cyclones outside the tropics, that inside the tropics there must also be found occasional wind systems with shifting directions, counterclockwise in the Northern Hemisphere, and that hurricanes only form along these lines. Such wind shifts, often called a *shear line*, indeed exist, near the surface and most frequently along the equatorial trough of low pressure. Here, since only the shearing motion of Figure 23a is observed, the relative vorticity will be cyclonic and the absolute vorticity will be greater than just the earth's rotation alone. It is for areas of cyclonic or positive relative vorticity that one inspects weather charts as a mark of potential locations of storm formation.

Now it is not quite evident why such areas should be specially favorable, since in Figure 22a we were able to create a cyclonic vortex from the earth's rotation alone. But it must be remembered that this development was forced by introducing contraction of the starting envelope. In nature, a mechanism for the contraction must be found—a subject that still eludes full comprehension.

Here we return to the point made above, that absolute vorticity is enhanced over the Coriolis parameter in areas with positive relative vorticity. A fundamental theorem, due to Carl-Gustav Rossby, stated that

Eq. 4.2
$$\frac{1}{\zeta_a} \frac{d}{dt} \zeta_a = -\operatorname{div}(V)$$

Here ζ_a is defined as the absolute vorticity; d/dt denotes the change with time, as one follows a particular mass of air; and the division by ζ_a on the left side indicates that the change is on a percent basis. On the right side is the horizontal divergence from Chapter 3; Equation 4.2 shows our major reason of interest for div (V). From the equation, absolute vorticity will increase with convergence and decrease with divergence. If we further consider the surface layer of air and recall Equation 3.1, we can also say that increasing absolute vorticity will be coupled with ascending motion, decreasing vorticity with descending motion. Thus, where absolute vorticity increases, deep cloud masses as shown in Figure 7 *may* (not *must*) be able to develop releasing condensation heat. This is helpful when we wish to consider the condensation mechanism an energy source for hurricanes. As may be noted right away, the larger the area over which convergence occurs and scattered cloud masses appear, the lower their potential for generating storms will be. Concentrated, solid cloud masses, in contrast, are a favorable indicator.

On synoptic charts, one often finds that cloud masses as seen by satellite are situated where areas of cyclonic relative vorticity are observed near the earth's surface. This need not mean that a vorticity increase is taking place at map time, but that it has taken place some time before, while air was moving from low to high vorticity in conformity with Equation 4.2. These observations place us in the position of locating the potential areas of hurricane development on each day.

Balanced Flow. The difficulty in finding a mechanism for contracting the envelope sketched in Figure 22a is compounded by the fact that air by and large seeks to move in a "balanced" state. From Newton's laws of motion, acceleration will be proportional to the force exerted on any given mass. But, when there are several forces, they may cancel, so that acceleration will be zero. In horizontal motion in the atmosphere, air is accelerated only by the *pressure force*; that is, the pressure difference over a certain distance accelerates air from high to low pressure. However, air practically never moves directly across isobars toward low pressure; other forces interfere. For the present, we shall largely ignore the friction force exerted by the earth's surface, which retards or decelerates air. But, we may note that surface air in the trade-wind areas, for instance, tends to move toward lower pressure at a small angle of just the right size, so that acceleration by the component of the pressure force downwind precisely cancels the frictional force opposing the motion (Fig. 24a; see also Fig. 10). Through such balance the trade winds manage to keep moving at constant speed; they are known as the steadiest wind system of the globe.

The most important force is provided by the earth's rotation (component B in Figure 21). As air starts to be accelerated toward lower pressure, the earth underneath and the low-pressure therewith rotate counterclockwise, so that the air aiming at low pressure misses the center and arrives on its side: when moving northward in the Northern Hemisphere, it will arrive to the east of the low pressure; when moving southward, to the west. The relation between the pressure force and the earth turning can be so finely tuned that the air does not move toward lower pressure at all but along the isobars (Fig. 24b); as in Figure 24a there will then be no acceleration. A wind balanced in this way is called *earth-turning*, or *geostrophic*. Really, a balance of forces in the Newtonian sense is not involved and would not be seen by an observer in space. However, to an observer fixed on the surface of the earth it appears that the air has been deflected to the right—eastward for northward motion and westward for southward motion. He may imagine, therefore, a force as pictured in Figure 24b as having accomplished this right deflection, a concept solidified by the physicist Coriolis of the nineteenth century and hence called *Coriolis force*. The magnitude of this force is fV, f being defined in Equation 4.1 and V being wind speed. Thus, the force increases with latitude and with wind speed. It balances the pressure force almost completely in middle and high latitudes, greatly hindering the type of contraction shown in Figure 22a. One used to think that the contraction should be easier to accomplish in the tropics, where the Coriolis force becomes small at small values of f. However, because

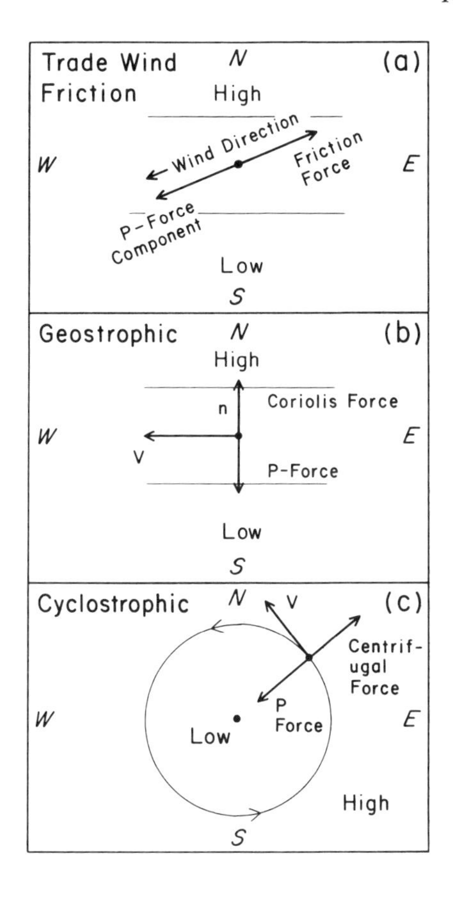

Fig. 24. Models of three balances of forces in horizontal motion.

the pressure force normally decreases with the Coriolis force, contraction is not facilitated.

Figure 24a and b show balance of forces in straight flow. In contrast, Figure 24c shows the balance in the hurricane with strongly curving flow. Once formed, the winds revolve around the low-pressure center at great speed; at small distances from the core the Coriolis force may be neglected. The pressure force strongly accelerates the air toward the center. But, in trying to reach the low pressure, the air is forced to turn counterclockwise in the Northern Hemisphere. This turning again may take place at such a rate that in actuality the air does not approach the core but revolves around it at constant pressure, in the case of a symmetrical storm. Such a wind is called *cyclostrophic*; comparison of geostrophic and cyclostrophic winds shows at once that the same principle is involved: the rotation—of the earth in the one case and of the revolving ring of air in the other—takes place at such a rate that the air always misses the low pressure center! Its speed remains constant. The acceleration corresponding to the pressure force in the hurricane is the *centripetal acceleration*. But on the surface of the earth an observer has the sensation of a centrifugal force. The situation can be likened to the experience of turning a sharp corner. In this case the pressure force is absent and, for balanced motion around the curve, we must compensate by leaning sideways, so that a component of gravity acts perpendicular to our bodies. In case of car travel, the balance of forces is readily accomplished by tilting the roadbed as is always done in constructing circular or oval auto race tracks.

In summary, we have encountered strong evidence that air is not easily accelerated in the sense that its speed changes. On the contrary the atmosphere contrives any number of ways to hold such acceleration to a minimum and to operate entirely through the centripetal acceleration. The formulas for the balanced winds described are the following.

Geostrophic case:

Eq. 4.3
$$fV = \frac{1}{\rho}\frac{dp}{dn}$$

where the left side symbolizes the Coriolis force and the right side the pressure-gradient force, p stands for pressure, ρ represents air density, and the coordinate n points from low to high pressure for conformity with the next equation.

Eq. 4.4
$$\frac{u^2}{r} = \frac{1}{\rho}\frac{dp}{dr}$$

Here r is the radius of the cyclone, positive outward from the center, and u is rotating velocity. The centripetal term u^2/r typically contains the square of the velocity.

Thermal Wind. Equations 4.3 and especially 4.4 are required to introduce another subject of great importance for the hurricane, namely, a way to deduce the temperature from pressure and wind fields. Finding such a method will be of particular benefit for analyses of low-latitude weather systems. Temperature gradients in the deep tropics are so small as to be well within range of instrumental error in reading temperature differences, especially from varying types of balloon sondes.

Consider first the geostrophic case at constant latitude. We can ignore variations of density for this discussion; in actual computations they can, of course, be readily included. Looking downwind in the Northern Hemisphere, pressure drops from right to left across the wind from Equation 4.3—Buys Ballot's law. For V increasing upward, the difference between the high pressure to the right and the low pressure to the left must also increase with height. Thus, the slope of isobaric surfaces becomes greater with height, and the vertical distance between isobaric surfaces must be greater to the right than to the left of the wind, sketched in Figure 25a.

Now, increasing vertical distance between two surfaces of constant pressure denotes lower density since density = mass/volume. The mass between two constant-pressure surfaces is a con-

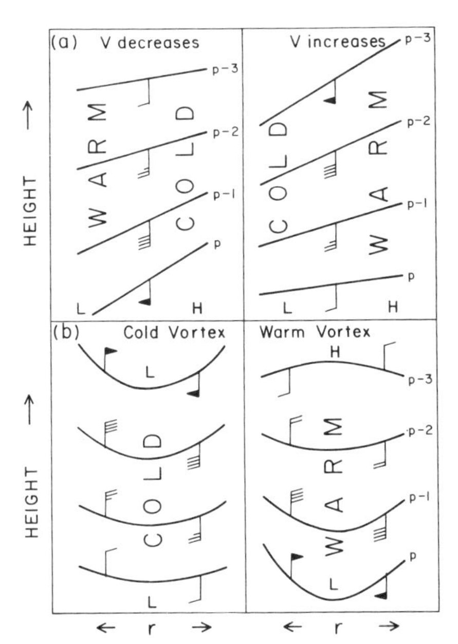

Fig. 25. Vertical wind shear in relation to distribution of cold and warm troposphere temperatures in Northern Hemisphere. *Top*, straight flow; *bottom*, vortex flow.

stant by definition of pressure. When the constant mass is spread over a larger volume—here only vertical distance is considered—the density must decrease. Following the ideal gas law, here applicable,

Eq. 4.5
$$p = R^* \rho T$$

where T denotes temperature and R^* is a characteristic constant of the gas; p and ρ have already been defined. If we vary this expression at constant pressure or, as in our problem, between constant-pressure surfaces, the percent changes of temperature and density are inverse, so that

Eq. 4.6
$$\frac{\rho'}{\bar{\rho}} = -\frac{T'}{\bar{T}}$$

where the overbars indicate general mean values and the primes deviations. Thus, when density becomes lower, the temperature will become proportionally higher.

We can now conclude that in Figure 25a temperature must be higher to the right than to the left of the wind direction in order for the slopes of isobaric surfaces to change slope with height, as needed for geostrophic balance when wind increases upward! We can form a thermal analogue to Buys Ballot's law: When wind speed (at constant direction) increases with height in the Northern Hemisphere, temperature will be higher to the right than to the left of the wind, looking downstream. When the wind decreases with height, temperature will be higher on the left than on the right.

We have achieved the objective of deducing the temperature field from pressure and wind fields, indeed from the wind field alone. Given the wind at any altitude, the pressure gradient is computed from Equation 4.3. Then the temperature gradient is determined from the pressure gradient at various heights—a procedure that we need not follow in detail quantitatively. The wind, the highest quality upper-air observation made in the tropics with available instrumentation, has made it possible to deduce the important features of the temperature field indirectly. The method here described is used widely in practice, though some care must be exercised, especially in curving flow.

For an assessment of the temperature field in strongly curving flow, let us make a parallel analysis with Equation 4.4. An inspection of Equations 4.3 and 4.4 shows at once that the same chain of reasoning applies, except we must consider changes of u^2 rather than V with height at any radius *r* from a hurricane center or from any weaker center as long as Equation 4.4 remains reasonably correct; *i.e.*, we need not include the Coriolis force. The magnitudes of Coriolis and centrifugal forces in hurricanes are compared in Table 5. For a tropical revolving storm in either hemisphere we can state directly: When the rotational speed u increases upward, the core of the storm will be cold relative to its outskirts. This statement is of immense value in judging the structure of tropical weather systems concerning their capability of acting as hurricane starters. For a hurricane to develop, the air inside an area of low pressure and cyclonic vorticity must be warm; the vorticity must decrease upward, and so must the low pressure, preferably going over into high pressure in the high troposphere and facilitating outflow from deep convective clouds (Fig. 25b).

We can push our insight one step further. If any rotational circulation is to have a lifetime beyond a few hours—hurricanes may last a week or more—balanced flow and thermal wind must be valid to a high degree of approximation. Otherwise the temperature field would not hold together, and the system would collapse quickly or never form—the normal case. If the temperature field is to hold together and the storm is to remain alive, a vertical shear of

Table 5. Comparison of the Magnitudes of Coriolis Forces (fu) and Centrifugal Forces (u^2r^{-1}) in Hurricanes, Expressed as 10^{-4} m sec^{-2} for Latitude 20°

u (*mps*)	*Coriolis force*	*Centrifugal force*					
		Distance from center (km)					
		20	*40*	*60*	*80*	*100*	*150*
10	5	50	25	17	13	10	7
20	10	200	100	67	50	40	27
30	15	450	225	150	112	90	60
40	20	800	400	267	200	160	107
50	25	1250	625	417	314	250	167

the rotating wind component must exist in order to make maintenance of the right kind of temperature field possible. This means that the storm must possess a mechanism to create the correct wind shear. The mechanism for the mature storm will be discussed in Chapter 6. But, for assessing the quality of the initial weather systems we should base part of our judgment on their vertical structure.

Initial Development: Energetic Constraints

The balanced state described cannot be completely valid, or else large storms and rain could never occur. An "imbalance" also exists, though it is small compared with the balanced part of motion, certainly in the initial stages of a hurricane, and therefore it is often difficult to identify. One may approach the matter from dynamic or energetic standpoints; here the energetic approach has been chosen.

Given $K = V^2/2$ as the kinetic energy per unit mass and d/dt as the time change following this mass, as in Equation 4.2, the generation of kinetic energy or energy of motion is given by

Eq. 4.7 $$\frac{dK}{dt} = \text{Generation by pressure forces minus friction}$$

In Figure 24, top, a balance between pressure and frictional forces shows that the velocity of the trade wind remains constant. A corresponding diagram can be prepared whereby the generation by pressure forces balances the dissipation by friction, so that kinetic energy remains constant (Fig. 91). If the two opposing terms are not quite equal, increase or decrease of kinetic energy results. Friction in most instances *responds* to other actions; it normally detracts from kinetic energy and, near the surface itself, is known to be proportional to V^2. Our principal interest, therefore, lies with the energy-generating term.

Potential Energy Release. In case of frictionless flow

Eq. 4.8 $$\rho \frac{dK}{dt} = -V \frac{dp}{ds}$$

where the coordinate s points downwind and the minus sign indicates that a gain of energy will result if pressure decreases along s, or along the wind. From a fundamental knowledge of physics, we know that energy in all its forms is conserved. Hence, when energy in one form—here energy of motion—becomes larger, a supply of another energy form must be available for conversion. In one of the simplest physical systems, potential energy may become free or available from the coexistence, side by side, of warm and cold air on the two sides of the wind-shear line in Figure 23a. Such coexistence denotes a state of potential energy raised above an arbitrary reference level such as the solid surface of the earth, which serves well for this purpose. Because of what we saw from Equations 4.5 and 4.6, the center of gravity does not correspond to the lowest possible position when cold and warm air lie side by side. The latter would be obtained when the cold air lies below the warm air. Thus, when cold air starts to sink below warm air, the center of gravity is lowered and potential energy is freed for conversion to energy of motion.

A strong case can be made for release of potential energy as a principal trigger in hurricane development. There are always cold air infusions from higher latitudes into the tropics, most effective in narrow channels, as depicted in Figure 40. Presence of such an upper-atmosphere channel alongside, but not above, an area with cyclonically rotating winds at the surface is strongly indicative of intensification when the cold air aloft starts to sink. Note, however, that the cold air must not be situated at the surface, like a cold front in middle latitudes, and must not enter the tropical center. Destruction of any initial warm core and immediate collapse of the winds result from such entry, as often happens.

Release of Latent Heat. As noted in the discussion of Figure 5, most release of latent heat of condensation in ascending masses of clouds is used to re-create the mean tropical atmosphere. Only the small temperature difference between this average atmosphere and the ascent with condensation in Figure 5 would be available for a direct energy infusion to an area with cyclonic winds through a small rise in temperature. In later stages of development this rise does occur and indeed plays a significant role. But experience shows that it cannot be used as a starter for two reasons.

1. Rising cloud masses tend to mix with their surroundings. They entrain air in the middle atmosphere from outside, especially dry air. Through such mixture the temperature of the rising cloud-forming air approaches that of the mean atmosphere and is no longer warmer. It may rain heavily for a long time, as much as 20 inches (50 cm) in a day; but no wind system develops unless there is release of potential energy—warm air rising and cold air sinking. In the great majority of rain events this does not happen. Flooding may be extensive, but there will be no trace of a hurricane.

2. The ascent of air to the high troposphere (12 km) takes place mainly in cumulonimbus clouds or, with preference, in groups of such clouds. The area covered by such groups is small, perhaps 50 km across. They are called *mesoscale cloud systems*; this scale is intermediate between that of single clouds and large storms of 1,000-km extent.

In areas with cyclonic winds at the surface, the groups occur in close proximity. Each is trying to become large and thus the groups compete with each other. As a result, none succeeds in attaining overall control; in the end they all die. No energy release has been achieved toward unification of all groups into a single revolving vortex. The air thrown up in the tall towers sinks again in the spaces between cloud groups. For an effective starter we need an initial favorable flow and thermal arrangement, for release of potential energy to occur. Once such a field has come into existence, we can ask how release of latent heat can be brought effectively into play.

Based on this discussion, we shall now proceed to look at a small but representative part of the great variety of possible hurricane-generating synoptic systems in the tropics and try to identify which of them may be regarded as promising seedlings. Indeed we find a number of such areas around the globe in the tropics on every day; they extend over several hundreds of kilometers, sometimes up to 1,000 km. Out of the large total, not more than perhaps 1% to 2% become tropical cyclones; but we must look at all of them on every day.

Synoptic Weather Systems

Equatorial Trough of Low Pressure. The most probable place for the initiation of hurricane seedlings is the equatorial trough, where during the summer rainy season low pressure and cyclonically shearing winds coexist in many areas of the tropical belt on most days. The low-pressure trough migrates, following the sun with a lag of 1 to 3 months, greater in the Northern than in the Southern Hemisphere. Excepting the Indian Ocean monsoon area, the north-south migration of the trough is less than that of the sun (45°), ranging from as little as 5° to 10° latitude to as much as 20° to 30°.

Over western Africa, in spite of monsoonal seasonally reversing winds, the latitude of the trough zone ranges only from about 5° N along the Gold Coast in winter to 15°–20° N in summer. The trough zone slopes strongly southward with height from its surface position, indicated near 17° N in the example of Figure 26, top, to the 600-mb (4 km) position near 10° N (Fig. 26, bottom). The strongest westward-moving weather systems are found mainly between these latitudes. The lower monsoon winds from southwest reverse to strong east winds beginning at about 1.5 km. High speeds of 50 kt and more are attained at 600 mb. Thus the east wind increases strongly with height. We may conclude from the last section that temperatures are higher in the north than in the south; temperature differences in this region are large enough that they are also apparent from direct temperature measurements.

A strong rain system in Figure 26 is located between the surface and 600-mb trough positions near longitude 5° E. Such centers travel westward at variable speeds, from 15 to 35 kt. The very fast ones, usually associated with strong wind squalls at the surface, normally last no longer than 1 day. Many of the slower ones, with an average frequency of about one in 5 days, live to reach the Atlantic Ocean. Figure 27 contains an example of such a system off the African coast near the Cape Verde Islands. This type of vortex may then cross the Atlantic Ocean toward the Caribbean. By this time it would be embodied in the trade winds and would no longer be part of the equatorial trough zone that slopes southwestward

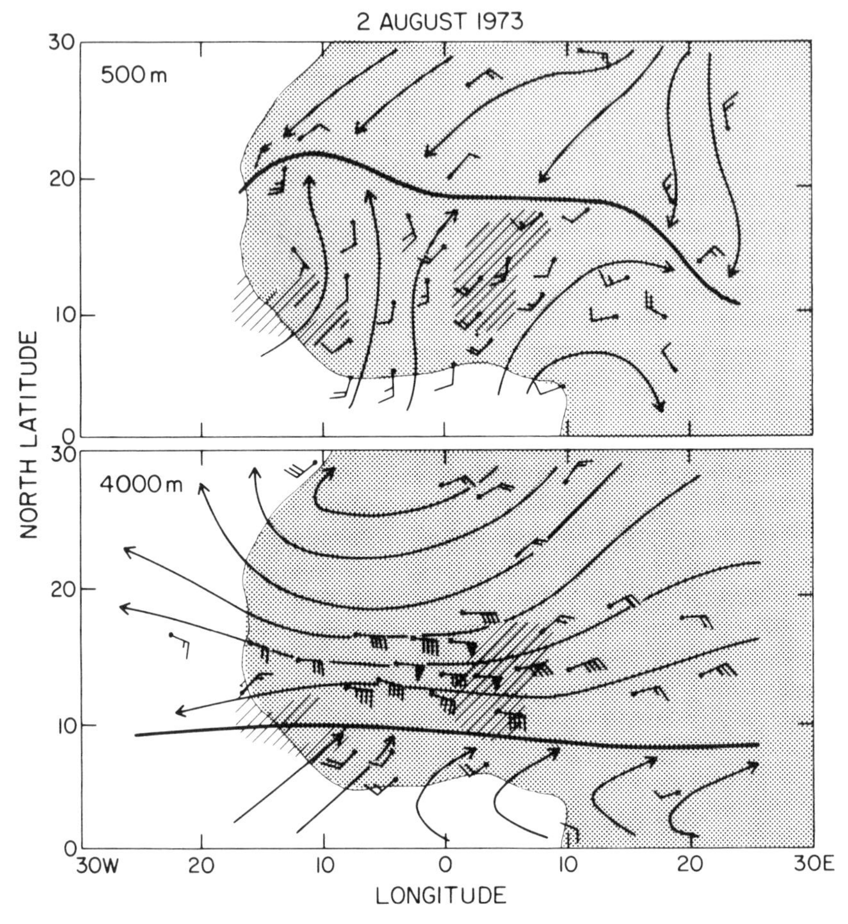

Fig. 26. Winds over West Africa at 500-m and 4,000-m altitudes on 2 August 1973. *Heavy line*, equatorial trough; *hatching*, principal satellite-seen clouds. Winds drawn for direction from which wind blows: *short barb*, 5-kt winds; *long barb*, 10 kt; *heavy triangular barb*, 50 kt.

toward South America (Fig. 10). The occasional hurricanes developing from such vortices have been called, not unreasonably, "Cape Verde" hurricanes. They give the longest advance notice of a potential hurricane threat—a whole week, the time normally needed to cross from eastern to western Atlantic.

Over South America, the equatorial low-pressure trough in the mean remains at a very low latitude, almost at the equator, in the

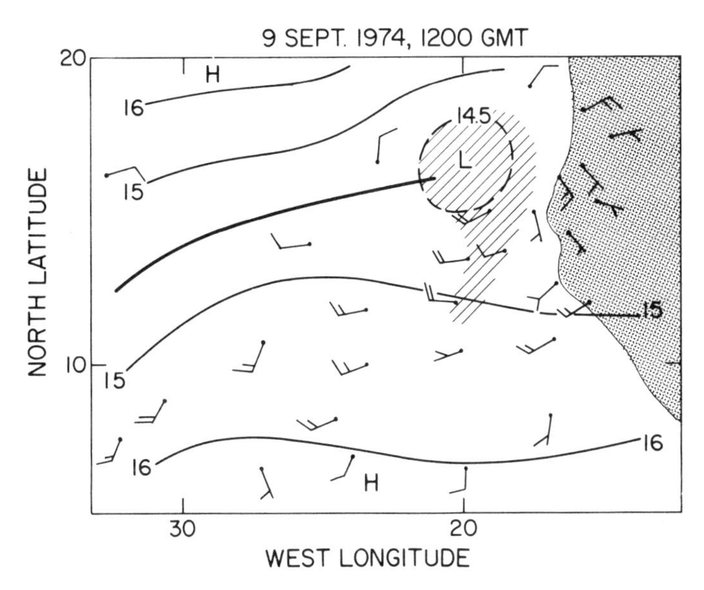

Fig. 27. Surface winds and pressures (mb above 1,000 mb) for 9 September 1974. *Heavy solid line*, equatorial trough; *hatching*, principal satellite-seen cloud.

rainy summer season (Fig. 10). However, the zone may be stimulated to travel northward across Venezuela by the waves arriving from the east and by large trade-wind surges from the Southern Hemisphere. The latter can be followed from satellite-cloud displacements, seen for August–September 1972, in Figure 28. As much as 5 days' warning can be obtained from the photographs of the impending entry of a surge from the south into Venezuela and possibly into the Caribbean. The equatorial low-pressure trough and a sharp cyclonic wind-shear line then appear in central to northern Venezuela. Occasionally, cyclonic centers form along the east-west wind-shear line.

Such centers, while over land, cannot become hurricanes. But they can produce very heavy rains leading to flood situations when they approach the eastern slope of the Andes. A composite wind analysis of several cases indicates a distinct center of rotation near the surface, with strongest winds to the right of the direction of motion (Fig. 29)—typical for hurricanes as well. The center weakens and degenerates into a wave structure already in the layer

750–650 mb. We conclude that with cyclonic winds decreasing, upward temperature should be warmest near the center and become colder on the outside. Passage of such centers into and across the Caribbean has at times been followed by growth to hurricane strength. Moderate precipitation, as much as 10 mm per 3 hours and 50 mm for the whole storm, has been computed for the average case. A maximum storm will release an estimated 100 mm, or double the rainfall from the average storm, into the rivers of the Andes—potentially a flooding case.[9] This information is helpful in designing dams and spillways for containing floods.

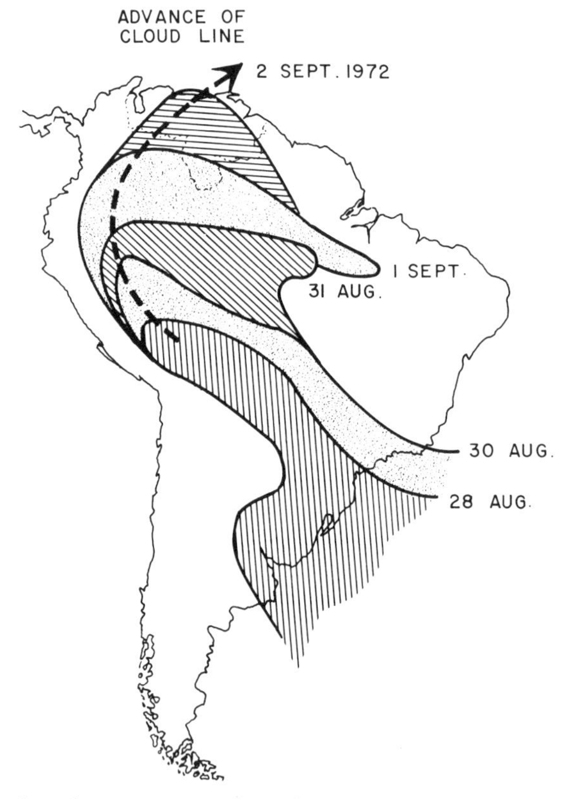

Fig. 28. Northward progression of satellite-seen cloud mass advancing from southern South America to Caribbean, 28 August–2 September 1972. The western edge follows the main cordillera of the Andes. H. Riehl, "Venezuelan Rain Systems and the General Circulation of the Summer Tropics: I. Rain Systems," *Monthly Weather Review*, CV (1977), 1402–20.

Trade Winds. In contrast to the equatorial trough zones, synoptic weather systems such as waves passing through the trade winds move in a regime in which the wind field is normally divergent at the surface (Fig. 12). The waves may be invisible at the surface except for the rain they produce. Whereas in the equatorial trough zone we find low-level winds with component from west overlain by upper easterlies, the opposite structure is normally encountered in the trade-wind areas. The amplitude of tropical waves

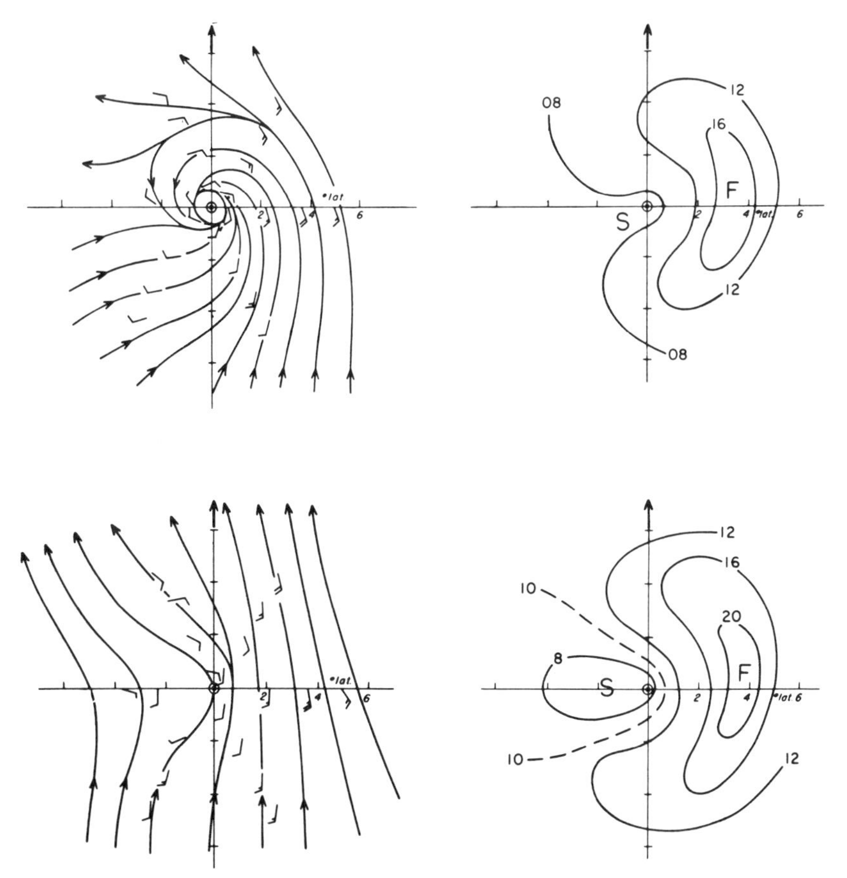

Fig. 29. Left, streamlines and *right*, wind speeds (kt) composited for Venezuelan cyclones with respect to direction of motion (*arrow upward*). *Upper*, layer 950–850 mb; *lower*, layer 750–650 mb. H. Riehl and H. R. Byers, "Computing a Design Flood in the Absence of Historical Records," *Geofisica Pura e Applicata*, XLV (1960), 215–26.

often is greatest in the layer 600–500 mb. They are often joined to troughs in the upper troposphere, as shown in a composite diagram for six waves at the 200 mb level, in August 1969 (Fig. 30). Rate of motion is westward at an average of 14–16 kt (7–8 mps); but sometimes it may be slowed to 10 kt or less. There are two preferred rain areas: north and east of the trough axis near latitude 20° and west of the axis near latitude 10°. But the favorable low-level structure of Figure 29 is *not* present initially. Rather, since the wave amplitude increases upward, the waves possess a cold core, at least to 400 mb—quite the opposite of what is wanted for hurricanes. A strong transformation must occur before a hurricane can be borne. An incipient center will tend to be situated 300–500 km from the cold trough, the potential energy reservoir—to the west in the southern portion and to the east in the northern portion.

Figure 31 shows a variant of such waves in time-section form

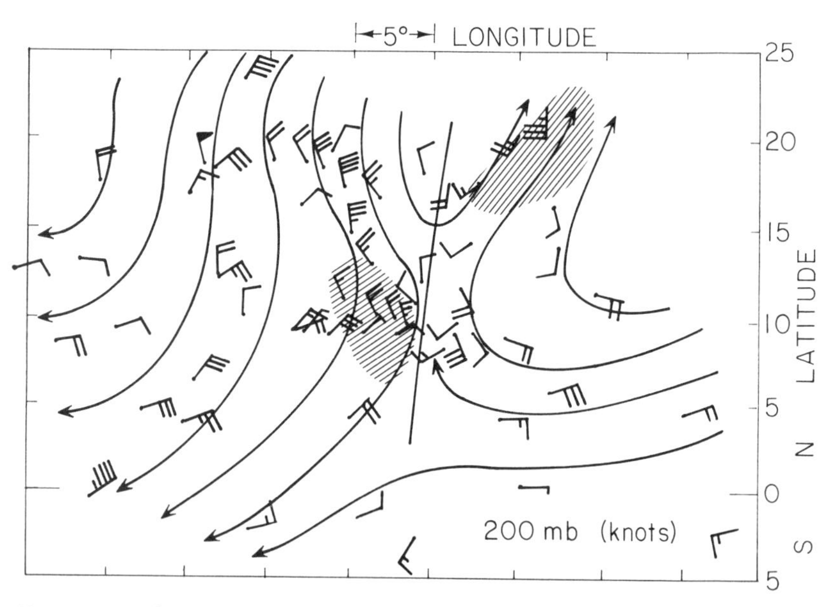

Fig. 30. Winds at 200 mb composited with respect to high tropospheric trough for five cases in August 1969 just prior to the onset of rain episodes in northern Venezuela. *Hatching*, preferred satellite cloud positions. H. Riehl, "Venezuelan Rain Systems and the General Circulation of the Summer Tropics: I. Rain Systems," *Monthly Weather Review*, CV (1977), 1402–20.

passing over the upper-air station of the small island of Grand Cayman in the western Caribbean (19° N, 82° W). As shown by the wind arrows trough amplitude is largest near 500 mb, decreasing both downward and upward. From the wind-shear and direct-temperature measurements the air is coldest near and east of the trough up to 500 mb; a reversal occurs higher up, a frequent case. The rain area, located close to the trough line, is situated largely in the cold air. It is not readily apparent how heating from precipitation can start a hurricane circulation in this setting. With potential energy highest near the trough, sinking of cold air would destroy rather than develop the wave. Such are the difficulties in judging hurricane formation, and it is readily seen, as in this instance, that usually nothing happens.

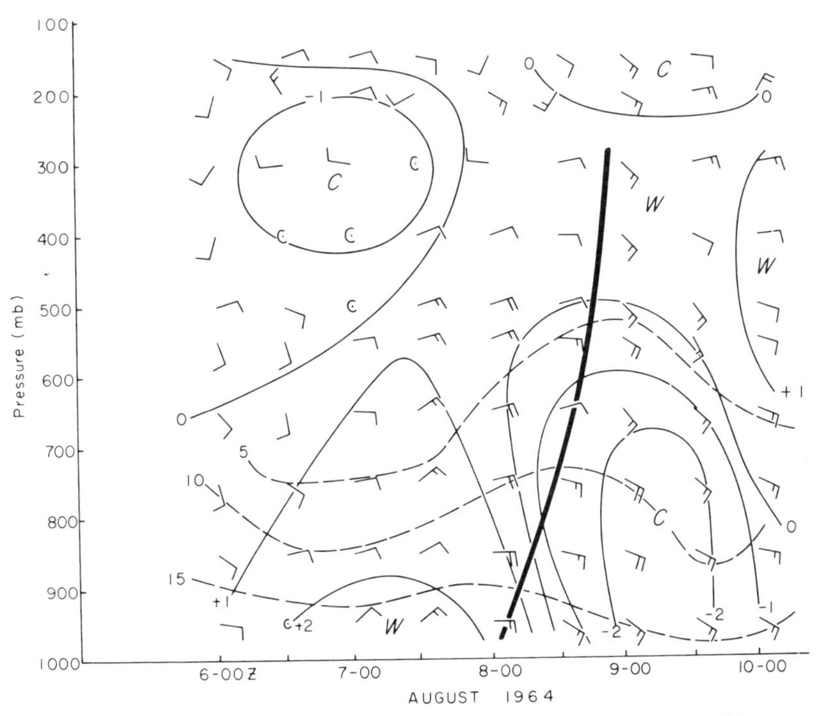

Fig. 31. Vertical cross section of wave moving westward in western Caribbean at Grand Cayman (19° N, 81° W) in August 1964. *Heavy line*, axis; *solid curves*, temperature deviations (°C) from mean at each level; *broken curves*, specific humidity deviations (g/kg).

Figure 32 presents a far more suitable arrangement. A strong upper-air trough with a cold core and high potential energy passed through the center of the United States as a small low-pressure center entered into the southwestern Gulf of Mexico. With the meeting of these two systems and the subsequent rapid sinking of the cold air over the southern United States, an initial energy source was supplied to the tropical center. The center indeed flared up and formed hurricane Audrey, which moved north-northeastward to strike the western Louisiana coast, one of the strongest and most damaging June hurricanes on record. We shall continue to encounter this interplay between weather systems of different areas in the later chapters.

Old Fronts. Fronts moving into the tropics quite often become stationary with a general east-west orientation. The temperature con-

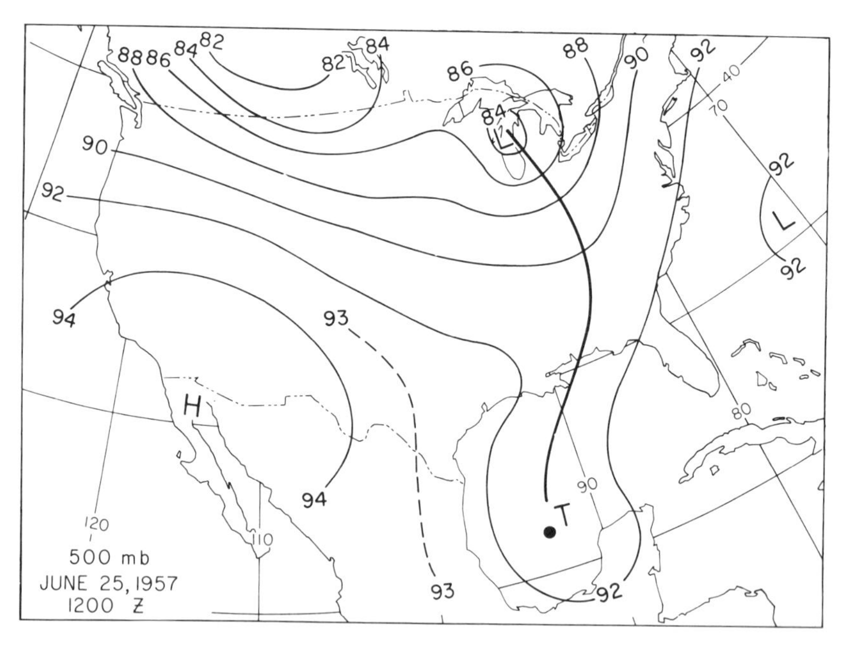

Fig. 32. Heights of the 500-mb surface for the United States and Gulf of Mexico at the time of inception (T) of hurricane Audrey, 25 June 1957. 500-mb heights in tens of meters above 5,000 m.

trast between the air on both sides vanishes, but the cyclonic shear zone remains, often at a rather high latitude in the summer season. A preferred region is the Gulf of Mexico and the adjoining part of the Atlantic Ocean. The shear zone may have a tendency to strengthen and extend upward in the atmosphere. We see a gross and quite rare case in Figure 33. At the surface, three low-pressure centers are indicated: one in the Pacific off the Mexican shore, one in the central Gulf, and one just east of Florida. In those earlier times, upper-air information in no way rivaled that of the present day. But there is a clear indication of a strong cyclonic shear line at 3 km (700 mb) extending across the three centers (Fig. 34). The middle troposphere must be relatively cold to permit the shearing circulation to increase upward. A trough passing eastward

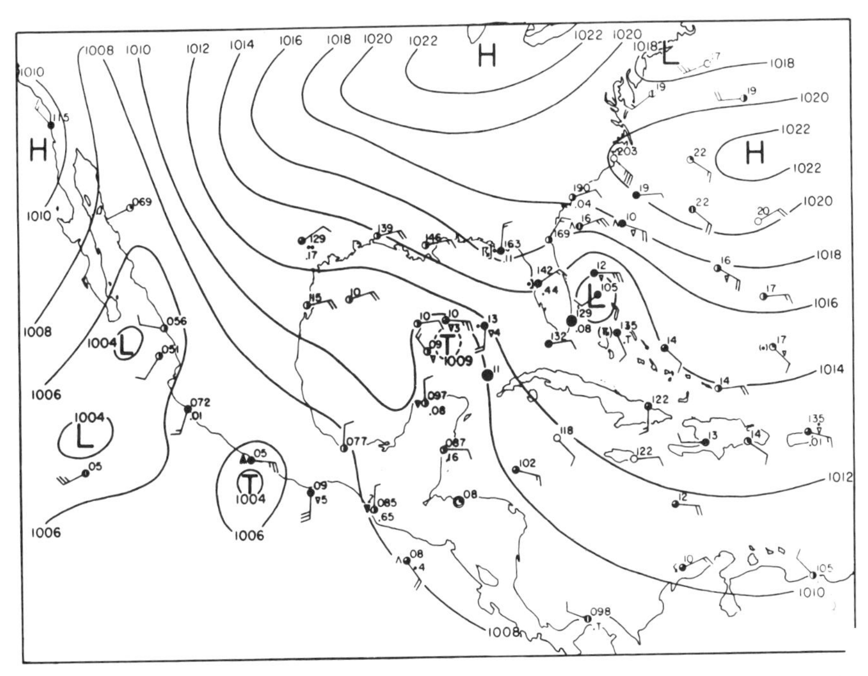

Fig. 33. Surface map showing three tropical low-pressure centers aligned northeast-southwest, 11 September 1941. H. Riehl, "On the Formation of West Atlantic Hurricanes, Part I," Miscellaneous Report 24 (Department of Meteorology, University of Chicago, 1948), 1–64.

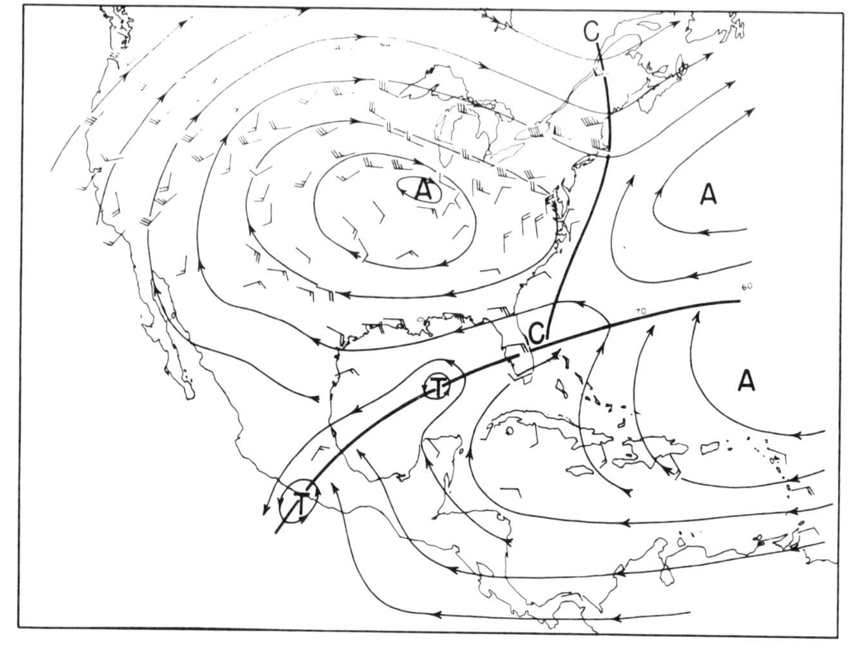

Fig. 34. Winds at 4,000 m for the same period.

in the United States, as in Figure 32, first touched off the surface center in the Gulf and then the one in the Atlantic, again indicating a sequence leading to the brink of a potential hurricane. But, the conditions of June 1957 evidently were not met. The Gulf center went northwestward and entered the Gulf coast as a weak tropical storm; there is no record of the eastern low intensifying at all. Organized sinking of cold air did not take place. The rain areas connected with the surface cyclonic centers extended upward into cool air, at least to the 700-mb level shown.

Evolution of Vortex Stage

While the display of potential hurricane-forming cases around the globe could be continued at great length, enough has been indicated for the reader to see that all situations shown had a cyclonic wind field and that they occurred at least 10° latitude away from the equator. The potential energy distribution was much more variable, being at times very favorable, as in Figure 29, and at others quite unfavorable, as in Figures 32 and 34. Nevertheless, because hurricanes can spring forth from any of these types of seedlings, we must find additional general or particular criteria to isolate the rare instances with live birth.

1. Some criteria are inherent in the physical structure of the geographical setting. As Figure 5 shows, the mean tropical summer atmosphere without temperature inversion must prevail. Coincidentally, surface ocean temperature must be high enough so that actual mass ascent following the mean tropical atmosphere curve can take place.

2. Two external mechanisms of the upper air will aid in enabling release of condensation heat to make a contribution toward establishing and intensifying a warm-core vortex.

a. Presence of an anticyclone in the high troposphere centered near 200 mb will establish an initial pressure gradient facilitating outflow from cumulonimbus clouds. Relative motion of surface disturbance and upper high toward each other—so that the high becomes superimposed—is definitely favorable. An outflow anticyclone left over from a previous hurricane is preferred but not necessary. In some dynamical studies it has been postulated that such an anticyclone must be "dynamically unstable"; *i.e.*, air dislocated from its center must accelerate outward. This hypothesis, though attractive, has not been verified by observations.

b. As seen earlier, unless there is vertical shearing of the horizontal winds in the environment, there will be an inadequate source of energy to give direct support to the production of kinetic energy required for the growth of a cyclone. However, assuming that cumulonimbus convection is able to follow ascent paths warmer than the mean tropical atmosphere, there are impelling reasons why the wind shear at the core of the disturbance needs to be minimal if there is to be significant development. The falls of pressure at the surface depend upon the increase in mean temperature at the center of the disturbance. If there is appreciable vertical shear of the horizontal wind in the disturbance core itself, the heat in the lower portion will be conducted in a different direction or with a different speed than that in the upper level. It will become more difficult to establish a central warm core through a deep layer.

Further, given a vertical column with positive temperature anomaly and a pressure fall at its base, the temperature excess over the surroundings may be sufficient to produce high pressure or greater heights of isobaric surfaces in the outflow layer, thereby strengthening the upper anticyclone postulated above and initiating its character as an *outflow anticyclone*. One can say that, through increasing temperature differences between inside and outside, the whole field of potential energy available for conversion to energy of motion is enhanced.

Thus we arrive at a double requirement of vertical wind structure: first, small shear—though large enough to prevent inhibiting the internal life cycle of large clouds—in the convective core; second, increasing shear in the environment, as one moves away from the center to a distance of roughly 500–700 km, where, from the external wind structure, potential energy may be released to accelerate the movement of air into a developing vortex in the lower boundary layer and away from the vortex at the top of the cloud system.

Earlier concepts that envisaged the need of a completely shearing environment, analogous to higher-latitude cyclones, for the development of a disturbance or pressure depression into a tropical storm or hurricane must be revised. While slow development is conceivable in a nonshearing environment, significant or rapid growth from depression to storm and hurricane stage is most likely to occur when the environment adjacent to the core of the depression becomes strongly shearing with increasing distance. Thereby the requirement of balanced shear and temperature fields, discussed earlier in this chapter, is met, guaranteeing coherence and prospect of sustained life for any incipient vortex.

3. Possible negative-feedback mechanisms have not received much treatment in the literature, but they may be mentioned at least briefly. The cloud mass of an intensifying low-pressure center most commonly does not overlie the area of lowest pressure but is situated to its side. Especially in the case of the equatorial trough it

has been noticed, and more recently confirmed from satellite photography, that vortex formation directly within the low-pressure trough is rare. Most often it occurs at a distance of some hundreds of kilometers on the poleward side, that is, at a higher latitude. One must then ascribe the density decrease of the atmosphere, leading to the low pressure, to descending air motion alongside the heavy cloud masses—a motion later intensified to become the hurricane eye.

4. High-level cold air moving against the flank of a rain area will serve as an external starting mechanism only if it actually sinks. If it does not sink but spreads over the convective area at high levels, the effect is reversed, causing an incipient storm to die; even mature hurricanes have been destroyed by high-level cold air invasions of their center.

In the following description of three rather different situations, the transition from the initial to the vortex and hurricane stage was successful. The latter transformation is described further in Figures 74 and 75.

Examples of Vortex Evolution

Hurricane Camille. The great hurricane Camille of 1969, herein followed from birth to maturity to late stages (see Figure 78a for the track), had the usual humble beginnings. Cyclonic flow with low pressure was followed across the Cape Verde Islands on 3 August and thereafter tracked across the Atlantic. On 10 August no more than a weak cloud mass advancing from the Atlantic could be followed on the satellite photograph; low-level winds over the Lesser Antilles were strongly easterly. However, the cloud mass made contact with the outflow from a previous hurricane situated 10° longitude east of Florida.

On 11 August winds at 850 mb were still easterly (Fig. 35), except for a suggestion of a northward surge by the equatorial trough zone over Venezuela, which could not be tracked back in time to Brazil, as in Figure 28. The connection of the cloud mass with the hurricane farther north had broken off again. A series of satellite photos shows that it had become an isolated, rather compact, rectangular cloud block between Puerto Rico and the Lesser Antilles and that it broke into two circular masses late in the day. Possibly the high-tropospheric events pictured for the next day were starting then, but information over the western Atlantic unfortunately was too sparse for analysis.

A spectacular change became evident on the next day, 12 August (Fig. 36), when the earlier diffuse system had consolidated into a very well marked wave at the 850-mb level. A broad cloud shield extended northwestward, as seen by satellite, and streamline amplitude had increased greatly, indicating strong cyclonic vorticity over and south of the island of Hispaniola. Pressure-height falls of 20 m (2 mb) had occurred in the center of the wave. This marked and rapid development can hardly be assigned to any spontaneous upheaval but must be related to the influence of large external circulations. Fortunately, high-tropospheric data for that day are ample for preparing 200-mb charts and other relevant charts.

At 200 mb (Fig. 37) the low-level wave, moving westward, had

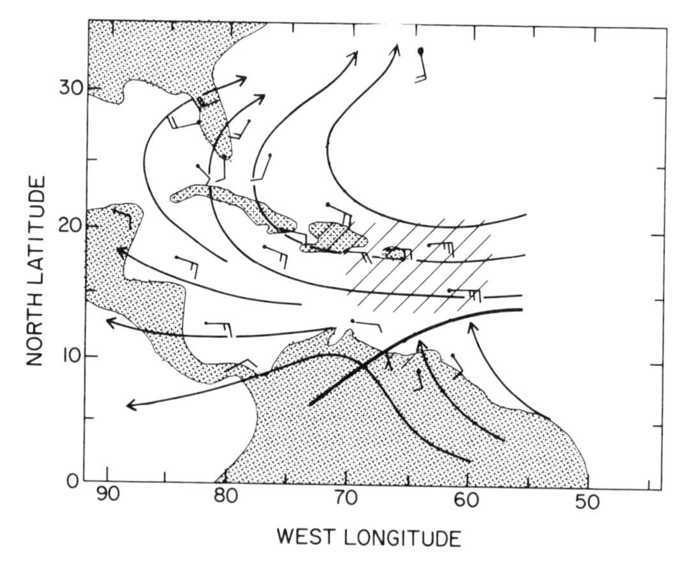

Fig. 35. Winds at 850 mb for the Caribbean, 11 August 1969. *Solid lines*, principal wind-shift lines from southerlies to trades in north; *hatching*, main satellite cloud mass.

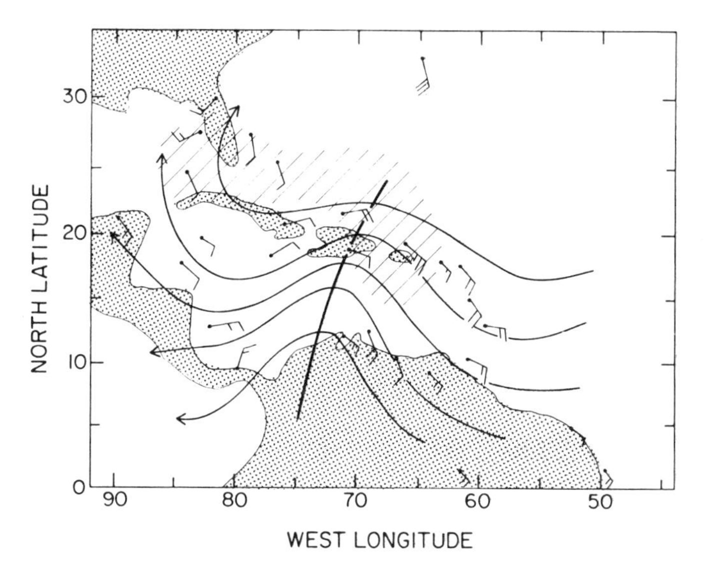

Fig. 36. Same as Figure 35 for 12 August 1969, 1 day later.

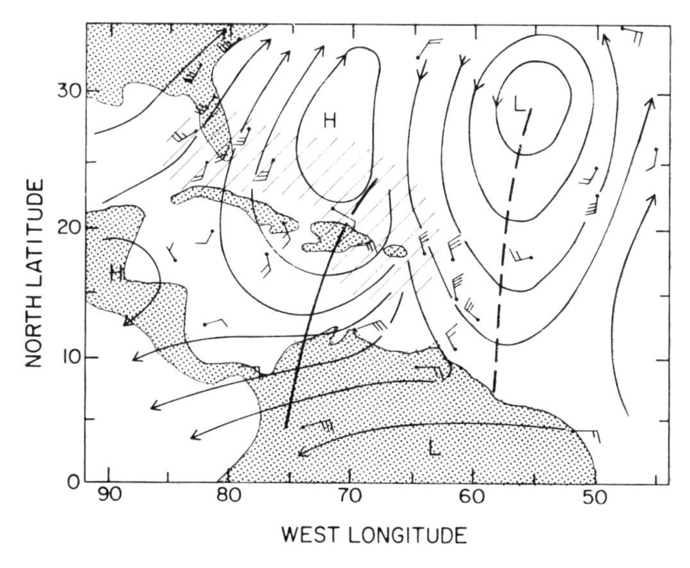

Fig. 37. Winds at 200 mb for 12 August 1969. *Heavy line*, axis from Figure 36; *broken line*, 200-mb trough axis.

moved under the outflow anticyclone of the previous hurricane, which was left behind even though the convection had died out, as is often found. Such superposition is always favorable, because an initial high-tropospheric wind and pressure-height field is advected and can serve as a mechanism for outflow from any deep convection. More important yet, a large trough intrusion into the deep tropics with cold air aloft is found to the east. No doubt intensification of this mid-Atlantic trough had started 1 to 2 days earlier, but there is no means of verification.

One can now picture the intensification process as follows. With a strengthening east-west pressure-height gradient at 200 mb, the northerly flow is accelerated (Fig. 38), leading to divergence at the western end of the accelerated current and convergence at the eastern end. Then the model of Figure 39 applies. The convergence in the east leads to sinking motion and thereby the release of potential energy transported southward along the western part of the North Atlantic trough. At the surface, pressure rises relative to the western side near the lower tropospheric wave axis, above which divergence has been initiated. In principle the model does not differ at all from that of simple heat engines, applied on various scales of motion in the atmosphere.

Strengthening mass convergence is implied at the surface from the inflow. Increase of mass circulation through a center may be a precursor of intensification; especially in a full hurricane it denotes further strengthening to extreme intensity. A property of the organization of inflow is consolidation of interior ascending towers into a set of solid bands, or a wall of ascent, that grow together and no

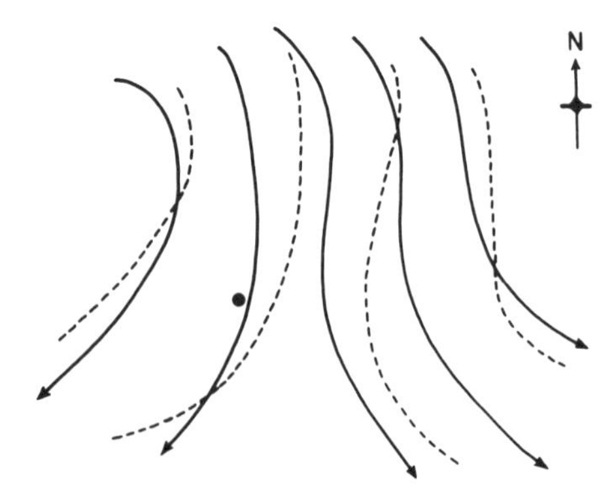

Fig. 38. Model of streamlines (*solid line*) and contours (*broken line*) at 200 mb during the early phases of deepening of tropical cyclone (*black dot*). H. Riehl, "A Model of Hurricane Formation," *Journal of Applied Physics*, XXI (1950), 917–25.

longer compete. This phase, the last one leading from initial vorticity concentration to initial cyclone, is the least easy one to see clearly physically. But, it is the stage best caught by numerical modeling. In the dark confines of the black boxes the consolidation into an organized upward current takes place quite readily, and

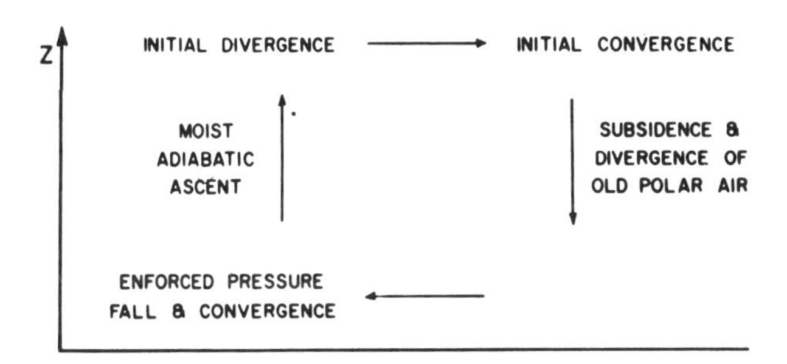

Fig. 39. Model of vertical circulation during early phases of hurricane formation. H. Riehl, "A Model of Hurricane Formation," *Journal of Applied Physics*, XXI (1950), 917–25.

with it a general warm mass of air in the high troposphere arrives and also starts to warm the middle layers of the troposphere. In Figure 40 we see cyclonic circulation decreasing upward with warm air in the middle of the system, just as was demanded in the discussion early in this chapter. The highest temperature coincides with satellite cloud mass, and thus the heat of condensation finally enters as an organized factor in the maintenance and further growth of the still-incipient center. Since the cloud-forming currents still must be lifted, *i.e.*, increase their potential energy, for the release of latent heat to occur, the previous energy cycle of potential-into-kinetic energy of the early stages (here, before 12 August) has now been supplanted by the cycle of *latent heat to potential energy to kinetic energy*. However, a vast cold reservoir must remain present as condenser so that the engine is able to operate efficiently and not be killed by its own heat production. The further history of Camille is examined in this light in Chapter 6.

Hurricane Edith (1971). For an initial synoptic system to become a vortex and eventually a hurricane, it appears necessary that large-scale energy transformations take place as described in Fig-

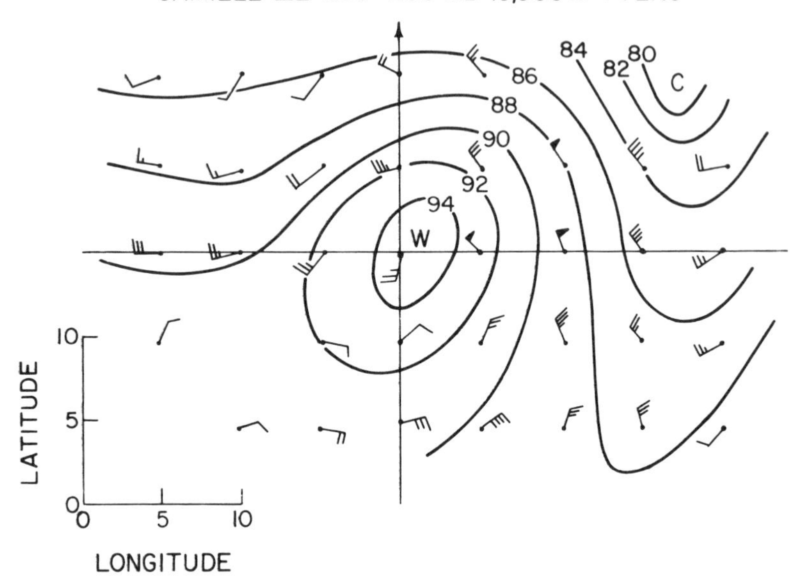

Fig. 40. Vertical shear of the horizontal wind 850–200 mb and thickness for the same layer (tens of meters above 10,000 m) in coordinates following the forming of hurricane Camille, 12–14 August 1969 in the western Caribbean. *Lower left,* relative distance in latitude and longitude; *W,* warm center; *C,* cold center.

ure 39. However, the precise geometry is irrelevant and does vary widely. Figure 32 shows an arrangement conducive to development in the northeastern part of the trough in Figure 30. The higher the latitude of the initial vorticity center, the more likely it is that one will find this latter sequence. At very low latitudes, even near the southern border of the Caribbean, the potential energy source often consists of an isolated mass of cold air at high levels, marked as a sharp trough or shear line, preceding an incipient hurricane. We see such a shear line in Figure 41 on 7 September 1971, when the surface center of what became hurricane Edith passed the island of Curaçao with 1,001-mb pressure. This was very close to the South American coast, the lowest latitude possible in the Caribbean. During the ensuing day the cold center, whose temperature was 3°C less than that of the environment, sank and disappeared completely, together with the 200-mb shear line. At the same time pressure fell to 982 mb. Edith had become a hurricane; rapid deepening continued to 943 mb, indeed a spec-

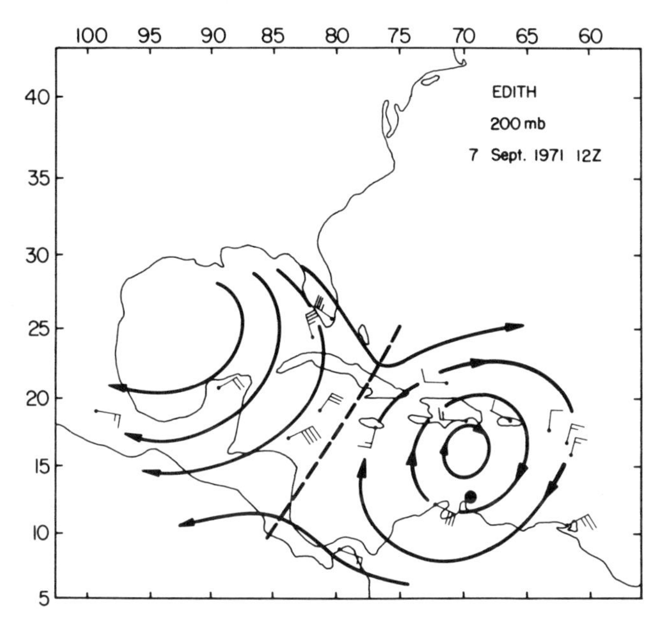

Fig. 41. Winds at 200 mb for the Caribbean on 7 September 1971 prior to formation of hurricane Edith (*heavy dot*, initial center).

tacular growth rate, when the storm entered Nicaragua and Honduras. We see again that it is the physical mechanism, not the precise geometry, that matters.

Cyclone Tracy. This survey will be concluded with an account of the information given by the satellite NOAA-2 in the case of the Australian hurricane Tracy that on Christmas Day of 1974 destroyed the port of Darwin, the principal city on the Australian north coast (12° S,131° E). The available data consist of satellite photography and of equivalent radiation temperature, that is, the black-body temperature corresponding to the total long-wave radiation measured by the satellite. Radiation temperature is averaged over 1° squares (110 × 110 km). Thus the details of the very highest clouds in the forming hurricane are not given, but the information is entirely adequate for following the general evolution. Discussion is mainly confined to temperatures with threshold value of −30°C (when clouds reaching to 300 mb [10 km] may be expected), −40°C (250 mb), and −50°C (200 mb, over 12 km). The series of seven pictures from 20 to 26 December is reproduced in Figure 42.

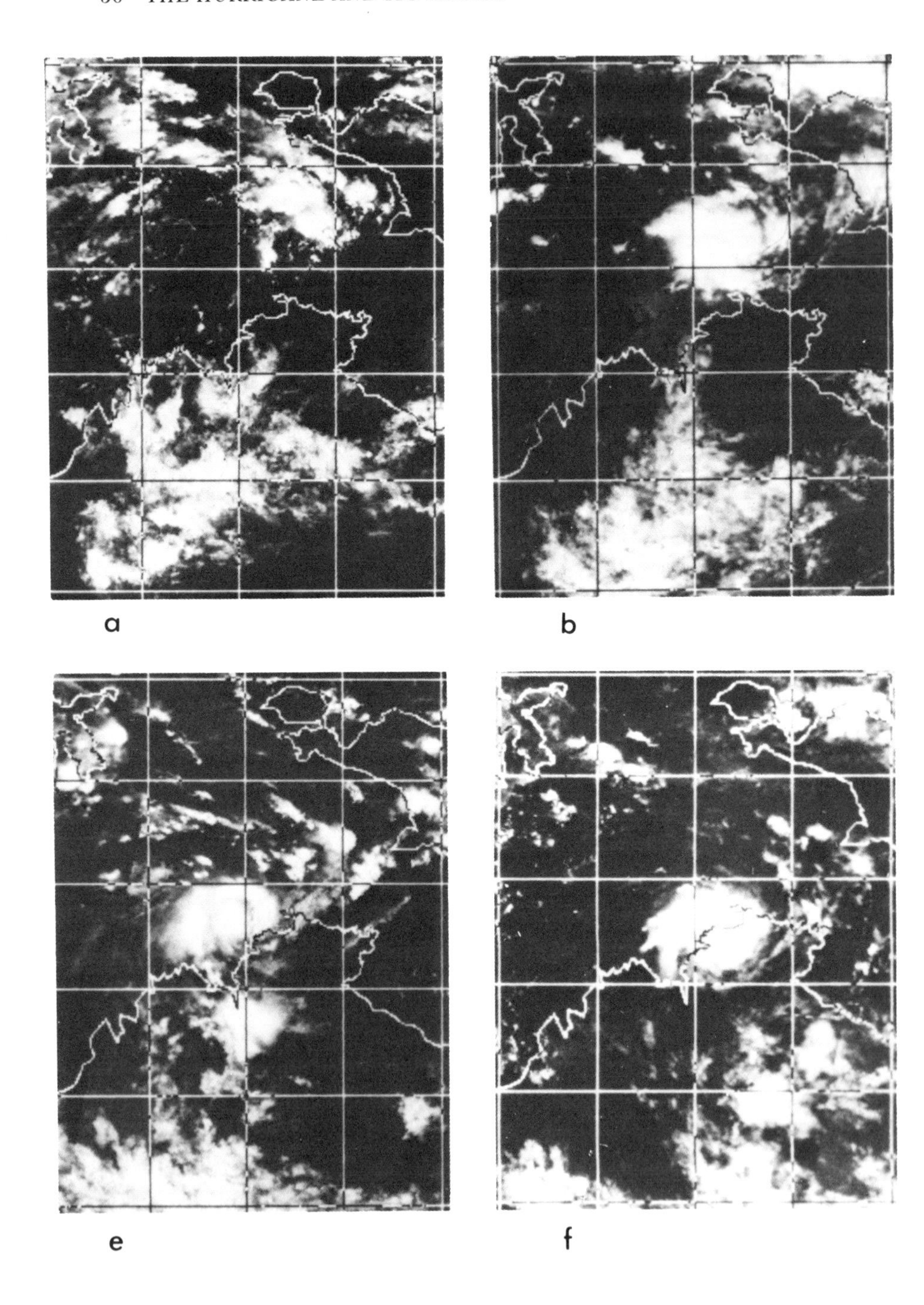
a
b
e
f

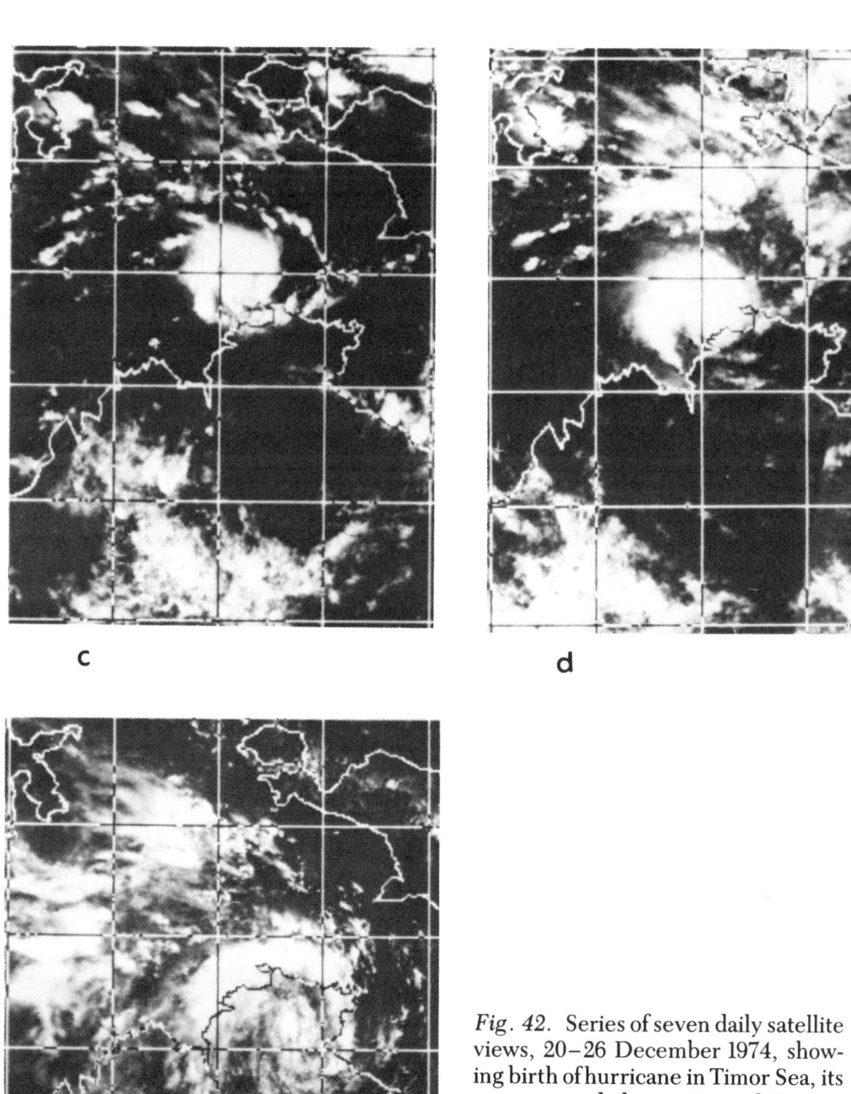

Fig. 42. Series of seven daily satellite views, 20–26 December 1974, showing birth of hurricane in Timor Sea, its overpass and destruction of Darwin on 25 December, and weakening of vortex over land on 26 December.

On 20 December one sees a cloud array over the area south of New Guinea, as on many days, with temperatures barely reaching −30°C in places. Cold temperatures of −50°C, found only near the high mountain range of New Guinea, indicate orographic convection. From the satellite data no special event is indicated.

The situation first becomes dangerous on 21 December, when the cloud mass drifts slowly southwestward from New Guinea across the Coral Sea and assumes a fairly circular pattern. One more day's travel would bring it close to the Australian coast. No temperatures from the satellite are available for this day.

Actual motion from 21 to 22 December is southwestward but very slow, at a rate of only 5 kt. Marked intensification is evident from the firm contraction of the center, a hard core and suggestion of an eye, and some low cloud lines to the north, leading cyclonically (here clockwise) toward the core. Ambient cloudiness remains suppressed. Radiation temperature now has a marked core with less than −50°C in six 1° squares, the coldest being −56°C. This cold area above the contracting cloud mass clearly supports the notion of cyclone development gained from the picture.

In the next 24 hours, to December 23, the slow southwestward motion continues. The center looks more uniformly covered with cloud, and there is a large outbreak of ambient convection to the north—another synoptic weather system. Though this second system might well have interfered with the forming hurricane, temperatures indicate that this was not the case. These attained their lowest value of −60°C—quite unusual—in 2 squares; temperatures below −50°C prevailed in 12 squares, both morning and evening. The temperatures were reported near 0900 and 2100 local time. Usually, convection over the ocean is strongest in late night and early morning; coldest temperatures are reported at the 0900 overpass, while at 2100, temperatures 10°C to 20°C warmer are the rule. On 23 December, however, the cold core of −60°C persists into the evening, a sure danger signal.

The second synoptic system, whatever its nature, is short-lived; ambient clouds have again disappeared on 24 December. Slow motion of the cyclone, now a hurricane, continues with direction turning to south. Dramatic intensification is suggested from the picture. There is an inner hard, circular core, almost separated from the outer cloud envelope by a nearly cloudless ring with a radius of about 100 km. Lowest radiation temperature is −56°C to −57°C, still very cold and suggestive, both morning and evening, in ten 1° squares.

Recurvature toward southeast continues on 25 December, when the hurricane enters land, now with a slight increase in speed. This is the period of the destruction of Darwin. The cold radiation temperatures persist in the morning, but all values below −50°C have disappeared at the time of the evening observation, suggestive of weakening of the hurricane core over land.

The final photo on 26 December shows the typical decay of the hurricane core over land, treated in Chapter 10. The cloud mass widens out and assumes a large-scale spiral form. Along the band extending northwestward from the south, the center next accelerates to reach southern Australia 24 hours later.

This history of cyclone Tracy proves the immense value of satellite information to deduce when the contraction postulated in Figure 22a and later is operative and when it is failing. The slowly recurving path must be predicted from conventional observations. But one of the most essential features of development, contraction into a narrow ring, together with information about strengthening or weakening of the circulation, is brought out by the satellite. Such a revealing mass of information is not readily supplied at a glance by any other data systems.

REFERENCES

1. G. E. Dunn, "Aerology in the Hurricane Warning Service," *Monthly Weather Review*, LXVIII (1940), 303–15.
2. H. Riehl, "On the Formation of West Atlantic Hurricanes, Part I," Miscellaneous Report 24 (Department of Meteorology, University of Chicago, 1948), 1–64.
3. J. Malkus and H. Riehl, "On the Dynamics and Energy Transformations in Steady-State Hurricanes," *Tellus*, XII (1960), 1–20.
4. R. H. Simpson, *et al.*, "Tropical Disturbances in the North Atlantic, 1968," *Monthly Weather Review*, XCVII (1969), 240–55.

5. K. Ooyama, "A Dynamical Model for the Study of Tropical Cyclone Development," *Geophisica Internationale*, IV (1964), 187–98; H. L. Kuo, "On the Formation and Intensification of Tropical Cyclones Through Latent Heat Release by Cumulus Convection," *Journal of Atmospheric Science*, XXII (1965), 40–63; S. L. Rosenthal, "A Circularly Symmetric Primitive Equation Model of Tropical Cyclone Development Containing an Explicit Water Vapor Cyclone," *Monthly Weather Review*, XCIX (1970), 643–63.
6. W. M. Gray, "Hurricanes: Their Formation, Structure, and Likely Role in the Tropical Circulation," in D. B. Shaw (ed.), *Meteorology over the Tropical Oceans* (London: Royal Meteorology Society, 1978), 155–218.
7. R. H. Simpson *et al.*, "Atlantic Tropical Disturbances of 1967," *Monthly Weather Review*, XCVI (1968), 251–59.
8. Simpson *et al.*, "Tropical Disturbances in North Atlantic," 240–55; P. J. Hebert, "Intensification Criteria for Tropical Depressions of the Western North Atlantic," *Monthly Weather Review*, CVI (1978), 831–40; J. C. Sadler, "Role of the Tropical Upper Troposphere Trough in Early Season Typhoon Development," *Monthly Weather Review*, CIV (1976), 1266–95; J. Simpson *et al.*, "A Study of a Nondeepening Tropical Disturbance," *Journal of Applied Meteorology*, VI (1967), 237–54.
9. H. Riehl and H. R. Byers, "Computing a Design Flood in the Absence of Historical Records," *Geofisica Pura e Applicata*, XLV (1960), 215–26; M. Garstang and A. Betts, "A Review of the Tropical Boundary Layer and Cumulus Convection: Structure, Parameterization, and Modeling," *Bulletin of the American Meteorological Society*, LV (1974), 1195–1205.

PART III

The Hurricane Event

CHAPTER 5

The Life Cycle

Normal Stages of the Life Cycle

Formative Stage. During surveillance of initial vorticity centers, one watches for one or more of the following developments:

1. Unusual falls of pressure over 24 hours by 2–3 mb or more in the center of the vorticity concentration and also in its outskirts, in the case of the equatorial trough as much as 500 km away.

2. Asymmetric strengthening of wind, and appearance of gale force in one sector. In the trade region, southeast winds will usually begin strengthening while winds with components from the west, always a danger signal, will appear where normally only east winds blow.

3. In the vicinity of the equatorial trough and along old fronts, formation of large elliptical or circular wind envelopes, which replace straight shear zones; contraction of the envelopes into smaller radii, as in Figure 22a, is a definite signal for alert.

4. A concurrent indication in satellite photos of isolation of a marked cloud mass from other nearby cloud areas; *i.e.*, a clear ring, or at least an arc, begins to surround a forming hurricane. Appearance of spiral bands may be another early but not universal precursor.

Immature Stage. Wind force first increases to hurricane strength, usually at a distance of perhaps 50 km from the cyclone center (now well identified), and pressure falls intensify. In the Caribbean, centers in the immature stage may be very small but strengthening rapidly at the same time. A center may still pass between two islands in the Lesser Antilles with 100-kt maximum wind speed, while neither island experiences more than 25 kt. Intensification may occur very rapidly, in just a few hours. Near the equatorial trough, in contrast, contraction of a cyclonic envelope may proceed with monotonous slowness over several days, as much as three, rendering early warning of at least gale-force, if not hurricane, winds rather easy.

One might add that there are no hard and fast rules about the immature stage, just as there are no rules without exception in anything pertaining to hurricanes. Some Atlantic storms have taken as much as 3 days to consolidate, even in midseason; there are instances of double storms trying to form and interfering with each other, so that neither can develop a closed circulation. On the other hand, some equatorial-trough formations in the West Pacific, notably early and late in the season, go through the immature stage more typical of formations in the Atlantic.

If an incipient center is in view of a radar set, it is worthwhile to follow closely the behavior of any bands and especially to watch for the sudden push of a band from the outside well into the central area. On the satellite photo the immature hurricane is marked by a small but very hard core and is enveloped by some fluffy-appearing clouds. Tropical cyclone Tracy of December 1974 was one excellent case illustrating generation of the small and very hard core from 23 to 24 December 1974 (Fig. 42). After landfall the picture transformed to the larger-banded structure typical of a late stage over land with a dying core.

Mature Stage. Duration of the mature stage is most variable; it has ranged from hours to as much as a month! On the average the mature stage occupies the longest part of the life cycle and most often lasts several days. Central speed and pressure need not exceed those of the immature stage. But the circulation widens out, and in moving storms, hurricane winds may extend several hundreds of kilometers from the center to the right of the direction of motion. A well-formed inner ring of maximum wind encircles the eye at a variable distance averaging perhaps 50 km. At this stage, heating from convective clouds furnishes the largest amount of energy for hurricane maintenance; the cyclone is warm cored. Pressure gradients are therefore largest at the surface, and conventional surface-pressure analysis furnishes an excellent tool for locating the general area of a hurricane. Of course the precise center position, vital for forecasting, may not be given by this source; sharper and much more reliable tools are satellite, radar, and aircraft reconnaissance.

The mature stage furnishes the widely known hurricane descriptions. One treats the storm with a "steady-state" model, in which there is no change with time. We shall employ this concept in the next chapter. But again it should be emphasized that this is a simplification with a purpose. In reality the hurricane undergoes substantial variations on the time scale, from hours to days; pressure in the center may go up and down by more than 30 mb over a day or two. Nevertheless, the whole hurricane envelope holds together remarkably well, indicating a basically stable vortex; thus the steady-state treatment is not without realism.

Decaying or Transformation Stage. Nearly all hurricanes weaken substantially upon entering land, because they lose the energy source furnished by the underlying ocean surface. The decay is especially rapid where the land is mountainous. Some hurricanes do die out over sea, and this event can be related to their moving over a cold ocean current or being invaded by a surface cold air mass behind a cold front or by a cold center at high levels moving over their top—markedly different from Figure 32.

There is a great variety of terminal stages. A hurricane still over sea or just entering land may arrive out of the tropics at the moment when, from the mid-latitude circulations, a cyclone is about to be born. The hurricane then furnishes the nucleus for the development, which can become exceedingly intense and rapid. This late hurricane stage is perhaps the most dangerous of all stages in the cyclone's life cycle, especially as forward speed may change from a leisurely 10–15 kt to a sudden 40–50 kt with the inherent

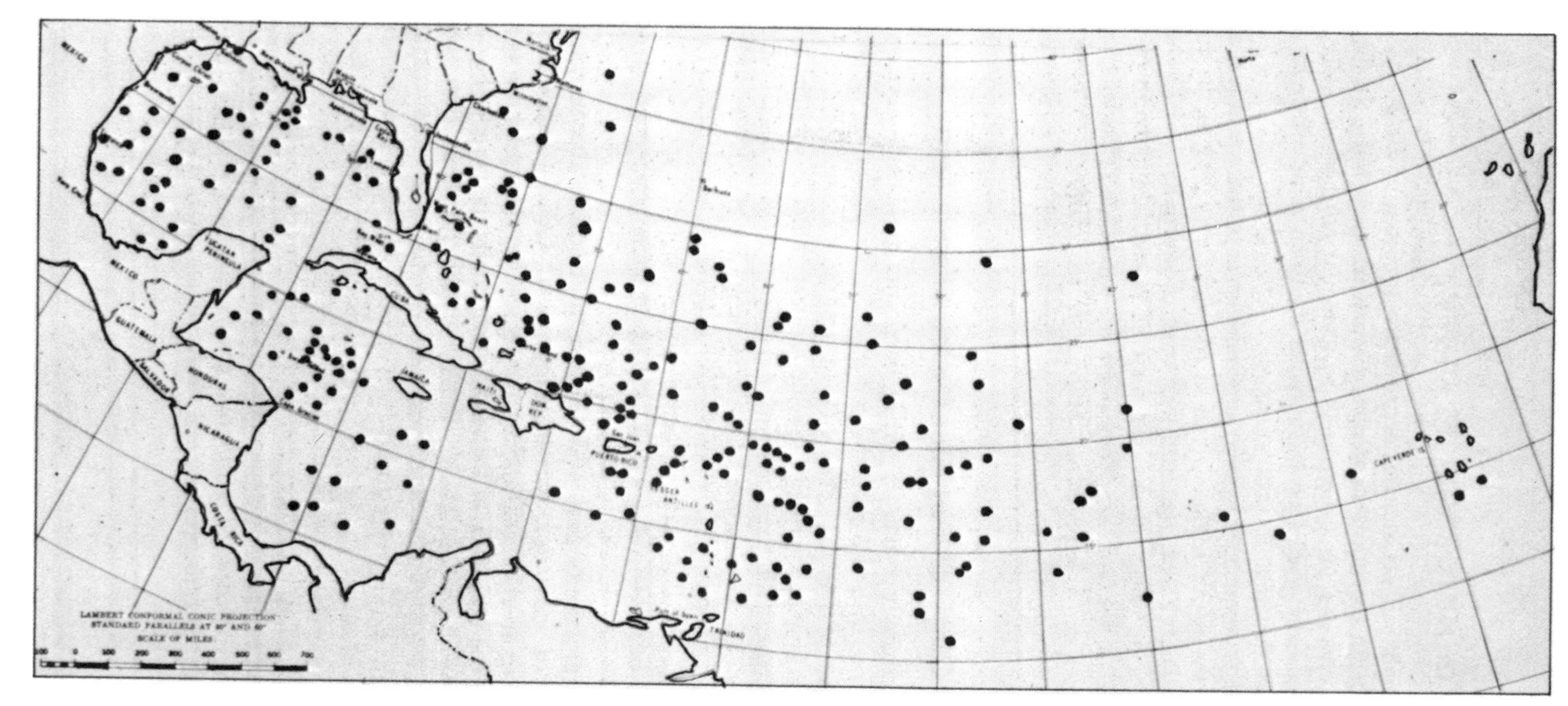

Fig. 43. Locations where tropical cyclones reach hurricane intensity, 1901–1957. Gordon E. Dunn and Banner I. Miller, *Atlantic Hurricanes* (Baton Rouge: Louisiana State University Press, 1960), 38.

threat of early landfall and of hurricane wind and rain catastrophe on land.

Favored Areas and Seasons of Hurricane Development (see Fig. 1)

In the preceding chapters we have found:

1. Hurricane development is rare between 5° N and 5° S latitude, where the vertical component of the earth rotation is very weak or zero (Fig. 21).

2. Formation is largely restricted to those portions of the oceans that do not have trade-wind inversions or persistently dry upper layers. Rather, the mean atmosphere of Figure 5 should be found.

To these points should be added:

3. Since the warming of the summer atmosphere and ocean does not stop at the solstices but extends 2 to 3 months beyond this date, highest hurricane frequency is delayed into late summer and early autumn, a displacement much more marked in the Northern Hemisphere than in the Southern.

4. Although there are warning voices, the suggestion is widely advanced that the warmer the ocean, the more likely it is that there will be a strong hurricane. It is an old notion that, the hotter the fire, the more rapidly the water in the kettle above it will boil. Over the oceans one observes that hurricanes rarely develop when surface water temperature is below 26.5°C.[1] At this temperature, the overlying low atmosphere normally is conditioned so that its undisturbed ascent—ascent without mixing with the surrounding atmosphere encountered on the way up—will be slightly warmer than that of the environment (Fig. 5). Such conditioning can, and sometimes will, occur at lower temperatures.

A value of 26.5°C to 27°C is generally a valid threshold, but it appears to be just that and no more. Figure 43 shows the location where incipient disturbances first attained hurricane strength in the Atlantic region, according to an analysis by Gordon Dunn.[2] The threshold of 26.5°C is well outlined. But, whereas temperature of the water rises all the way westward to the Gulf Stream and the Gulf of Mexico, one does not find that the dots in Figure 43 become more crowded in these areas. Rather, they are fairly uniformly spaced, suggesting that mechanisms other than warm ocean temperatures must be the principal hurricane starter. In contrast, it is well established that hurricanes weaken markedly and even disappear when moving over water with temperatures substantially below the threshold value, particularly cold ocean currents.

In gross outline, the foregoing holds well around the globe; on account of regional geography and varying dynamic factors, however, there are also notable deviations. A survey of the main regions of formation follows; but all details cannot be explored in their complexity, nor can the number of storms be given with any accuracy. Earlier estimates, particularly in regions with sparse ship traffic, have been badly upset by the satellite evidence, which shows that some areas have far more tropical storms than indicated by old statistics. Annual number of hurricane-strength cyclones may now be estimated as seventy-five to one hundred, almost double old guesses but still small compared to the number of cyclones outside the tropics.

North Atlantic and East Pacific. These two areas, separated narrowly by Central America, should be considered a single complex. In the Atlantic portion the seasonal variation in the strength of the subtropical anticyclone dominates the picture. The equatorial trough appears infrequently and weakly in the western Caribbean area in May and June, so that there is a slight maximum of storms mostly moving north or east of north from inception. This regime extends into the Gulf of Mexico.

In midsummer the anticyclone strengthens, and the trade winds overlie most of the Atlantic in July, a poor hurricane month. With the advance of the equatorial trough in the eastern part of the ocean, the Cape Verde storms make their appearance in August and September, normally guided all the way across the ocean by the strong anticyclone. Late summer and early fall bring a resurgence of the equatorial trough in the western Caribbean. Interaction between tropical vorticity concentrations and waves in the westerlies farther north multiplies with the coming of autumn and leads to frequent repetitions of the pattern of Figure 32 until late in October or even November.

Many of the Atlantic systems—hurricane-strength or weaker—cross Central America and regenerate there to move west-northwest along the Pacific shore until they encounter the cold California current and die. Other formations occur along the equatorial-trough zone, well marked in the eastern Pacific, under circumstances still not fully known. Rarely do hurricanes in dying stages reach California, western Arizona, and Nevada, though if they arrived there at the height of the dry summer season, they would not be welcome. The agricultural economy, notably grape and raisin production, is geared to a growing cycle with dryness at that time of year. Combined mean annual frequency, subject to much variability is assessed at above twenty, with the eastern Pacific furnishing the majority in most years. A few of the East Pacific hurricanes travel far west to the south, or even close to the Hawaiian Islands, and may reach the West Pacific.

Western North Pacific. Here is the area of the world's greatest tropical storms, locally known as *typhoons*. Continents like South and Central America do not hinder full development. The world's largest cyclones observed there—"supertyphoons"—may occupy areas as large as the whole United States. Annual frequency is well above twenty. Although the maximum is in September, no month of the year has been entirely free of hurricane-force cyclones. Those of midwinter are found mainly below latitude 10° N, crossing from the Pacific over the southern Philippines into the South China Sea.

Bay of Bengal and Arabian Sea. Just like the western Caribbean, these two parts of the North Indian Ocean have a double season: during the advance of the summer monsoon and during its retreat (May–June and September–October). On occasion, the season is even later in the Bay of Bengal along the southern coast of India. At the height of the season, when the equatorial trough is farthest north, large cyclones with heavy rain but only weak wind systems occur—a point to remember. The transition seasons, when there is interplay between upper troughs extending from Siberia into the tropics, are the main severe cyclone seasons. In the history of tropical storms some of the greatest disasters causing loss of life have occurred in the northern Bay of Bengal where the coastlines converge and storm surges can grow to formidable portions on the flat and very densely populated shores.

Australia and Surroundings. The isolated continent of the Southern Hemisphere may experience tropical cyclones from three sides: the Pacific Ocean in the east, the Coral Sea to the north, and

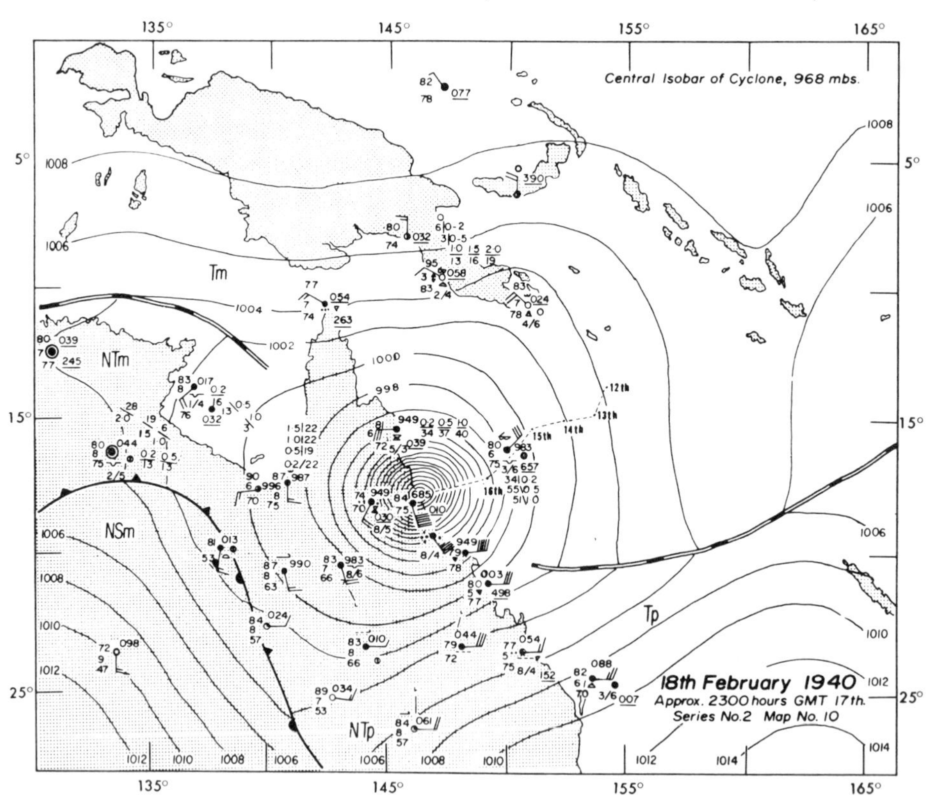

Fig. 44. Hurricane from Pacific arriving on Australian east coast. Gordon E. Dunn and Banner I. Miller, *Atlantic Hurricanes* (Baton Rouge: Louisiana State University Press, 1960), 68.

the Indian Ocean in the west. Cyclones from the Pacific are rare and mostly recurve southward. But they may assume huge proportions, about half the size of the continent, and strike the east coast with great vigor (Fig. 44). In the north, the main part of the equatorial-trough zone overlies Indonesia; in most years the trough appears to be prevented from entering the continent deeply by strong subtropical westerlies farther south. Hurricanes in the Coral Sea are rare and small, but they can be devastating (Fig. 42). West coast cyclones often originate from large, weak rain systems that have made their way westward across Australia from the Gulf of Carpentaria and have intensified upon reaching the Indian Ocean. They may hug the coast, remaining nearly stationary or on slow irregular paths, for days before either moving away toward west or curving back into the continent between 20° and 25° S. In instances of prolonged proximity to the west coast, cloud physics aspects may assume an unusually important role, with large-scale introduction of a continental nucleus spectrum from the continent into the hurricane rain areas (Fig. 45).

South Indian and Pacific Oceans. Over the wide expanse of the South Pacific Ocean hurricane-type cyclones have from time to time visited all the different island groups. Because less literature is available on this large ocean than on any other, it is very difficult to form a reliable picture of the origin of these storms.

In the Indian Ocean, the trade-wind temperature inversion typical of the eastern parts of oceans evidently does not hug the Australian west coast consistently, in contrast to North and South America. But, an analogue to the Cape Verde hurricane is found along the equatorial-trough zone extending westward from Indonesia. Here, just as in the South Pacific, the satellite now observes many hurricanes, including cyclone pairs, whose existence had been largely unknown. However, a well-known area of concentrated cyclone occurrence is situated at the western edge of the ocean and affects the islands of Mauritius and Madagascar and occasionally the African coast. Seasons are comparable to those in the western Atlantic and the main season of the western Pacific.

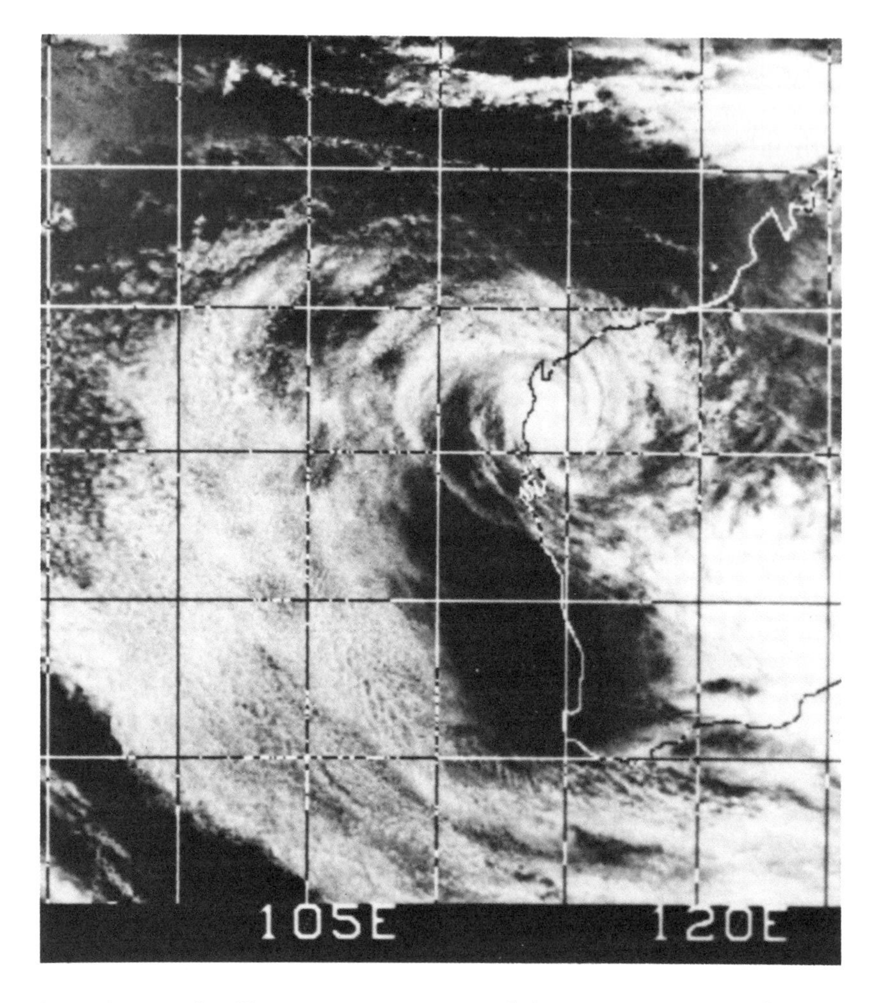

Fig. 45. Example of hurricane entering Australia's west coast. Note *outflow* spiral cloud bands, counterclockwise in the Southern Hemisphere.

South Atlantic Ocean. In the South Atlantic Ocean no hurricane has ever been reported. Many hypotheses have been offered: the ocean is too small; the ocean is too cold; the trade winds don't warm up enough; the equatorial trough hardly, if ever, penetrates south of the equator; strong upper westerlies are present at all times of year including summer. The reader is invited to believe in the hypothesis of his choice, but our most honest statement is that we just don't know. Table 6 provides a summary of hurricane areas and seasons of occurrence.

Table 6. Tropical Cyclone Climatology (All Intensities)

Region	*Season*	*Max. months*	*n**
North Atlantic	June–Nov	Aug–Oct	8
Northeast Pacific	June–Nov	Aug–Sept	(15)
Northwest Pacific	May–Dec (some in each month)	July–Oct	22
Bay of Bengal	May–June, Sept–Nov	same	6
Arabian Sea	May–June, Oct–Nov	same	2
Australia Area	Dec–Mar	Jan–Feb	5
Southwest Indian Ocean	Dec–April	Jan–Apr	(5)
Southwest Pacific Ocean	Dec–Mar	same	

*All frequency estimates are very uncertain.

In Figure 46 we see the frequency of the day of hurricane formation in 3-day running totals for August to October 1900–1957. Evidently, there are no preferred calendar days for hurricane formation; the irregularities in the curve shown are due to insufficient data.

Climatic Trends

On the time scale of a few years, significant changes of frequency of landfall along any coastline may be brought about by shifts of preferred track location, which are related to changes in the upper-air circulation outside the tropics. Considering longer time scales, the influence on secular trends of population increases and long-distance ship and air traffic must be recognized. Australia's cyclone statistic typifies the slow rise in frequency related to population growth. Frequency rose from five to ten per 2 years around 1910, when a statistic was started. It then rose twenty to twenty-five around 1960 and later, clearly showing that the early counts were

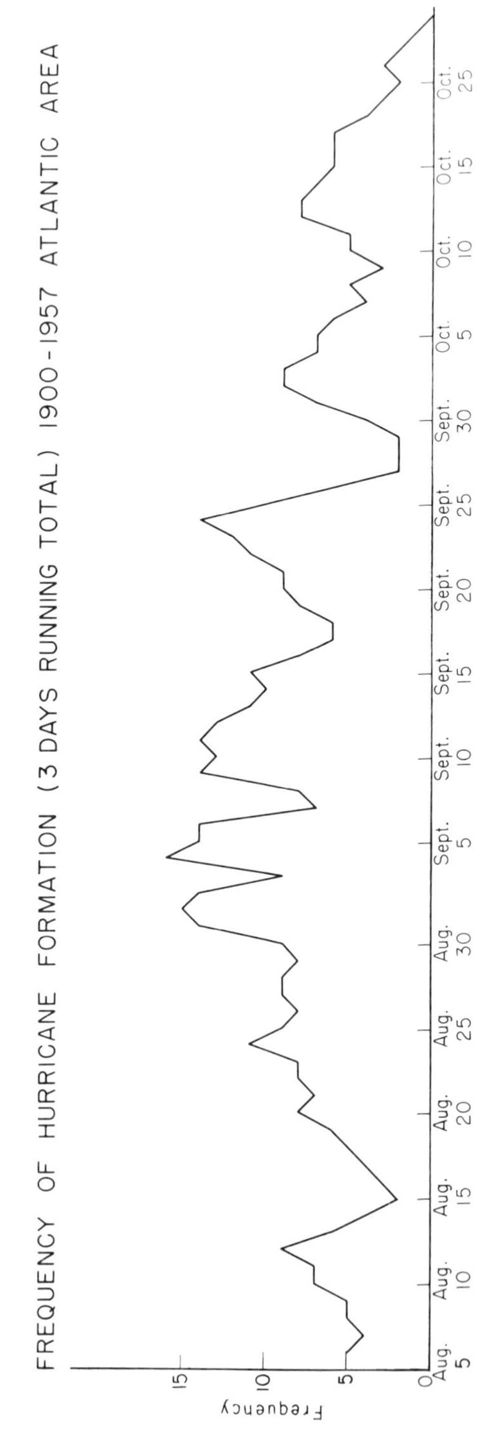

Fig. 46. Frequency of hurricane formation (3-day running totals) 1900–1957 for the Atlantic Ocean area.

low. Here a tropical cyclone is defined by a maximum wind greater than 63 km per hour.

Only the Atlantic appears to give reasonably reliable frequency information. Frequency for hurricanes and tropical cyclones has averaged about eight per year from 1885 to 1975. From Figure 47, below-average frequencies predominated from 1895 to 1930. If one adds the yearly deviations from the average cumulatively, a marked low point is apparent at the end of the 1920s. Then in a single year, from 1930 to 1931, hurricane frequency rose to a level well above average and remained there until the 1970s, when renewed decline commenced. The decline, however, was felt even earlier in the number of landfalls on the United States coasts (Fig. 134). Annual frequency averaged six from 1895 to 1930 and ten from then on to recent years. This is a swing of four per year, or half the yearly average for the whole period—a very large change that cannot be dismissed as a random feature and that is crucial in hurricane prediction and assessment of landfall probability.

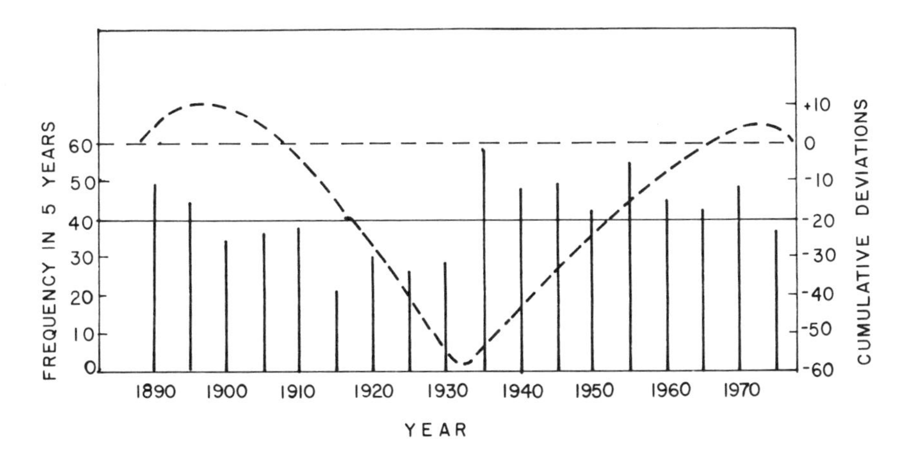

Fig. 47. Sums of hurricanes and tropical storms for the Atlantic area, 1887–1977, in frequencies of 5 years. *Broken curve*, cumulative deviations of the sums from *solid horizontal line*, which represents the mean of forty per 5 years.

Hurricane Tracks

Steering. As well recognized in the nineteenth century, many hurricanes tend to follow an approximately parabolic path around the subtropical anticyclones over the oceans (Fig. 10). Winds from these large systems in which tropical cyclones are embedded were thought to "steer" the cyclones. According to this concept the hurricanes first move westward at low latitudes in the prevailing easterlies. Near the western margin of an ocean their path is turned northward or southward depending on the hemisphere. As they leave the easterlies and come under the influence of the westerly winds of high latitudes, they recurve toward northeast in the Northern Hemisphere and southeast in the Southern Hemisphere. They then retrace the same ocean in which they started at a higher latitude and, unless they strike land, may even go into the polar regions. Early writers went so far as to compute parabolas that any particular hurricane would follow for a week or more, given its starting position. If such an "ideal" situation prevailed, there would be little need for the forecaster! However, this old and ingrained idea turns out to be a gross oversimplification that breaks down in light of actual hurricane situations. In addition to a great variety of interactions between tropical storms and middle-latitude weather systems—the main problem in predicting recurvature—the hurricane controls its own destiny at least to a small extent and, in the case of large and severe storms, even to a considerable extent.

Poleward Component of Motion. Common to most hurricane paths, and also to paths of cyclones outside the tropics, is a tendency to move steadily toward higher latitudes; correspondingly, most anticyclones trend equatorward. The difference in Coriolis parameter on the north and south sides of a cyclone is thought to be chiefly responsible for this dynamic component of displacement. Rare deviations of hurricane paths back into the tropics were once thought to be associated with decay. But there have been many gross deviations, typified by hurricane Anita in the western Gulf of Mexico during September 1977 (Fig. 78a). It deepened from 963 to 926 mb on the last day of its journey on a course toward the southwest into Mexico. Thus the above "rule" has had to be abandoned.

Steering Seen as Vortex Interaction. The idea of steering can be derived from the concept of two bodies rotating around each other. There will be a common center of gravity around which the rotation takes place, as in the case of the earth and moon. Now the mass of the earth is so large compared to that of the moon that the center of gravity actually lies inside the earth. We only observe the rotation of the moon around the earth; refined instrumentation is required to notice the moon's effect. By analogy, if the subtropical

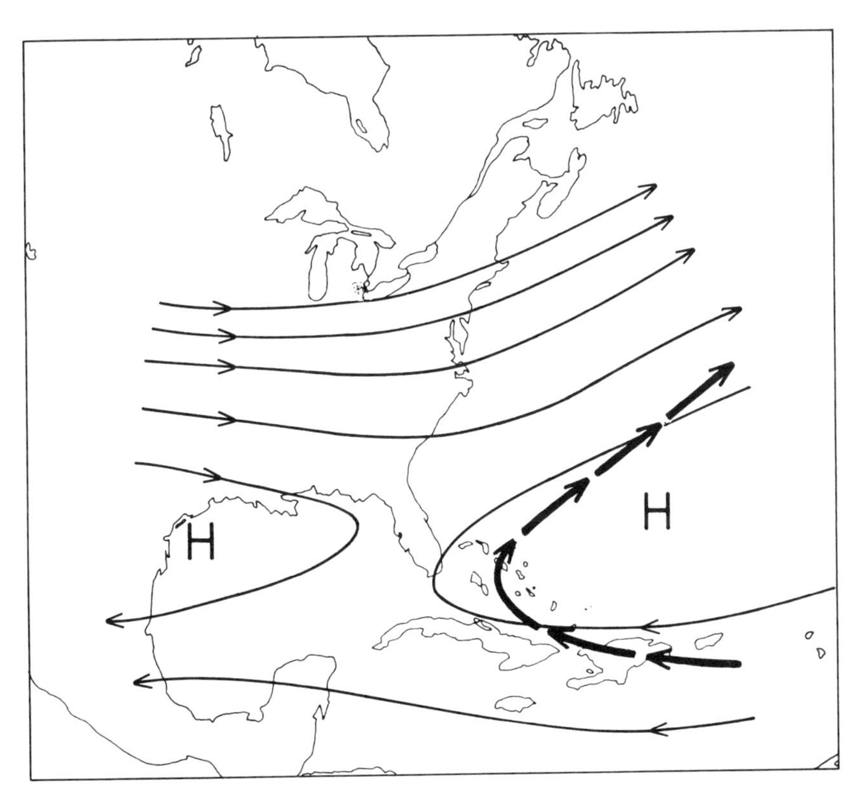

Fig. 48. "Normal" quasi-parabolic hurricane track sketched with 500-mb streamlines that indicate a flat westerly pattern not extending into the tropics.

anticyclone is large (some thousands of kilometers across) and a hurricane small (hundreds of kilometers wide), the center of gravity lies close to the anticyclone center, which moves very little while the hurricane revolves around it (Fig. 48).

However, with growing hurricane size, the picture changes as the storm begins to exert an increasing influence upon its surroundings. Outstanding are the occasional Pacific supertyphoons, which may assume dimensions of the size of the United States or Australia. They then become as large as the subtropical anticyclones; as a second factor, their much stronger rotation rate,

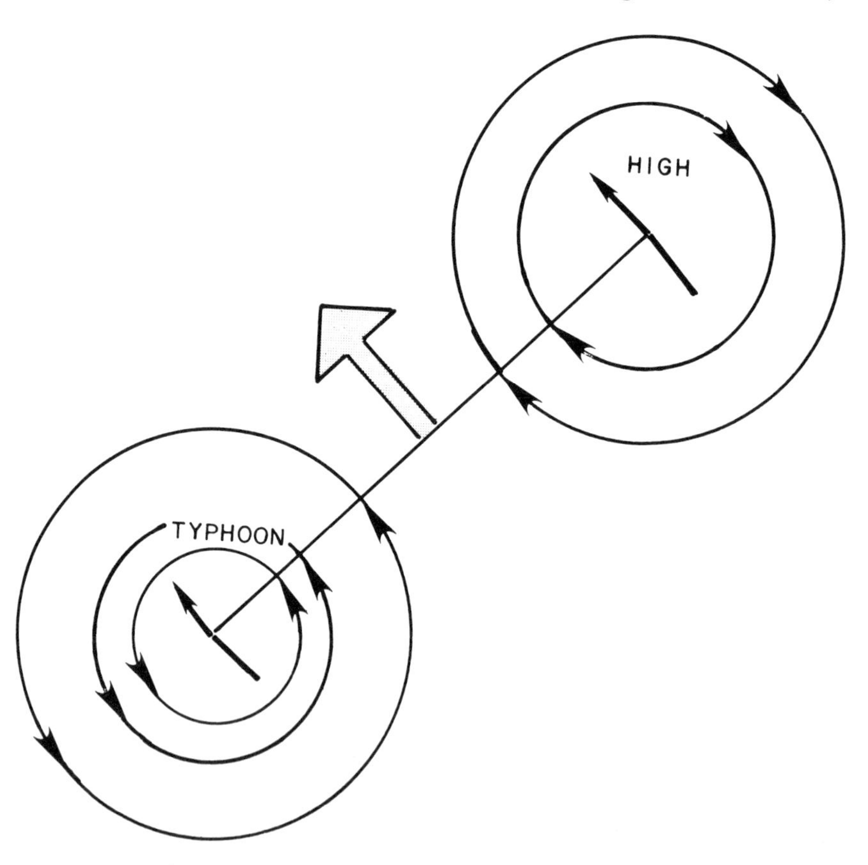

Fig. 49. The influence of Pacific supertyphoon and of subtropical anticyclone upon each other and resulting motion of midpoint between them.

compared to that of the anticyclones, enters the picture. Pacific typhoons at times have "captured" their subtropical anticyclones and moved with them on long straight paths directed west-northwest over half the Pacific to make entry into China in the latitude range 25° N–35° N (Fig. 49). One further observes that often hurricanes which undergo a late secondary intensification stage refuse to recurve on schedule but trend northward to westward, throwing up a large upper anticyclone on their eastern side (see description of hurricane Camille in Chapter 7).

Hurricane Pairs. An interesting variant of vortex interaction is that of two hurricanes impinging upon one another. Experience shows that hurricanes must be less than 1,500 km apart (15° longitude) for interaction to become effective. And one storm cannot be much larger than the other, or else the small one will simply rotate quickly around the large one and lose intensity in the course of 2 days. When interaction occurs around a common center of gravity roughly midway between hurricanes, their rotation around each other is counterclockwise in the Northern Hemisphere and clockwise in the Southern Hemisphere. They also tend to approach each other, and this increases the rate of rotation to as much as 45° in a day or more.

Forecasting becomes especially difficult under these circumstances. The forecaster must assess the rotation rate, but the midpoint on the axis connecting them may become displaced in response to other outside circulations that impinge on the hurricane pair. Most of the time the midpoint tends to move toward higher latitudes (Fig. 50).

Internal Oscillations. In addition to the foregoing, plus the uncertainties in hurricane fixes that always remain a problem, some storms undergo noticeable oscillations in track direction over periods of some hours. These cannot be ascribed to random errors in positioning. The rather involved theory was most thoroughly taken up by T. C. Yeh.[3] We see a spectacular example in Figure 51. Here is another unwelcome problem for the forecaster who has to decide whether sudden changes in track are really indicative of the general storm motion or whether they are just to be taken as wiggles on an otherwise constant track.

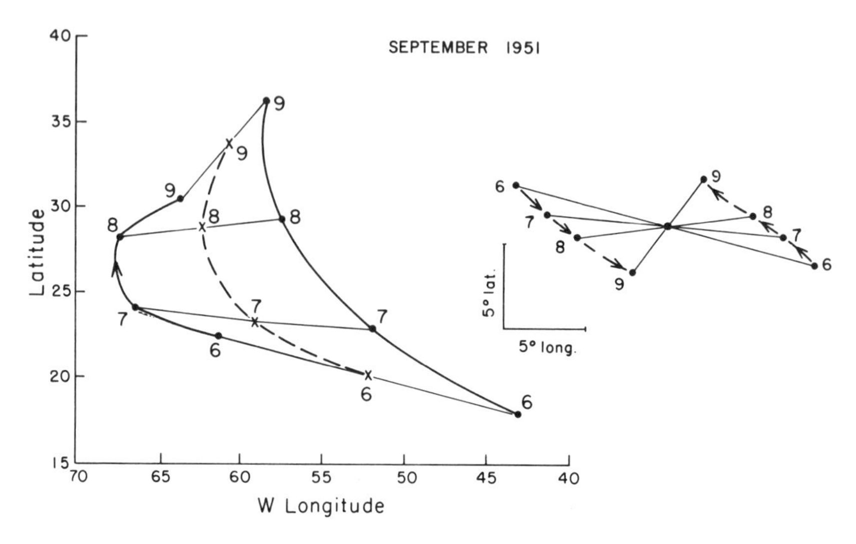

Fig. 50. Tracks of two hurricanes, with *broken line* denoting midpoint in Atlantic Ocean, 6–9 September 1951. *Right,* relative motion with respect to midpoint showing acceleration of rotation and mutual attraction; *lower left,* distance scale.

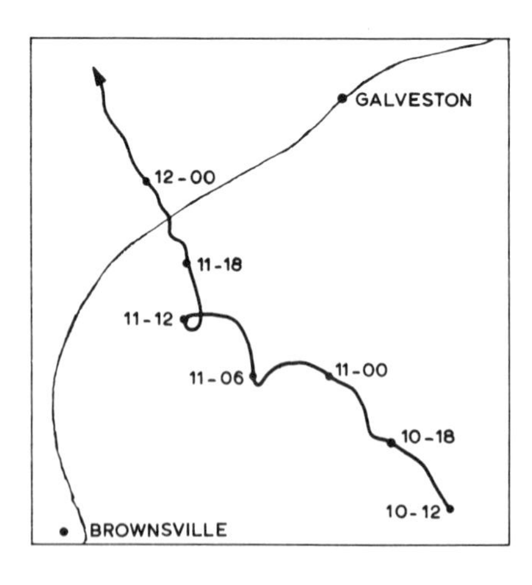

Fig. 51. Wobbly approach of hurricane Carla from Gulf of Mexico toward Texas coast, indicative of internal oscillations, 10–12 September 1961.

Seasonal Track Variation. A very broad and generalized picture of where tropical cyclones form and where they move is drawn in Figure 1. Superimposed on these mean tracks is a variety of seasonal and regional features. The subtropical anticyclones are most pronounced in the middle of the hurricane seasons of all major oceans. At that time the longest westward tracks are encountered (Fig. 52). We observe a distinct divergence of tracks near longitudes 60° W–70° W in the Atlantic. About half the cases turn north and northeast. The other half continues straight into the western Caribbean and the Gulf of Mexico to make landfall there. From Figure 52, we see that a crucial decision faces the forecaster when a hurricane reaches latitudes 15° W–20° N, longitude 60° W. Most storms do pass through this critical area. Where will the hurricane decide to go? Sometimes the decision is easy: when the anticyclone is far west of its average position, the hurricane cannot recurve, and must strike land. When the anticyclone is far out in the eastern part of the ocean, the hurricane is likely to recurve without landfall. But, there are many intermediate and complex cases that do not offer an easy solution.

In contrast to midseason, the westerlies are found farther equatorward in the early and late months. The belt of easterly winds is much narrower, and thus forming hurricanes very quickly come into contact with the westerlies. Any parabolic-track feature may be lost completely; the storms may start moving directly toward north-northwest or north and then quickly acquire a component toward east (Fig. 53). But here again one finds exceptions. The high-pressure ridge of the upper troposphere lies east-west near 10° N over Southeast Asia and the South China Sea area in the northern winter, forced into this position by the flow of the strong upper jet stream around the Himalayas and onward toward Japan. First, persistence of this ridge enables tropical storms to form even in midwinter in this area, though rarely. Second, the zonal upper easterlies enforce straight westward tracks through the Philippines and the South China Sea.

In the Arabian Sea and the Bay of Bengal hurricanes arise with the advance and retreat of the summer monsoon, as noted. These cyclones often travel very slowly over periods of several days to a

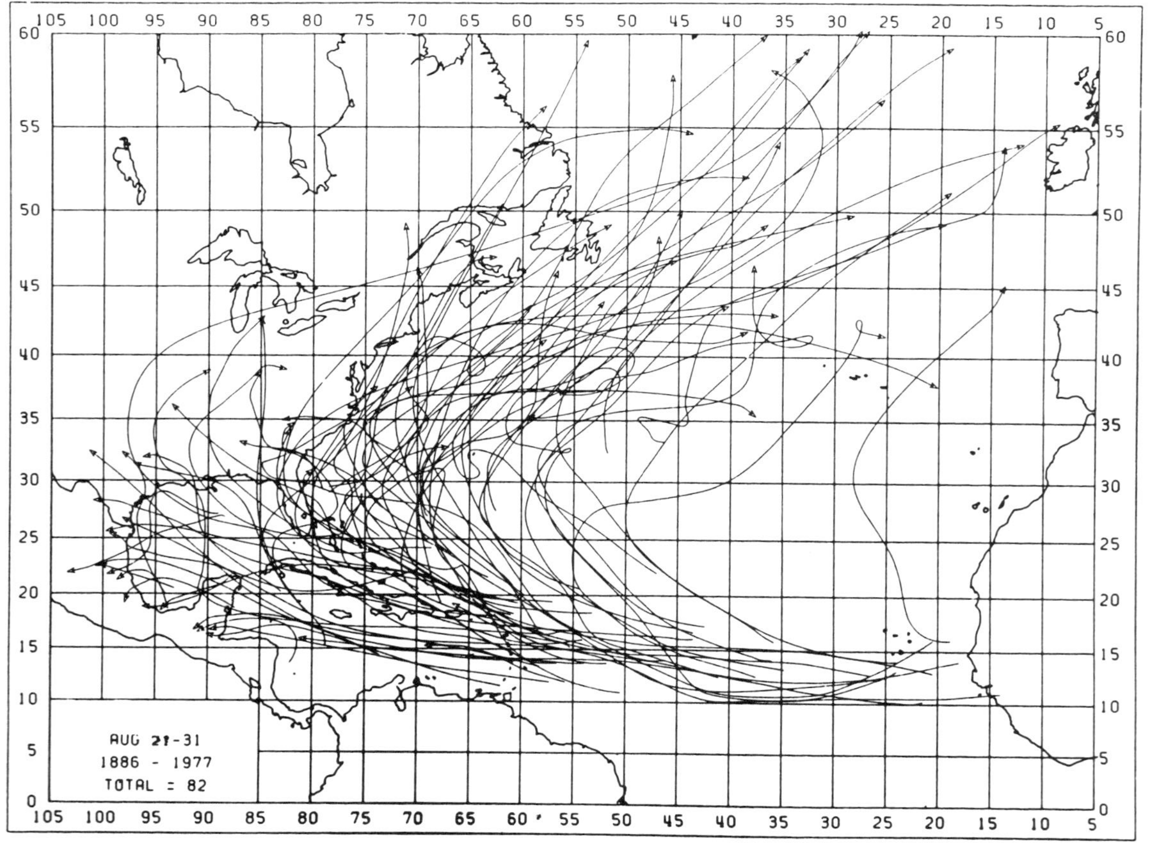

Fig. 52. Tracks of hurricanes for the period 21 to 31 August, 1900–1978. C. J. Neumann *et al*., *Tropical Cyclones of the North Atlantic Ocean, 1871–1977* (Washington, D.C.: NOAA, Government Printing Office, 1978), 161.

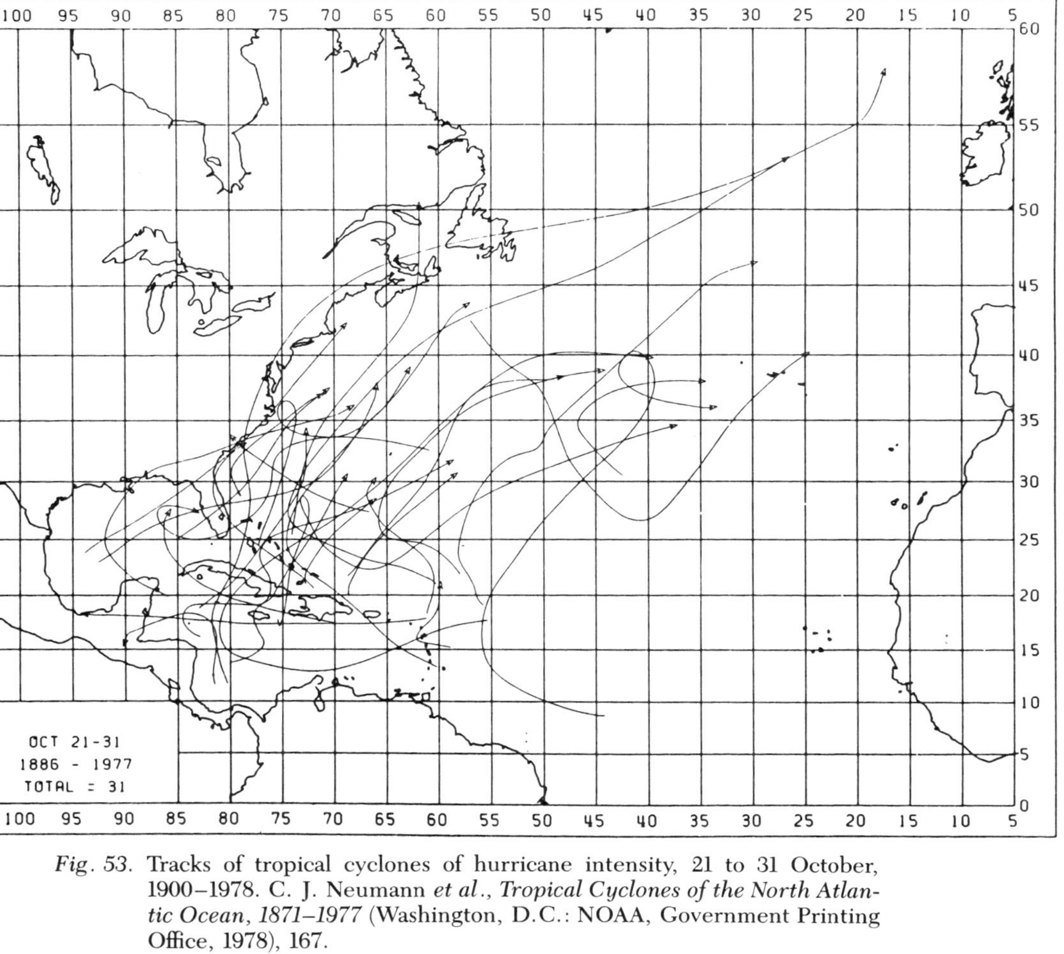

Fig. 53. Tracks of tropical cyclones of hurricane intensity, 21 to 31 October, 1900–1978. C. J. Neumann *et al.*, *Tropical Cyclones of the North Atlantic Ocean, 1871–1977* (Washington, D.C.: NOAA, Government Printing Office, 1978), 167.

week and then pass on rapidly. Predominant tracks are toward northwest in the Arabian Sea and north to northwest in the Bay of Bengal. But marked and sudden deviations occur, with much impact on the dense populations, at landfall. Arabian Sea cyclones may swing northeast to the Indus Valley. In the Bay of Bengal the worst disasters are at the head of the bay; but occasional landfall near Madras from east to northeast is not unknown and has led to some severe catastrophes. Satellite observations appear to be of great value in analyzing and tracking these systems.

For a longer-range outlook, often needed, the 3-day vector resultant path has been computed for the Atlantic tracks in terms of initial latitude and direction of motion, using the numbering scheme of Table 7. All seasons are represented in this table. The result is quite straightforward (Fig. 54). At the lowest latitude and with most westerly tracks, the 3-day vector remains west-northwest at moderate speeds. With more northerly path orientation, the 3-day path will also turn north and northeast, somewhat more slowly. At high initial latitudes the trend toward the northeast is stronger, and speed becomes higher again after recurvature. Because the error ellipses for the climatological motions are quite large, especially at initial high latitudes, the illustration can only serve for first rough orientation.

Table 7. Class Numbering Scheme for Figure 54

		Current direction toward	
Latitude	*270°–300°*	*310°–340°*	*350°–040°*
13°–18°	1	4	7
19°–25°	2	5	8
>25°	3	6	9

Terminal Hurricane Stages

The outdated proposition that a hurricane was finished upon landfall does not hold at all. Indeed, the variety of terminal stages is very great. Some have already been mentioned, and others will be discussed in subsequent chapters. Here, it will suffice to recount some of the most common varieties.

1. A hurricane on a parabolic path moves steadily into high latitudes without landfall and becomes an extratropical cyclone. This occurs in all oceans.

2. A hurricane moves onto cold ocean current and dies. This is most prominent in the eastern Pacific.

3. A hurricane makes landfall, is blocked by an anticyclone, and weakens under influence of inflow of colder and drier air.

4. A hurricane makes landfall when an extratropical storm is about to form anyway. Rapid redevelopment occurs over land. Some of the great hurricane disasters evolve from this coincidence in timing. On other occasions, hurricanes have continued as strong extratropical cyclones all the way across the Atlantic Ocean to the British Isles.

5. The surface hurricane circulation weakens on landfall but the upper-air center carries on and will lead to very heavy flooding on westward to northeastward paths. Such flooding is difficult to predict and can lead to enormous catastrophes over large areas and whole river basins.

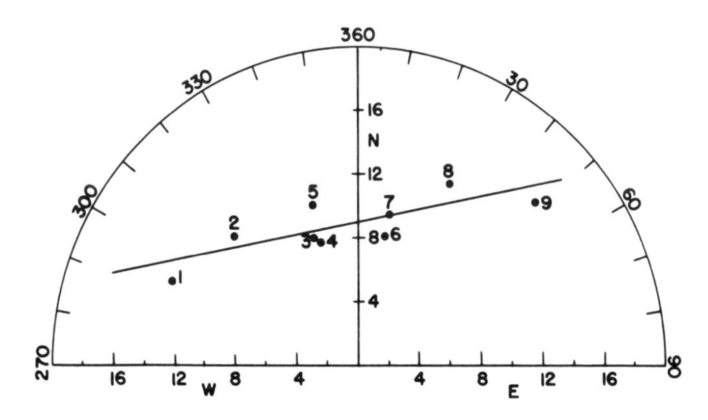

Fig. 54. Average 3-day tropical cyclone motion in Atlantic area for the nine categories listed in Table 7.

REFERENCES

1. E. Palmén, "On the Formation and Structure of Tropical Hurricanes," *Geophysica*, III (1948), 26–38.
2. G. E. Dunn, "Areas of Hurricane Development," *Monthly Weather Review*, LXXXIV (1956), 47–51.
3. T. C. Yeh, "The Motion of Tropical Storms Under the Influence of a Superimposed Southerly Current," *Journal of Meteorology*, VII (1950), 108–13.

CHAPTER 6

The Mature Hurricane

A clear demonstration of the hurricane as a working engine can be obtained by forming a composite, or a typical hurricane picture in all its facets from the ocean surface to the upper troposphere. This task has become possible to achieve through the many balloon soundings taken around hurricanes since the 1940s and through the invaluable penetrations of many storms by research aircraft of the United States. The latter method furnishes observations in the inner storm area, where balloons either cannot be launched due to high wind or are beaten back down by heavy precipitation. We begin with surface observations, of greatest value; they not only describe the hurricane's power but also permit inferences on the interaction between air and ocean, of vital importance for these storms.

Surface Structure

Pressure. The most reliable and widely used surface instrument yielding quantitative data is the barometer. Wind-measuring instruments are much more complex and are difficult to operate at sea, and at hurricane wind speeds they are often, if not usually, blown away. Waves on the ocean, until recent times, have been entirely a matter of qualitative judgment, though long swell, qualitatively, has provided advance warning. So it is the barometer that has always been carried by captains at sea and to which they have looked for their primary guidance. As Joseph Conrad relates in his famous tale "Typhoon," when the ship is in the eye of the storm:

> Captain MacWhirr had gone into the chartroom. . . . He groped for the matches, and found a box on a shelf with a deep ledge. He struck one, and puckering the corners of his eyes, held out the little flame towards the barometer whose glittering top of glass and metals nodded at him continuously. It stood very low—incredibly low, so low that Captain MacWhirr grunted. The match went out, and hurriedly he extracted another, with thick, stiff fingers. Again a little flame flared up before the nodding glass and metal of the top. . . . There was no mistake. It was the lowest reading he had ever seen in his life. Captain MacWhirr emitted a low whistle. . . . Perhaps something had gone wrong with the thing! There was an aneroid glass screwed above the couch. He turned that way, struck another match, and discovered the white face of the other instrument looking at him from the bulkhead, meaningly, not to be gainsaid, as though the wisdom of men were made unerring by the indifference of matter. There was no room for doubt now . . . the worst was to come, then—and if the books were right this worst would be very bad.[1]

The lowest barometric surface readings recorded are in hurricanes; and the lowest values known from ship reports and aircraft drop-sondes in eyes are in the range 860–890 mb, compared with standard surface pressure of 1,013 mb—lower by more than 10%. Could Captain MacWhirr have encountered so low a reading? Yet we can easily reach such low pressure all the time by just going up in the air or into the mountains by a little more than 1 km. Taking the elevator to the top of the Sears Building in Chicago, we arrive at the minimum pressure recorded in many powerful hurricanes. In the upper stories of that and other buildings the central pressure of moderate hurricanes prevails constantly, but no one is stopped from doing their daily work! The explanation is that the force exerted by pressure differences along the vertical (an elevator shaft) is balanced by the enormously powerful gravitational force; whereas pressure differences measured over much greater distances along the earth's surface are balanced (almost) only by the forces described in Equations 4.3 and 4.4

In the most severe hurricane that has struck the northeastern United States in the twentieth century, on 21 September 1938, lowest pressure was estimated as 950 mb when the center approached Long Island and the Connecticut coast from the south. This pressure still held at New Haven (41° N, 73° W) on the Connecticut shore (Fig. 55), passed over directly by the center. The barograph trace is particularly steep, because the hurricane moved very rapidly from the south with a speed of 55–60 mph (90–100 km/h); Figure 78d shows the track. In the last 3 hours before arrival, pressure fell by 40 mb, which is about the total pressure difference experienced in a strong winter cyclone. In terms of distance, the pressure gradient (pressure difference per distance) was 40 mb/250 km, showing the concentration of the extreme hurricane manifestations in a compact core even as far from the equator as latitude 42° N. Because of the compactness, traces such as that in Figure 55 are seldom obtained. Usually the center does not travel over a large weather station so directly! For comparison, the

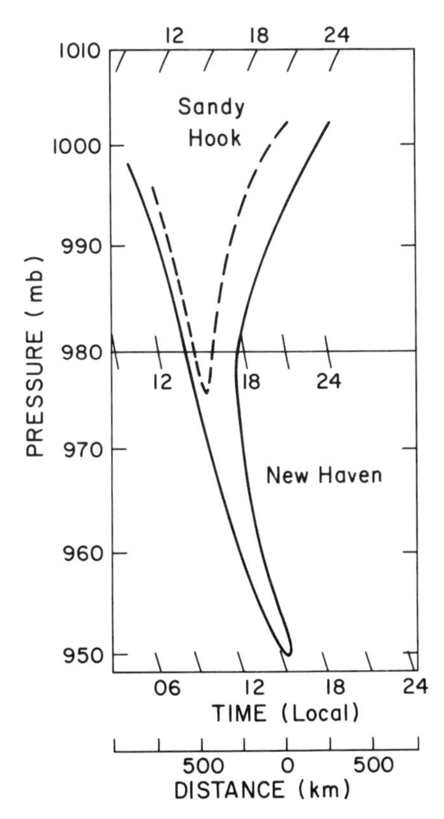

Fig. 55. Barograph traces at New Haven, Connecticut (41° N, 73° W) and at Sandy Hook, 120 km to the southwest in New York harbor, 21 September 1938.

trace at Sandy Hook on the southern end of New York Harbor, 120 km to the southwest, is added. Here, both fall and rise were as rapid as at New Haven, but the lowest pressure experienced was barely 975 mb—still near the record low pressure experienced there in big winter storms but a full 25 mb higher than central pressure! We see readily that, although distance inside the 1,000-mb isobar was 700 km (Fig. 85), the most active working part of the hurricane occupies only a few percent of the area enclosed by the 1,000-mb isobar, even though winds of 15 to 20 mps may still be blowing around that boundary.

A hurricane with 950-mb central pressure is always rated a severe storm. From one correlation between highest wind speed and central pressure, 950 mb corresponds to a maximum wind of 50 mps or 100 kt, averaged around the ring of strongest wind. Because, as the next section shows, winds are nearly always asymmetric about the center, a much higher speed in one quadrant of the storm and, of course, still higher short-term and gust speeds are implied. Attempts have been made to correlate the pressure difference between the inside and outside of hurricanes with the maximum wind speed, based essentially on Equation 4.4 but with allowance for friction. Figure 127 shows a correlation diagram that appears reasonably well borne out by observations.

The Wind Field. Because wind is a vector (carrier or bearer), it has direction as well as speed; they must both be considered. First a few models may be considered. Figure 56 affords a comparison of wind profiles with radial distance from the center, for maximum, moderate, and minimum hurricanes as well as for a "tropical storm" that has not, and may never, attain hurricane strength. In the latter case strongest winds are typically, though not always, at a greater distance from the center than in hurricanes in which the ring of strongest wind is by and large found at radii of 30–50 km (20–30 miles). Appendix C contains a rating of "damage potential" in terms of ranges of strongest wind speed.

When a hurricane lies embedded in what we may term a *steering current* of large scale, the speeds of the steering current and of the vortex are largely, though not completely, additive (Fig. 57). To

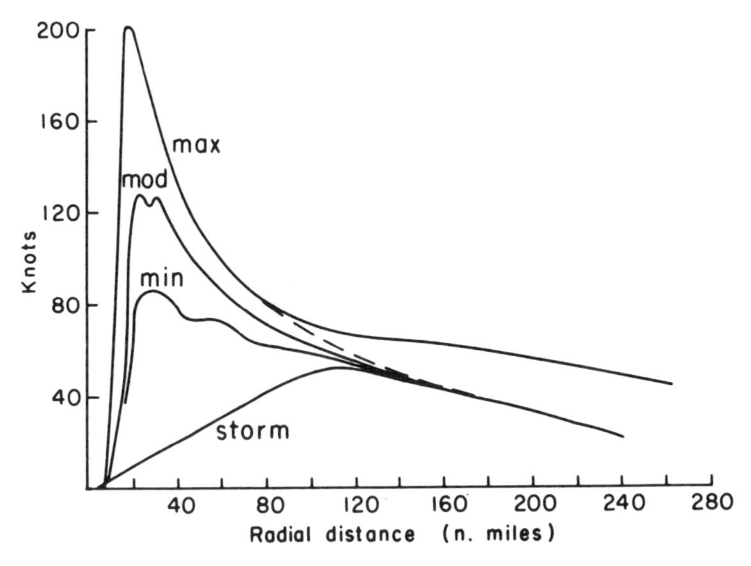

Fig. 56. Model of radial profiles of wind speed for three hurricane intensities and for tropical storm.

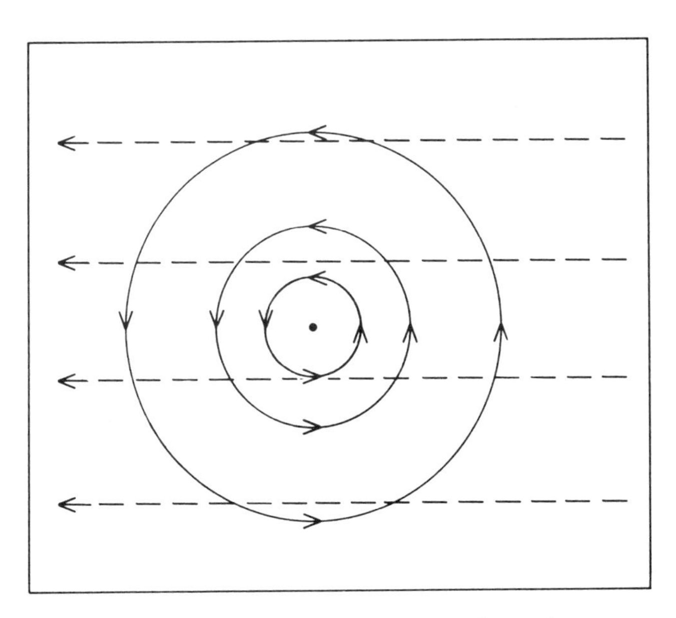

Fig. 57. Linear superposition of vortex circulation and straight steering current.

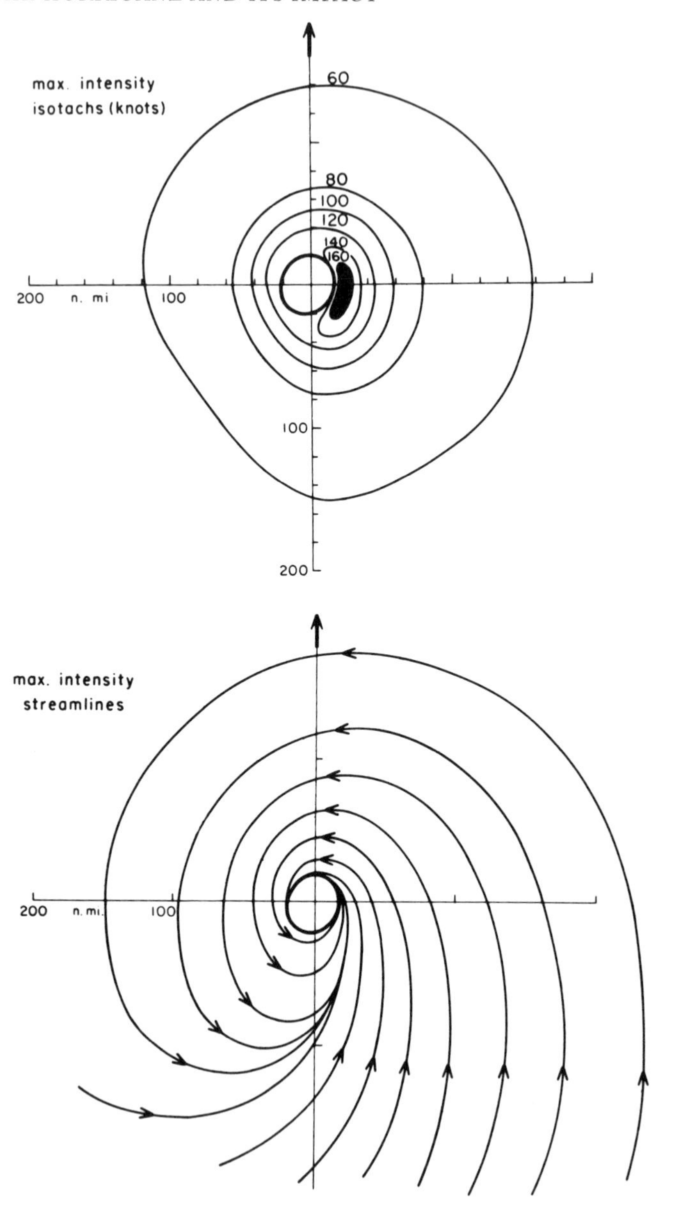

Fig. 58. Model of wind-speed distribution and streamlines for extreme hurricane, drawn with respect to direction of motion pointing upward (arbitrary).

the right of the direction of motion of the center, the direction of vortex motion and steering current coincide; on the other side they are opposed. Thus speeds are almost invariably higher to the right than to the left of the direction of motion in moving hurricanes, though with a few glaring exceptions (hurricane Celia of 1970).

Figure 58 is a model of a very intense storm moving at a speed of 20 kt (10 mps). Thus, speeds attain 160 kt (82 mps) to the right of the direction of motion, while they are held to 120 kt (62 mps) to the left. In the lower part of the diagram we observe the counterclockwise, or cyclonic, rotation for the Northern Hemisphere; it would be inverse for the Southern Hemisphere. The streamlines all spiral inward to the ring of strongest wind. The spiral is observed in all hurricanes. Figure 59 portrays typical speed distributions for moderate and minimal hurricanes. It is seen that the latter has hurricane-force winds only on the right side. Because of the spiraling streamlines, the stronger the winds, the stronger will be the circulation of mass through the central core. As a hurricane intensifies, mass transport through the central part increases. In tropical storms of less than hurricane intensity the ring of strongest wind usually is much farther out from the core than in hurricanes. Contraction of the ring, like contraction of the cloud envelope, serves as a warning signal of intensification.

Turning now to actual cases, we see in Figure 60 the large-scale wind field of a severe northeastward-moving hurricane in October 1944. Strongest winds are situated to the right of center direction; but winds of 30 mps (near hurricane force) extend over 500 km toward east and southeast—only 200 km to the northwest of the core. For wind distribution near the center, by far the best picture is obtained from aircraft traverses that are unhampered by the problems of measurement close to the surface. Figure 61 shows the wind distribution in hurricane Anita, moving westward in the central Gulf of Mexico, as taken from an aircraft of the National Oceanic and Atmospheric Administration (NOAA) Research Flight Center on 1 September 1977. The northern side is to the right of center direction, with winds of 130–140 kt (65–70 mps) over a band 20 km wide, whereas speeds barely attain 100 kt (50 mps) on the south side. The eye, with dropoff of wind to nearly calm condi-

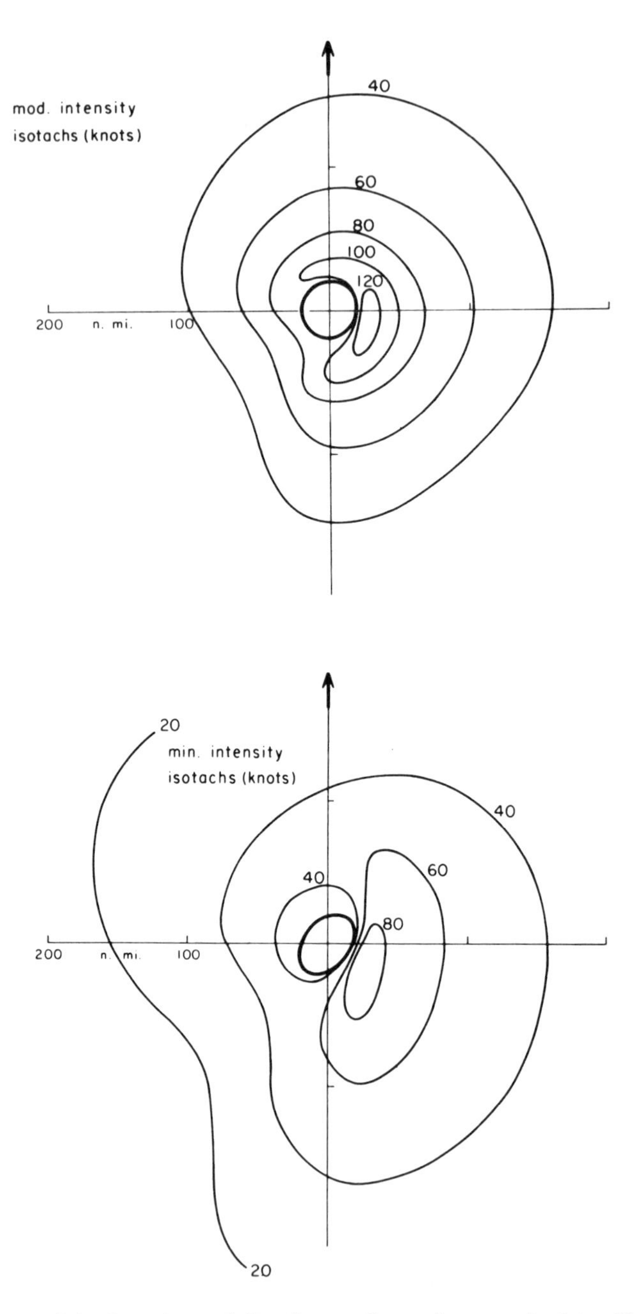

Fig. 59. Models of wind-speed distribution for moderate and minimal hurricane, drawn with respect to direction of motion.

tions at the center itself, is very small, with a radius of only 20 km or diameter of 40 km, in this rather formidable cyclone. Quite often eyes with a diameter twice this distance are observed in mature hurricanes, but in intense immature storms still smaller eyes have also been encountered.

When wind speed is plotted against radius on a logarithmic scale a virtually straight, sloping line is obtained outside the ring of strongest winds, drawn in Figure 62 for the winds to the right of the direction of motion. Wind distributions that can be represented by such a straight line are common and have been published from all hurricane areas. It may be noted that the flight was executed at a 3-km altitude (700 mb). The altitude should hardly affect total wind speed, which is nearly constant from the vicinity of the surface to perhaps 7 to 8 km (400 mb) height. But the spiral

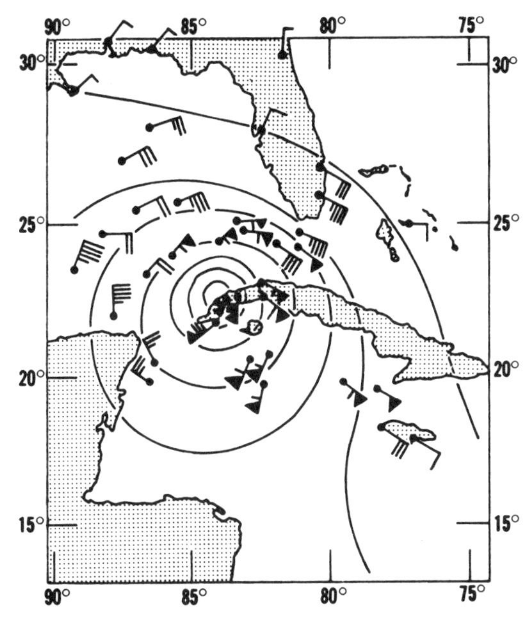

Fig. 60. Surface winds and isobars in the Gulf of Mexico-Caribbean area on 18 October 1944, 1,200 GMT. Winds drawn for direction from which wind blows: *short barb*, 5-kt winds; *long barb*, 10 kt; *heavy triangular barb*, 50 kt. Outer isobar, 1,016 mb; isobar interval, nearly 7 mb. Gordon E. Dunn and Banner I. Miller, *Atlantic Hurricanes* (Baton Rouge: Louisiana State University Press, 1960), 67.

shown in the model of Figure 58 and also evident in Figure 60 is no longer present, for at flight altitude, air no longer drifts inward. The whole mass flow toward the center tends to be confined largely to the lowest 1,000 m (100 mb).

In Figure 61 the lines representing equal height of the 700-mb surface, comparable to isobars on a surface-pressure map, are nearly circular. Their profile is very similar to that of Figure 55; the pressure gradient steepens inward quickly to the edge of the hur-

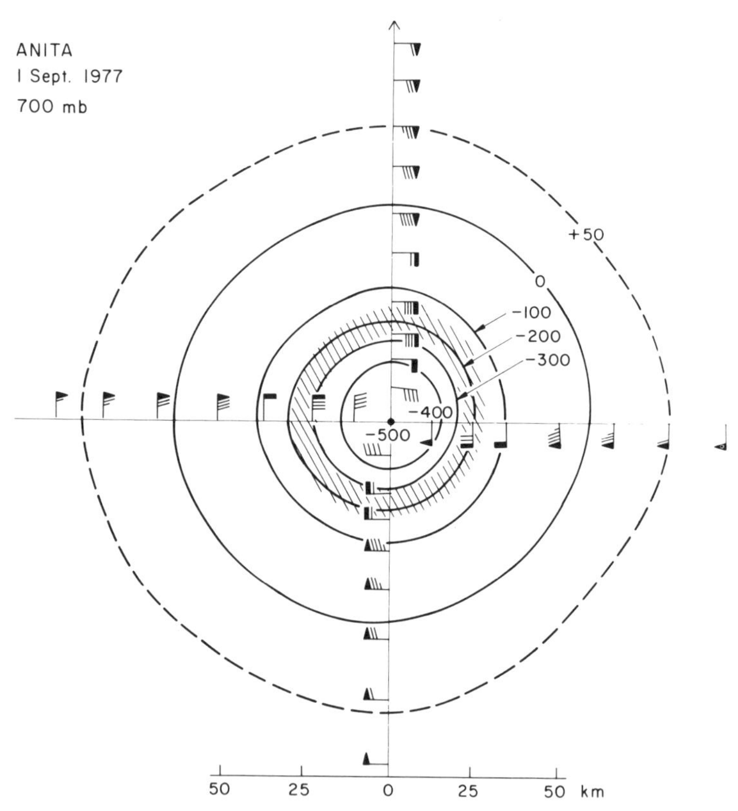

Fig. 61. Wind speed composited with respect to hurricane Anita in the Gulf of Mexico on 1 September 1977 (see Fig. 78 for position). *Short barb*, 5-kt winds; *long barb*, 10 kt; *heavy triangular barb*, 50 kt; *heavy rectangular barb*, 100 kt; *circular curves*, deviations (in meters) of the height of the 700-mb surface from standard height of 3,010 m; *hatching*, ring of maximum wind; *bottom*, distance scale. Courtesy of Robert C. Sheets, NOAA.

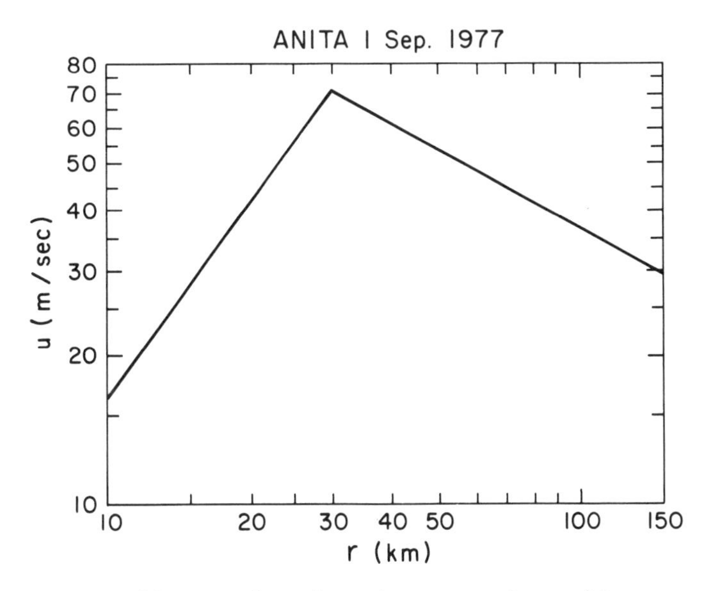

Fig. 62. Rotational (tangential) wind speed, u, averaged around hurricane Anita (Fig. 60) and plotted logarithmically against the radius.

ricane eye. Total height difference is equal to 50 mb, comparable to Figure 55; but the distance over which this fall occurred is little more than 100 km, showing greater concentration of pressure fall together with stronger wind speed in Anita than in the September 1938 hurricane.

The reader will note in Figure 61 that the relation between pressure gradient and wind along the north-south axis, which is perpendicular to the direction of storm motion, is not fully explained by Equation 4.4. The contours of the 700-mb surface are quite symmetrical, whereas wind speed is asymmetrical, with higher speeds to the north than to the south of the center, in conformity with the hurricane motion. Equation 4.4 is written strictly for a stationary system. For a moving center, the radius of curving trajectories, rather than the simple distance from the center, must be introduced. As the reader can verify quickly, the radius for the westward-moving center will be smaller for trajectories to the south than to the north of the center; they must curve more sharply. Introducing such a variable radius in Equation 4.4, the basic relation is again shown to be valid. For the same pressure force, winds must be weaker in the south than in the north.

Clouds. In earlier centuries hurricane warnings were largely issued by missionaries who, with sound training and high skill in observations, were able to develop a tradition in systematizing advance warnings of an impending storm. Among these indications was a wind blowing from unusual quadrants in the trades, especially from the west, or just a dying down of the trades. Particularly clear skies were another forerunner which, however, made it very difficult to convince the local populations that bad weather was in the offing. Various catastrophes involving much loss of life ensued from neglect of the warnings, even to present times. Satellite photos very definitely support the missionaries' observations. For instance, there was no outer forward cloud shield at all in cyclone Tracy on 24 December 1974 (Fig. 42) as it turned south-southeast toward the coast. The outer cloud bands were all rearward. In the experience of mariners the impression of a sudden appearance of hurricane clouds and winds is expressed by their

Fig. 63. Cloud line in hurricane off Palm Beach, Florida, on 17 September 1947 (see Fig. 78 for position). Note towering cumulonimbus with strongly shearing anvil. Photo from aircraft at 10 km, by R. H. Simpson.

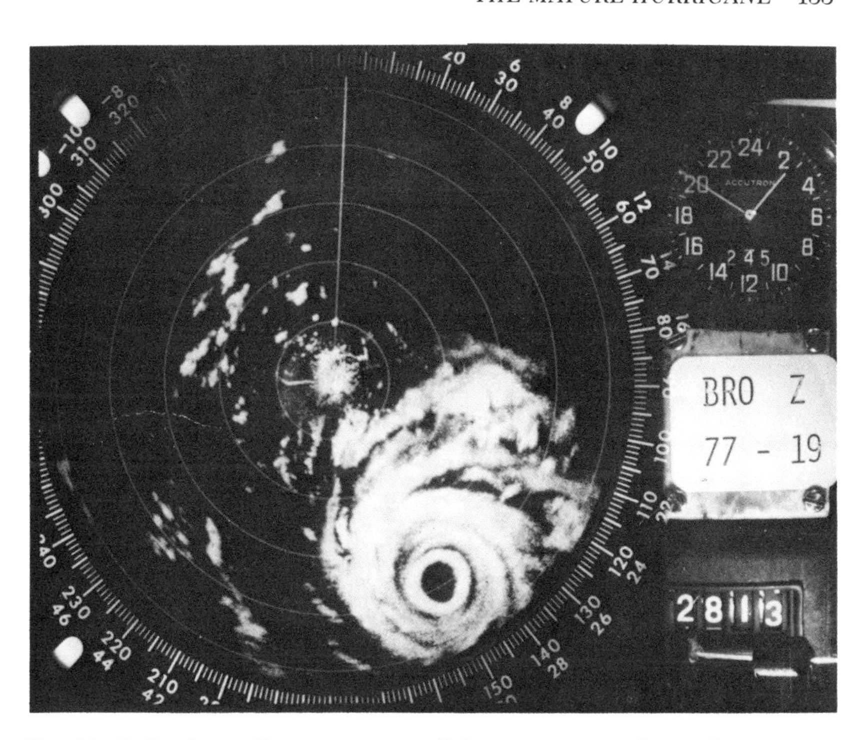

Fig. 64. Radar photo of hurricane Anita off the Mexican coast during the morning of 2 September 1977 (for position, see Fig. 78), taken from Brownsville, Texas (26° N, 97.3° W). Distance of hurricane center from radar is about 180 km. WSR-57 radar, plan position indicator (PPI) presentation. Range marks at 25-nm intervals (45 km). Courtesy of Neil Frank, National Hurricane Center, Miami.

term "bar of the storm," or the edge of the severe portion, so evident on 24 December 1974.

The clouds, especially at the outer edges, form long streets that then spiral inward. Occasionally, lucky photography catches an outer band, as in Figure 63, where a strange anvil from strongly shearing cumulonimbus reaches out over the top. Because satellites see hurricanes from above, the broader spiral bands on satellite photos often show outflow clouds that move clockwise in the Northern Hemisphere and counterclockwise in the Southern. Best portrayal of the inflow spirals of the low troposphere comes from observations by radar, which "sees" heavy rain. A good example is

Fig. 65. Application Technology Satellite III (ATS) photograph of hurricane Camille in lowest part of the Mississippi basin, 18 August 1969. Width of heavy cloud mass, or central dense overcast, is about 500 km. See Figure 78 for position.

hurricane Anita of 1977, seen by the Brownsville, Texas, radar (26° N, 98° W) from the north (Fig. 64) as Anita approached the Mexican coast on a track toward southwest (Fig. 78a).

With rare exceptions tropical hurricanes are overlaid by a cloud mass, as seen for Camille in Figure 65. The most intense part is situated off-center to the right of the direction of motion, which is toward north-northwest. Usually a central dot denoting the eye is visible, and one or two long curving arms extend from the central portions toward southeast and south and/or toward northeast, all occurring in reverse for the Southern Hemisphere. The equatorward branch follows the outflowing mass (Fig. 72) to a large extent.

Precipitation. In this setting the precipitation occurs that, subsequent to the initial formation, becomes the main driving mechanism of the hurricane. The upper outflow of cloud, mostly cirrus, can be counted as precipitation as well, since it also removes water or ice from the immediate hurricane environment. But the amount is estimated as small, perhaps one-tenth of the precipitation, and therefore is well within the error limit of any known method of estimating precipitation within the storm.

These measurements have proved very difficult. Most rain occurs in the high-wind area; it is estimated that at wind speeds above 50 kt (25 mps) not more than half the falling water is caught by the typical rain gauge. Further, local distortions such as buildings, trees, or hills near a gauge will introduce strong local factors. Much depends on how a hurricane moves with respect to a gauge (whether to right or left); how fast it progresses; and last, though of most interest, how intense and large is the circulation of mass through the hurricane. Under these circumstances widely different values of hurricane precipitation have been reported all over the world, ranging from practically nothing (in spite of hurricane-force winds) to as much as a whole meter (40 inches) and even more. It must be emphasized that many of the heavy rains are only ancillary to a hurricane (see Chapter 10) and that often they occur in very late stages when the hurricane moves outside the tropics. In a core area such as that depicted in Figure 58, the heavy rainbands are sometimes close to the center in the cloud eye wall, where wind also attains its strongest velocities. At other times the precipitation is rather widespread; it may lie outside the ring of strongest wind and extend over a radial distance of 100 to 300 km, especially to the right of direction of motion.

Under these highly confusing circumstances the best way to find out about hurricane precipitation is by numerical computation instead of by measurement, though in the future radar may provide a means for reasonable estimation. The computations have the further advantage that one obtains precipitation following the storm itself, which excludes variable rates of storm motion as a factor. From aircraft traverses with one, two, or, at most, three aircraft flying at different levels one can only establish, with some uncertainty, the precipitation in concentric rings around a center. The amount of moisture transport inward by the spiraling streamlines between two radii, say 50 and 100 km from the center, is computed from the aircraft data. The convergence between these rings forces ascent and condensation to occur (Chapter 3). If the rain essen-

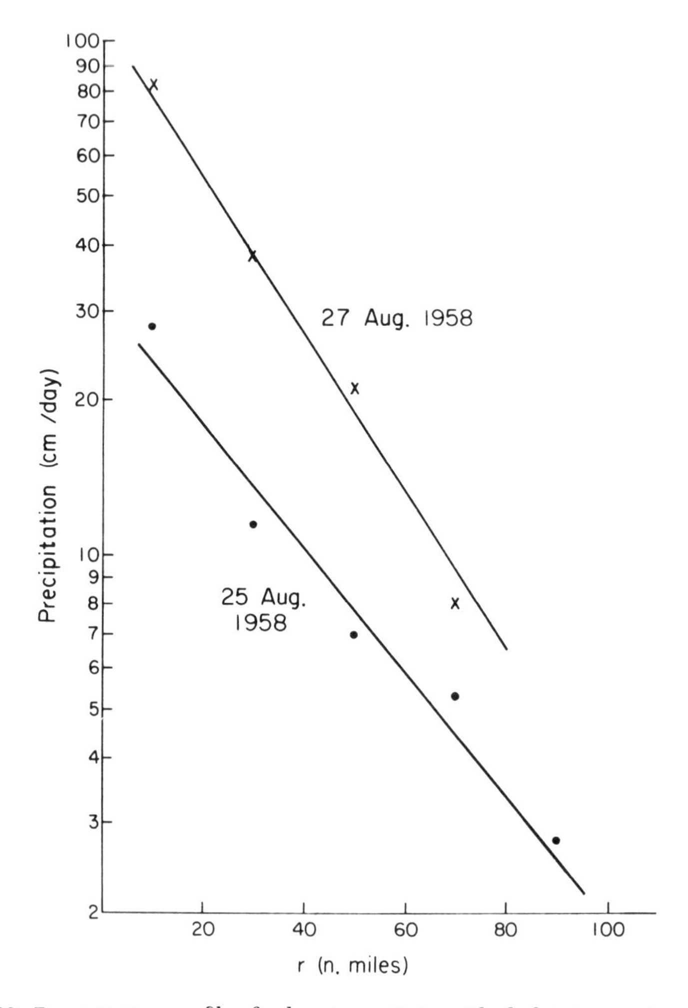

Fig. 66. Precipitation profiles for hurricane Daisy. *Black dot*, in growing stage, 25 August 1958; *x*, in mature stage, 27 August, off Florida's east coast. Logarithmic plot against radius. The precipitation is computed moving with the hurricane center and represents ground rain-gauge accumulation only if hurricane is stationary. H. Riehl and J. Malkus, "Some Aspects of Hurricane Daisy (1958)," *Tellus*, XIII (1961), 181–213.

tially falls out in the ring in which the water is condensed, if little or no water is lost by reevaporation on the way down, and if about 10% of the water or ice is transported away at high levels, then the residual must be the precipitation, assuming the hurricane itself does not change shape and intensity.

In spite of these many *ifs* computations in different storms have led to comparable results and give a rather clear picture of the precipitation distribution in principle. In Figure 66 we see precipitation profiles for hurricane Daisy for 1958, first, in the formative stage on 25 August and then in the moderate mature stage 2 days later. Precipitation triples over the 2 days, as the hurricane intensifies. Both profiles, as well as all others published, show a logarithmic distribution with radius, which emphasizes concentration of rain rate close to the center. Even on the first day, when aircraft intercepted the center just as winds were beginning to exceed hurricane force, rainfall in the very core was 30 cm/day; it ran to 80–90 cm/day on 27 August. Similar values have been found in other cases, including a composite average Pacific typhoon (Fig. 67), in which a rather low amount of mass inflow was used.

When the hurricane is moving along, of course, the extreme amounts do not occur at any fixed location on earth. One can lead a station through the computed rainfall pattern of a hurricane at the speed of hurricane travel and compute the amount collected in a rain gauge during the whole storm passage. The latter, on the average, takes 2 days, and if the path of the rain gauge is directly through the center, total rain for the event will be about 35 cm (14 inches), which is indeed large and may serve to furnish an average measure. If a storm halts in its track, however, locations near the center may indeed experience a full meter of rain and a flood catastrophe. This happened in October 1963 when a hurricane wandered aimlessly for 4 days through the mountainous area of eastern Cuba (Fig. 83). A remedial feature is that such very slow motion, especially partly over land, is not conducive to a hurricane's maintaining full force and a concentric shape, and thus central precipitation may diminish.

The hydrologist, in estimating potential floods and their magnitude, must bring to bear still another consideration; namely, while rainfall rate decreases logarithmically from the center, the area covered by rain increases with the square of the radius! Thus, depending on size and shape of a river basin and its eventual discharge downstream, the heaviest contribution of water may not be derived from a small central area but from a larger peripheral area. In Figure 66, for instance, the crosses on 27 August represent means of concentric rings 20 nmi apart. If we extend the curve outward to obtain another point in the interval 80 to 100 miles, Table 8 results. Clearly, the area between 35 and 110 km distant from the center makes the greatest contribution to water received at the ground and therefore will largely determine the amount of flooding to be expected. Even in the interval 110–150 km, total water depos-

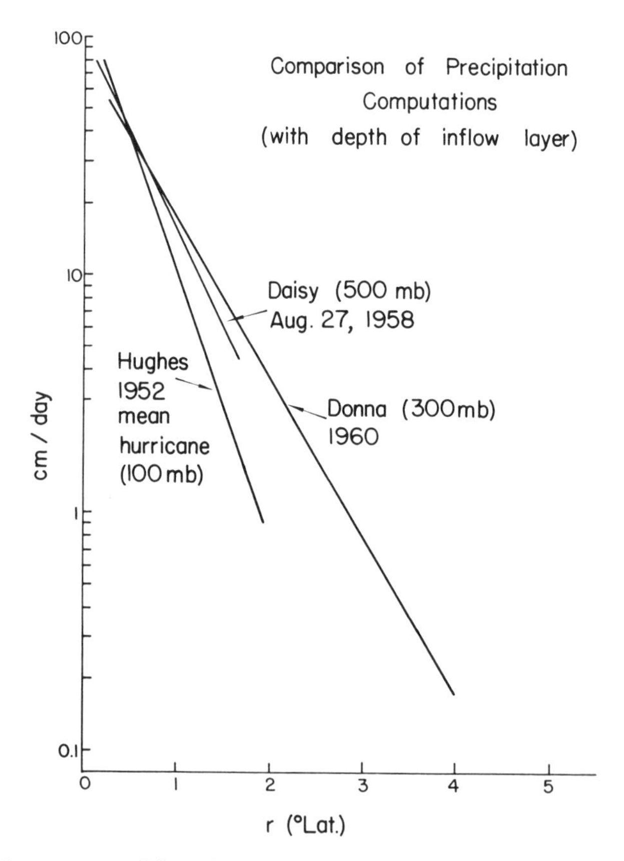

Fig. 67. Comparison of three hurricane precipitation profiles, plotted as in Figure 66. Hughes data from Pacific typhoons.

Table 8. Daily Volume of Water (m^3) from Hurricane Precipitation between indicated Radii (Following Center)

Radial interval (km)	*Area ($10^2 km^2$)*	*Rain rate (cm/d)*	*Water volume $10^8 m^3/d$*
0–35	37	80	30
35–70	120	40	48
70–110	220	20	44
110–150	310	10	31
150–190	440	5	22
			175
			= 17.5 km^3

ited at the ground is still as large as in the very core! Further, the grand total of over 17 km^3 in one day is no less than the total Colorado River discharge in a whole average year! Similar accounts have been computed for other hurricanes. To be sure, the Colorado runoff is near only 13% of annual average precipitation in the upper basin, which contains the mountainous source region. Most of the precipitation evaporates. But in a hurricane most water is likely to run off unless there has been marked antecedent drought. The comparison grossly highlights the flooding dangers and the problems of river engineers with hurricanes, as well as the devastation of parts of New England through flooding when in August 1955 one hurricane followed within 10 days on the path of another one for some distance.

We see at once that the message contained in Figures 66 and 67 depends very much on what aspect of the profiles the user considers most relevant to his problem. The viewpoints of meteorologist, hydrologist, and say, civil-defense worker all diverge strongly. For the meteorologist, who places the rainfall in context with the barograph traces such as those in Figure 55, clearly the total mass of water derived from the whole storm is of little concern. His problem is to judge the efficiency of the rainfall in providing energy to drive the hurricane; from all that has been said herein, this is not at all an easy or straightforward task. From the viewpoint of latent heat release (p. 35), the efficiency is 10% compared to normal 1%–2%. However, the efficiency of the pressure drop from outside to inside (Fig. 55) will rise to fully 50% for air that attains the maximum speed.

The Hurricane Eye. The pressure observation by Captain MacWhirr in Joseph Conrad's story was made in the typhoon center, where the extremely strong winds abate. Such abatement inside the ring of strongest wind is also evident in Figures 58, 59, and 61 as well as in all tropical hurricanes and cyclones outside the tropics. The center is revealed as a "singular point" (small area): pressure stops falling, wind stops blowing hard, rainfall ceases, clouds lighten or disappear so that the satellite photograph shows a central small hole, and the ocean waves are confused. Exhausted birds settle on ships' railings, and people come out of their houses and breathe a sigh of relief, though if they know what a hurricane is, they also realize that the second, and often strongest, half is still to come.

Eye diameters vary from 5 to over 60 km, depending on rate of storm propagation, the relative calm and quiet will last from minutes to hours at any location experiencing center overpass. In the case of New Haven in Figure 55 it was only minutes, since the hurricane was racing northward rapidly. Sandy Hook, on the western side, had no lull at all but experienced gradual turning of wind from north to northwest and west. Inside the eye, temperature may be almost or exactly as high as on the outside of the hurricane, in spite of extremely low pressure. People have often reported warm suffocating air in the eye, but this is dismissed largely as a psychological reaction to what has gone on before. In a few famous observations on record, temperature rose by as much as 10°C in the eye, and not just at noon, when if the eye is large, sudden sunshine could have a marked influence. Observations such as those at Manila and Formosa, in the immediate vicinity of mountains, have been interpreted as warm foehn-type, downslope-derived temperatures. It is not inconceivable, however, that on rare occasions very warm and dry air from higher levels does briefly break through to the surface. Usually, the lowest 1,500 m (150 mb) of the eye is very moist and often with a cloud in the middle, or even one or two little cloud spirals seen clearly from aircraft flying in eyes at higher elevations.

Fig. 68. Low-level view of banded spiral clouds in hurricane Beulah eye over the western Atlantic Ocean in September 1963. Photo by R. H. Simpson.

Figure 68 is a spectacular eye photo taken in the large hurricane Beulah of September 1961. A low, narrow stream of air visible through a chain of low clouds pours into the eye from the wall, dimly visible in the background. It then curls up in the cloud situated in the very middle. The inner edge of the wall cloud towers like a giant above the pleasant low-level picture (Fig. 69). At the top large ice falls cascade into the eye, and the striations down the wall portray the progress of descending motion over a vertical distance that may be 12–15 km deep. The ice streaks slant down counterclockwise. Ascent of the cumulus towers shows the reverse: leaning backward with height as wind speed diminishes upward.

Eyes, as seen by radar, are always in a state of unrest; they change shape, sometimes form double eyes, intensify and weaken. The rain bands from outside, carrying the strongest updraft, have a tendency to push into an eye which then becomes circular and very strong. After a few minutes the impulse subsides; the eye may

Fig. 69. Eye wall with downward slanting striations toward left (cyclonic) in hurricane Esther, 1961. Cloud top estimated above 12 km (200 mb). Photo by R. H. Simpson.

become deformed and weaken until the next band forces its way inside. Minor pressure oscillations accompany these changes; sometimes pressure is lowest at the eye center, sometimes at the rim with a flat pressure field inside. All these are turbulent features experienced in the center of a very violent and energetic event; they do not affect continuity in following a storm eye over hours or days. In fact, in spite of the many disruptive forces and motions present, hurricanes tend to hold together remarkably well.

Upper Air Structure

From earlier discussions, especially of initiating disturbances in Chapter 4, it has already become evident that the high troposphere clear to its top is heavily involved in the hurricane event. Thus the latter cannot be a system confined to the low atmosphere, say up to 3 km, as once believed by some meteorologists. As shown conclusively by Bernhard Haurwitz in 1935, such a constriction of the hurricane to low levels is quite unthinkable.[2] Consider, for instance, a top of a storm at 16 km (100-mb pressure), with surface pressure of 950 mb, as in Figure 55, and outside pressure of 1,010 mb. Then the pressure drop is 60 mb and, from the static viewpoint, must result from temperatures that are warmer inside the hurricane than outside. With the data given, the temperature difference between inside and outside is 8.4°C for the whole troposphere; if the vortex was confined to the lowest 3 km, the difference would have to be 50°C, a totally ridiculous value not obtainable by any known physical process. Our problem will be to see how to obtain approximately 8°C temperature difference; this is far from a small task!

Vertical Wind Structure. We begin with a description of how winds change with height in different parts of hurricanes. Such information has become available through numerous balloon sondes in the outer portion of hurricanes and through aircraft reconnaissance probes in the core. The air layer closely adjoining the ground will be treated in later chapters. One distinguishes between the wind component flowing around a center (tangential component) and the component blowing inward or outward (radial component). For an average large mature hurricane or typhoon, composited from many individual cases, the tangential wind increases inward from the 400-km radius shown (Fig. 70). At the 100-km radius (1°) hurricane force is attained, on the average, around the center. The tangential component is nearly constant, with height to 500–400 mb (6,000–7,000 m), and then decreases upward rapidly. At the outer 4° radius (400 km) the sense of rotation changes from cyclonic to anticyclonic (counterclockwise to clockwise in the Northern

Hemisphere) above 200 mb (12,000 m). At the 2° radius (200 km) cyclonic rotation is still present, although weak, and at the 1° radius (100 km) it still averages close to 20 m/sec, variable with storm size and intensity. From balloon soundings in the vicinity of hurricanes, one can assume safely that the circulation has wholly or very nearly disappeared at 100 mb (16,000 m), which is the base of the stratosphere. The principal feature shown by Figure 70 is that at high levels an outer anticyclonic ring of air encloses the inner cyclonic flow.

The radial wind is small compared to the tangential component and is difficult to present in composite form. Nevertheless, the mean values of Figure 71 give good general guidance for the outer parts of hurricanes. Inflow is largest very close to the ground, and decreases upward. Nearly all inflow is confined below 800 mb (2000 m), and there is a strong suggestion that the top of the inflow layer decreases to 1 km or even less quite close to the central cloud wall. However, in the whole outer region slight contraction must take place in the middle troposphere. Otherwise, the deep layer with nearly uniform cyclonic rotation could not exist, as explained very succinctly by K. Ooyama.[3]

In the upper troposphere, there must be outflow compensating for the low-level inflow if a hurricane is to remain steady: as much mass must leave as comes in. The outflow, from Figure 71 and more detailed evidence, is strongly concentrated around 200 mb and even as high as 150 mb in intense storms. The top of the outflow layer rises with storm strength.

Radial Mass Flow. Whereas the radial component (inward or outward) is larger at small than at large radii, this does not serve as an immediate indication of converging and diverging mass flow, the feature of most interest because it determines vertical mass flow and formation or dissolution of clouds and rain. The mass flow is given by the product: radial motion multiplied by the radius. In Figure 71 the inward flow is 4 mps at the 4° radius and 8 mps at the 2° radius. In each case the multiplication gives 16; *i.e.*, the mass flow is constant between the two radii. There is no net mass ascent (though it can well occur in individual quadrants). The conver-

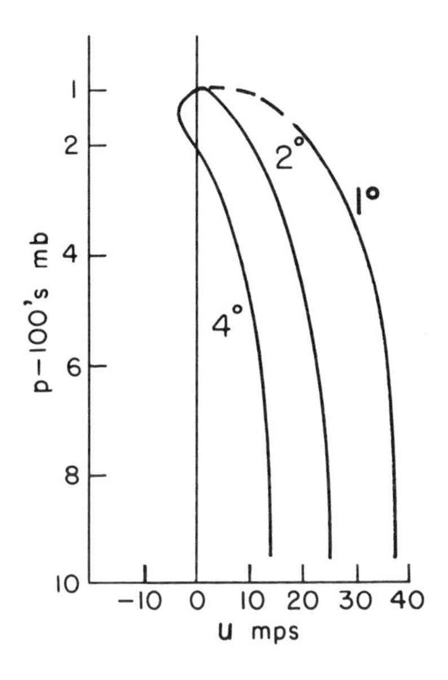

Fig. 70. Vertical profiles of rotational velocity, u, averaged around a composite typhoon at three radii.

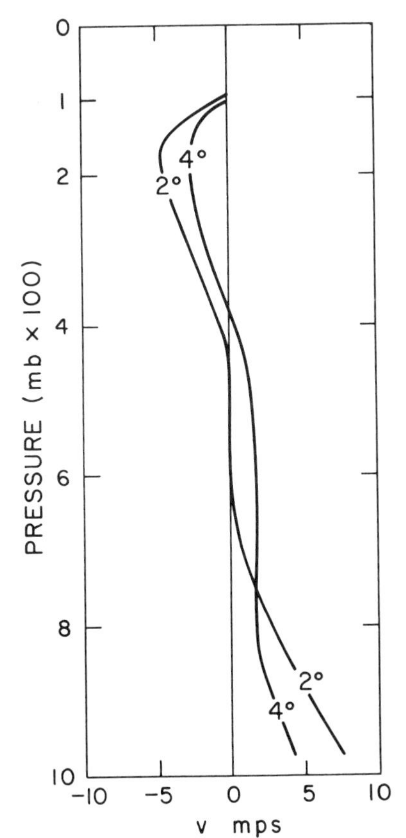

Fig. 71. The corresponding radial wind component, v, at two radii.

gence of mass leading to the severe inner cores and the cloud shields seen by satellite usually starts near the 2° radius and becomes very intense near the 1° radius (the bar of the hurricane). Similar reasoning of course holds for the upper outflow.

A composite view of the inflow layer in map form has been presented in Figure 60. For the outflow layer we again chose the intense hurricane Camille as an example (Fig. 72). All data are obtained from balloon sondes as the hurricane passed northwestward through the Gulf of Mexico at peak intensity with surface pressure near 900 mb. The inner cyclonic circulation is sketched but could not be observed, because no special high-level research flight was undertaken. From the balloon wind data, the clockwise flow spiraling outward from the core, mostly to the right of the direction of motion, is clearly evident. Some of the outflowing current exists into the westerlies farther north; a major portion turns toward southeast and south with increasing strength up to 50 kt and settles into the upper trough to the east. There sinking motion must occur, as evidenced by Figure 65; the cloud bands terminate

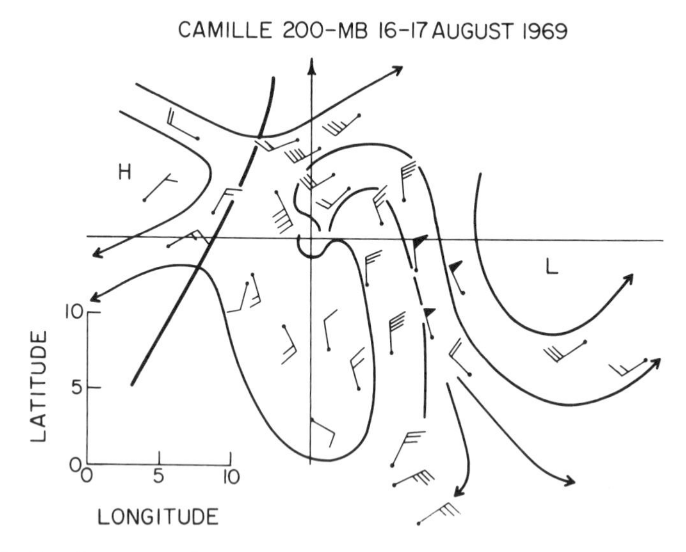

Fig. 72. Winds at 200 mb around the Gulf of Mexico and surroundings plotted with respect to hurricane Camille. *Center*, during period of extreme intensity, 16–17 August 1969; *lower left*, distance scales.

abruptly in that area. In contrast, a long outflow band keeps curving clockwise toward southwest, with strong upper winds of 30 kt (15 mps) and more at great distances from the center. Such outflow toward southwest, and eventually west, is particularly well marked in many West Pacific typhoons. Cessation of upper clouds and outflow also occurs west of the center, where a line of sharp cyclonic wind shear is situated—again a typical feature of the high-level flow.

Figure 73 affords a schematic view of the circulation through the hurricane in vertical cross-section form. We observe the strong low-level inflow to the core and corresponding outflow at high levels. The ascent within the core connects inflow and outflow layers. The eye wall is marked by heavy broken curves, nearly vertical up to the outflow layer and then sloping outward and becoming indistinct. Variability of this slope is regarded as one factor affecting surface pressure, since the slope is correlated with the depth of the warmest eye temperatures (curve D of Figure 74) in the upper tro-

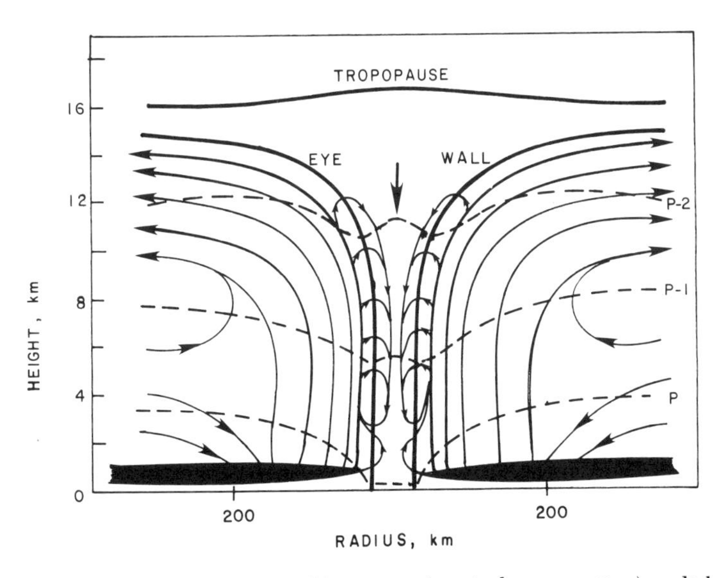

Fig. 73. Model of a mature tropical hurricane (vertical cross section), radial wind circulation versus radius, symmetrical case assumed. *Solid black*, surface inflow; *heavy lines*, tropopause and eye wall. Slope of pressure surfaces sketched schematically.

posphere. Air from the cloud wall mixes into the eye and descends. In the lowest 1.5–2 km, surface air also enters. The ejection needed for mass continuity must be located mainly in the middle troposphere.

The slope of surfaces on which pressure is constant has been added schematically with broken lines, to give an idea of the vertical change of pressure gradient. The lowest broken line, marked P, may be the curve of Figure 55. The middle one, P-1, has a slightly smaller slope on account of inward increase in temperature. But because of the central layer with very warm temperature, it is the highest curve, P-2, that shows only a weak outward rise. Beyond the 200-km radius, the curve actually slopes downward, locating a pressure-height maximum where, in Figure 72, inner cyclonic circulation changes to outer anticyclonic circulation.

A rise of pressure height has also been placed inside the eye, rather hypothetically. If surface pressure is flat in the eye, a small high-pressure center should be found at upper levels from the very warm temperatures present. Some means is needed to reduce the

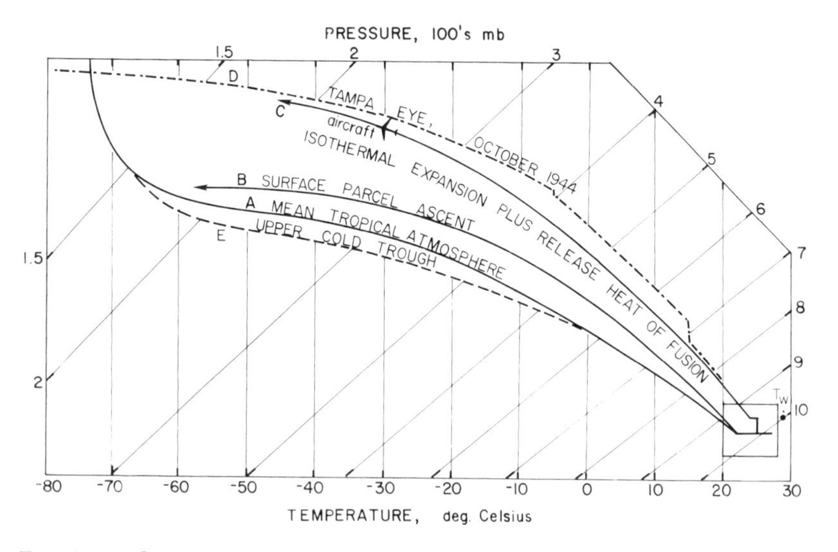

Fig. 74. Tephigram (thermodynamic chart) showing five types of vertical temperature distributions in and near a hurricane. *Aircraft*, high-level observation on *curve C*; T_w, ocean temperature.

strong winds of the air encircling the eye to the observed very low speeds inside the eye. From Equation 4.8 the most plausible mechanism is motion toward higher pressure.

Thermal Structure. In the mature stage, the temperature difference between the inside and outside of a hurricane, which permits the system to run, depends on the degree of interior warming that release of latent heat from condensation plus freezing (fusion) can achieve. The outside source of cold air aloft, discussed in previous chapters, will never amount to more than 2°C–3°C, compared to ten times this amount available for generating cyclones outside the tropics. As seen in Figure 5, the latent-heat release in question is only the amount that can provide a temperature increase above the temperature of the ambient air, which is already largely determined by successions of convective clouds during previous history. For reference, the broken curve of Figure 5, a good average of the ambient atmosphere, is shown again as curve A in Figure 74, this time all the way to the tropopause.

We have said that it is difficult for rising cloud towers to produce a temperature increase over the ambient atmosphere, because towers tend to intermingle with the cooler, and especially drier, environment described by the mean tropical atmosphere. Only if the contraction of air portrayed in Figure 22a is achieved does it become possible for the entrainment to be reduced over a substantial area, and thus temperatures in that area will take on the values of the undiluted moist ascent of Figure 5. The undilute curve is shown in full as curve B in Figure 74. Assuming this curve does become established, with the help of the starting mechanisms discussed, the whole troposphere will become warmer by about 2°C, the temperature increase being largest at middle levels around 500 to 400 mb. There, precisely, various investigators have observed the first temperature rise on the road to the hurricane or typhoon stage.

Since the ascent along curve B approaches the temperature of curve A near 200 mb, one may ask what effect the warming, with decreased density, will have on surface pressure, assuming the atmosphere above 200 mb remains undisturbed. The computed surface-pressure drop is 10 mb; hence, it will generate a central pressure of 1,000 mb, given 1,010 mb in the wider surroundings. At such a presure a tropical storm with whole gale force (50 kt wind) in one quadrant, pictured in Figure 56, can develop. The great majority of incipient vorticity centers that show signs of intensification level off at this pressure, which may be regarded as a threshold value. We remain short of explaining the pressures typical of hurricanes as shown in Figure 55. Still another energy source is required to shift the ascent in the core to higher temperatures than those of curve B.

Such an energy source is readily available at the ocean surface, through transfer of sensible and latent heat from the water to the atmosphere. Given a substantial area with gale-force winds, agitating the ocean and throwing masses of spray up into the air, the transfer will be enhanced greatly over normal values. Pertinent observations have been made by aircraft as well as exposed coastal stations and lighthouses offshore. The cloud base lowers from the normal height of 500–600 m to 100–300 m in the core of hurricanes. Air temperature stops falling when the difference between ocean and air temperature reaches 2°C to 3°C. Even in the eye of a typhoon with extreme surface pressure of 900 mb, as measured by R. H. Simpson in 1951, did surface temperature remain at 27°C, just as it did on the outside.[4] From expansion during passage from 1,000 to 900 mb pressure, the temperature should have lowered by 10°C in unsaturated air. If the condensation temperature is reached at a higher pressure (say 950 mb, on the outside), a dense cloud—a fog—should form on the ocean surface. But, this does not happen.

The history of suggested low-level processes is illustrated in Figure 75, an enlargement of the area marked by the square at the bottom of Figure 74. In the outskirts of the hurricane, surface air temperature falls to 26°C at 1,000 mb, where the difference between ocean and air temperatures becomes 2°C. Moisture, given by the dew-point line on the left, is assumed constant. Ascending air with these, the initial, properties from outside will attain condensation at 950 mb. The ascent of curve B in Figures 74 and 75 results.

Next follows the inward motion, at a constant temperature of 26°C, until the assumed pressure of 950 mb is reached in the eye-wall cloud. If the cloud base there is 150 m, the mixing ratio must increase along the slanting dew-point curve, which, for convenience, has been drawn as a straight line. If mixing ratio already increases farther out, for example, from evaporation of falling rain, a somewhat different diagram would have to be made. For the sequence of events assumed, the mixing ratio increases from 17.5 to

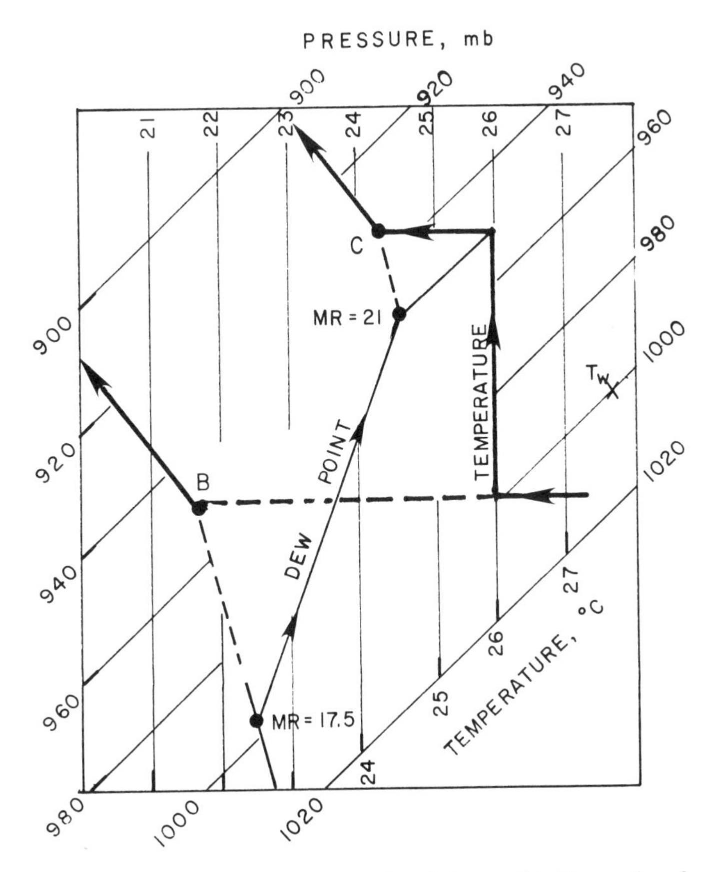

Fig. 75. Expansion of square in lower right of Figure 74. The paths of surface temperature and dew point following inflowing air near the surface are marked with *arrows* to 950 mb. The beginnings of ascents *B* and *C* from Figure 74 are marked. *MR* = mixing ratio.

21 g per kg. Then, upon ascent to the condensation level, curve C of Figure 74 is realized. Thus, through the mechanism of energy addition to the low-level inflow layer, it becomes possible for air to follow ascent paths different from, and warmer than, that of curve B. We note that curve C is one of these ascents; but there can be a whole family, depending on the pressure at which the air leaves the surface and begins to rise.

The latent heat of the inflow layer is increased by 10 joules/gm along the dew-point curve of Figure 75; the sensible heat rises by 4 joules/gm along the T = 26°C line. The ratio of sensible to latent heat increase is 4/10 = 0.4, larger by a factor of about 2 than the ratio of sensible to latent heat transfer from the ocean. Additional heat may be gained from turbulent downward transport or from dissipation of kinetic energy, especially in very intense storms. These are matters not yet clarified by observations; the difficulty in making measurements is extreme. Further discussion is contained in Riehl's *Climate and Weather in the Tropics.*[5]

From the foregoing the generation of very high temperatures in the upper layers of the hurricane, as seen in Figure 75, and of low-surface pressures can be explained. Of course, the two events occur simultaneously. As air columns begin to acquire increased energy in the inflow layer, atmospheric density decreases when the inflow air ascends vertically on a path such as that of curve C. Surface pressure must fall, given an undisturbed top, which may be as high as 100 mb. Aircraft traversing a hurricane center will encounter temperatures that, at first, increase very slowly and, in the end, close very quickly to the ring of strongest wind. This corresponds to the increasingly rapid fall of surface pressure, as in Figure 55, and also to the increasing slope of the height of constant pressure surfaces, seen at 700 mb in Figure 61. A profile of temperature at that level (Fig. 76) demonstrates the accelerating rise. Total increase in temperature between the 100-km radius and the edge of the eye was 4°C, in close correspondence to the difference between curves B and C at that level. At the time of the flight, Anita's pressures were in the range of making possible the ascent C in Figure 74.

The process of energy addition to the surface layer, in spite of its

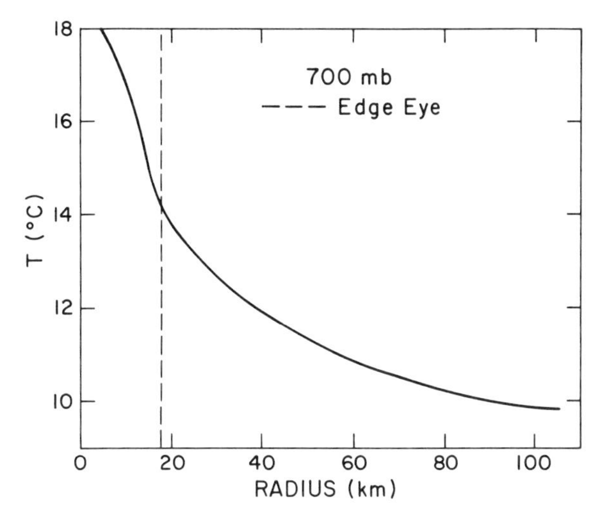

Fig. 76. Radial temperature distribution for hurricane Anita on 1 September 1977 (see Fig. 61).

great effectiveness, still cannot account for extremely low pressures of the 900-mb class; at least, no satisfactory mechanism has as yet been offered. Further, one often finds the lowest pressure, not in the rain area, but at the edge of and inside the hurricane eye, where clearing and subsidence prevail. Usually this happens when the eye is large. But in case of Anita (1977), with a small eye, the jump of temperature from the eye boundary to the center was as great as the whole rise outside (Fig. 76); sometimes the jump in temperature is even more dramatic. We must examine the thermal structure of the eye to find a viable mechanism to arrive at the lowest surface pressures observed.

Curve D, the warmest of all soundings shown in Figure 74, is an eye sounding made at Tampa, Florida, when the hurricane of Figure 60 reached that city, one day later, with a very large eye. Several stable layers attest to the sinking motion for the lower atmosphere. Moisture was low but not nearly as low as it would have been had the air in the eye all come from the stratosphere—an idea of early days. Observations suggest that the air is drawn inward from eye walls, as seen in Figure 69. Sometimes a whole ice fall cascades downward over the upper limit of the eye—a magnificent sight. The air drawn into the eye and descending there must also be evicted again. As suggested in Figure 73, it is incorporated into the cloud area again at rather low levels above the lowest layers where inflow and cloud formation prevail (Fig. 68).

Temperature in the eye is correlated with surface pressure, as may be expected. In a 900-mb storm, temperatures have been measured at 700 mb that are as high as those at the surface. This shows clearly why ascending air cannot produce such high temperatures. The atmosphere must become absolutely stable and ascent resisting for extreme temperatures of 25°C at 700 mb and of 18°C at 500 mb to occur.

Why do hurricanes so differ in their ultimate central pressure, which may range from about 985 to less than 900 mb pressure, and in their winds, which may range from 65 to 20 kt? A completely satisfactory answer cannot always be given, but some instances will be analyzed in Chapter 7. Curve E of Figure 74 indicates that even in a mature hurricane a meeting with an upper trough several degrees colder can substantially affect the temperature difference between inside and outside, especially in a hurricane of minimal intensity. There are many cases on record that support this suggestion.

The Role of Surface Friction

The discussion of the mature hurricane may be rounded off with a look at the stability of the whole hurricane machine, that is, its tendency to hold together very well in principle and its ability to persist for long periods of time. This subject was broached in Chapter 4 in connection with the thermal wind discussion. "If any rotational circulation is to have a lifetime beyond a few hours . . . balanced flow and thermal wind must be valid to a high degree of approximation. Otherwise the temperature field would not hold together, and the system would collapse quickly or never form—the normal case. If the temperature field is to hold together and the storm is to remain alive, a vertical shear of the rotating wind component must exist in order to make maintenance of the right kind of temperature field possible. This means that the storm must possess a mechanism to create the correct wind shear." In analyzing the transition from wave to vortex in the incipient hurricane Camille, we found that a mutually adjusted thermal wind and temperature field indeed evolved (Fig. 40). The persistence of such an arrangement has been found widely satisfied in mature hurricanes.

Still earlier in Chapter 4, in connection with expanding or contracting rings of air, as pictured in Figure 22, we related the changing area to the development of cyclonically or anticyclonically rotating winds. At that time we avoided making reference to a famous theorem, well known from the Middle Ages onward, that in the absence of any forces acting on expanding or contracting fluid, the latter will conserve its angular momentum. The latter is here defined as the mean wind speed around a circle multiplied by the length of the arm from the center, or the radius per unit mass. If we denote the rotating wind component by u (similar to its use in Equation 4.4) and the radius by r, as before, the angular momentum will be ur. Following the same chain of reasoning as in Chapter 4, this expression must be the angular momentum of the winds *relative* to the rotating earth. An observer in space will see the earth's own angular momentum in addition. Glancing back at Figure 21, this momentum, given as the rotation component B, will be $\omega \sin\varphi r x r$, using the definitions given with Equation 4.1. Therefore, introducing this definition, the component of the earth's angular momentum at any latitude may be written as $fr^2/2$. Adding this component and the relative angular momentum, we arrive at the definition of *absolute* angular momentum (Ω), always per unit mass.

Eq. 6.1

$$\Omega = ur + \frac{fr^2}{2}$$

Over short latitudinal distance one may apply this equation to a plane tangent to the earth at the latitude of the hurricane center, that is, a uniformly rotating disk; we shall adopt this simplification.

Glancing now at the model of Figure 73, Equation 6.1 should be most applicable to the upper hurricane outflow, even though turbulences may interfere somewhat. However, near the earth's surface the motion clearly does not take place without constraint from external forces. Surface friction, depicted in Figures 24 and 91, is present and acts against the wind. Thereby the friction *extracts* angular momentum from air to ocean, so that Ω decreases in the inflow. Figure 62 shows the type of wind profile typical for the inflow layer; use of 700-mb for comparison is permissible, since, from Figure 70, the tangential wind component is nearly constant with height in the lower atmosphere.

If we now plot a typical inflow profile on a linear scale against radius and add a computed profile from Equation 6.1 for the outflowing part of the circulation through the hurricane, we obtain the distribution of the tangential wind u with radius as seen in Figure 77. The outflow takes place at the reduced value of Ω, with which the low-level air arrives at its innermost radius—even though the rotational wind speed is at its maximum there. Thus, in the outflow, wind speeds are lower than in the inflow, increasing outward. We can again apply the principle stated in Chapter 4: when the rotational wind speed u decreases upward, the core of the storm will be warm relative to its surroundings. *A balanced,*

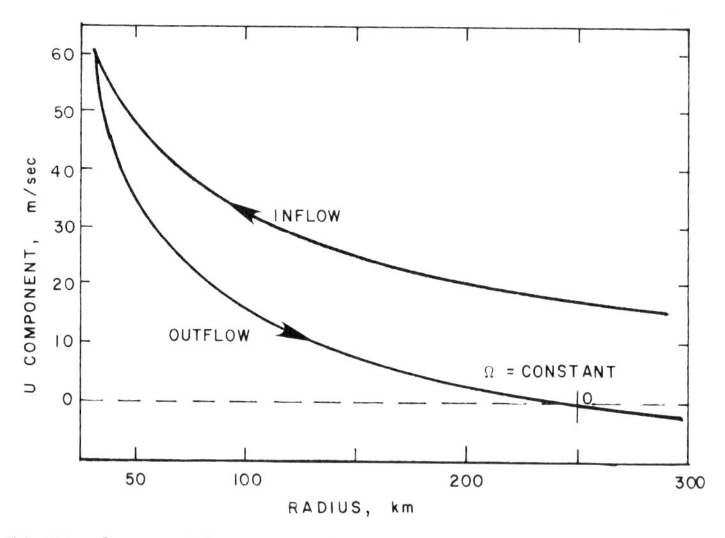

Fig. 77. Distribution of the rotational wind component averaged around a strong model hurricane (Donna of 1960 approaching Florida; see Fig. 78) in inflow and outflow layers. The symbol Ω denotes absolute angular momentum.

nondispersing temperature field can be maintained in steady state to the degree that surface friction extracts absolute angular momentum from the hurricane. This is one of the most fundamental statements we can make about the hurricane.

Pertinent values from Figure 77, and the calculated temperature field between inflow and outflow layers—determined using the thermal wind derivable from Equation 4.4—are given in Table 9. The calculated mean temperature field is added cumulatively inward and is compared with hurricane Anita (Fig. 76), even though a slightly weaker storm was used in the computation. It is seen that there is a very high degree of agreement, proving the close coincidence between observed temperature and that demanded for a steady-state hurricane.

Finally, we may take Figures 73 and 77 in conjunction to deduce the mechanism enabling the upper air to flow out of the storm cen-

Table 9. Vertical Shear of the Rotating Wind, Induced and Observed Thermal Gradient in the Troposphere

Rotating wind				*Temperature difference (°C)*		
				From shear		*Observed*
Radius (km)	*Inflow (mps)*	*Outflow (mps)*	*Difference*	*Between radii*	*Cumulative*	*Anita 1977*
30	60	60	0		3.7	3.7
				+1.7		
50	48	35	13		2.0	2.2
				+0.9		
75	38	24	14		1.1	0.8
				+0.5		
100	32	15	17		0.6	(0.6)
				+0.3		
125	28	11	17		0.3	
				+0.2		
150	25	8	17		0.1	
175	22	6	16	+0.1 (150–200)		
200	20	3	17		0	
250	18	0	18			
300	16	−3	19			
				Positive inward		Assumed equal values at 100 km

NOTE: Latitude 25°; layer depth from surface to 12 km.

ter. Inside the 200–250 km radius, the outflow is directed toward higher pressure. But the outward increase has been greatly diminished in the high troposphere, owing to the very warm inner temperatures, as noted before. Thus, the cyclostrophic balance of Figure 24 is not entirely fulfilled. Rather, the centrifugal force somewhat exceeds the pressure force; accordingly, there must be a net acceleration outward from Newton's laws of motion. Through this net acceleration the upper air is enabled to move outward, and its energy of motion decreases in accordance with Equation 4.8 until, from Figure 77, a radius is reached where the cyclonic goes over into anticyclonic flow. Here, the balanced flow again is fully in force, since we observe maximum height on isobaric surfaces at this radius. Farther outward, the outflowing air once again moves toward lower height.

The preceding, rather lengthy exposition of fundamentals concerning the mature hurricane shows things both simple and complex. Not everyone who picks up this volume will need to enter into these finer and more critical points. The reader may be quite content with the storm structure models and observations. But these pages are offered in what is hoped to represent a fair appraisal of the present state of knowledge and to offer encouragement to those curious about complex atmospheric processes and interested in going ahead to research further the secrets of the hurricane.

REFERENCES

1. Joseph Conrad, "Typhoon," in *A Conrad Argosy* (Garden City: Doubleday, Doran, 1942), 258–59.
2. B. Haurwitz, "The Height of Tropical Cyclones and the Eye of the Storm," *Monthly Weather Review*, LXIII (1935), 45–49.
3. K. Ooyama, "Numerical Simulation of the Life Cycle of Tropical Cyclones," *Journal of Atmospheric Science*, XXVI (1969), 3–40.
4. R. H. Simpson, "Exploring the Eye of Typhoon Marge," *Bulletin of the American Meteorological Society*, XXXIII (1951), 286–98.
5. H. Riehl, *Climate and Weather in the Tropics* (London: Academic Press, 1979), 455–58.

CHAPTER 7

Extraordinary Changes of Intensity and Path

In view of some of the preceding discussion and illustrations, the reader may have concluded that the range of intensities, sizes, and paths of hurricanes is so great that each storm must be carefully treated as an individual event. To a large degree this is true. Yet, Chapters 5 and 6 have shown that there are common elements of hurricane structure, mechanics, and life cycle and that there is some regularity in the seasonal variations of hurricane paths. Logically, when confronted with a hurricane situation, one starts out by considering it in terms of an average life cycle and a probable path that fits the majority of paths for the season. One tends to look for long tracks westward in August, contrasted with immediate northward to northeastward displacement early and late in the season. On the average, *hurricanes reach their greatest intensity shortly before or just at the recurvature point*. Once they acquire a reverse eastward component, their strength diminishes over water, as they gradually move toward higher latitude and toward a cooler underlying surface. For Pacific typhoons the decline statistically is found to be 50% over a latitudinal displacement of 15° past the point of recurvature in the climatic mean.

Beyond such general guidances, a sharp lookout is required for extraordinary turns in a hurricane's history, lest the unusual turn into an undesired surprise occurrence. For convenience, the four charts with storm tracks appear as Figure 78a through d; the corresponding information on pressures is contained in Table 10.

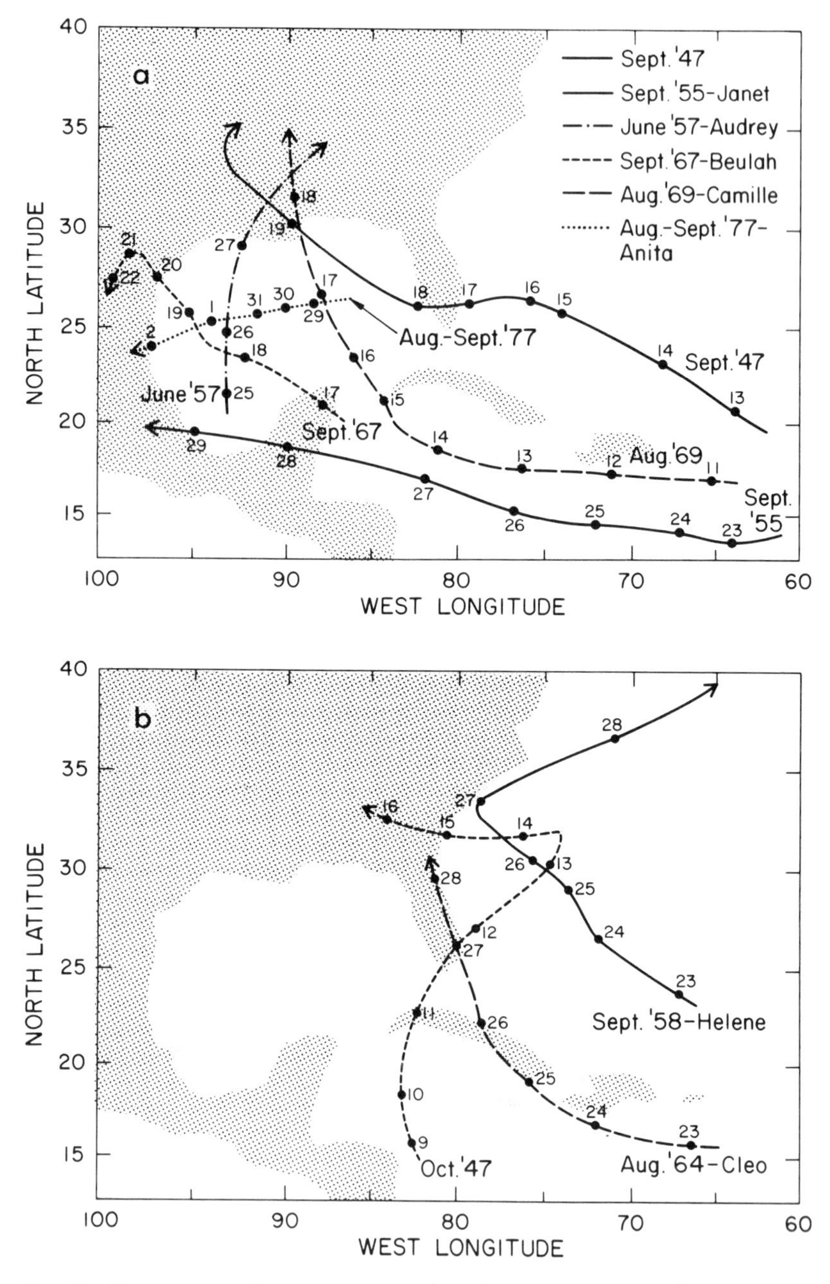

Fig. 78. Hurricane tracks: *a*, various modes of strong to extreme intensification; *b*, some odd recurving and nonrecurving paths; *c*, very strange paths; *d*, paths of large to extreme northward acceleration. Only relevant portions of the storm tracks are shown.

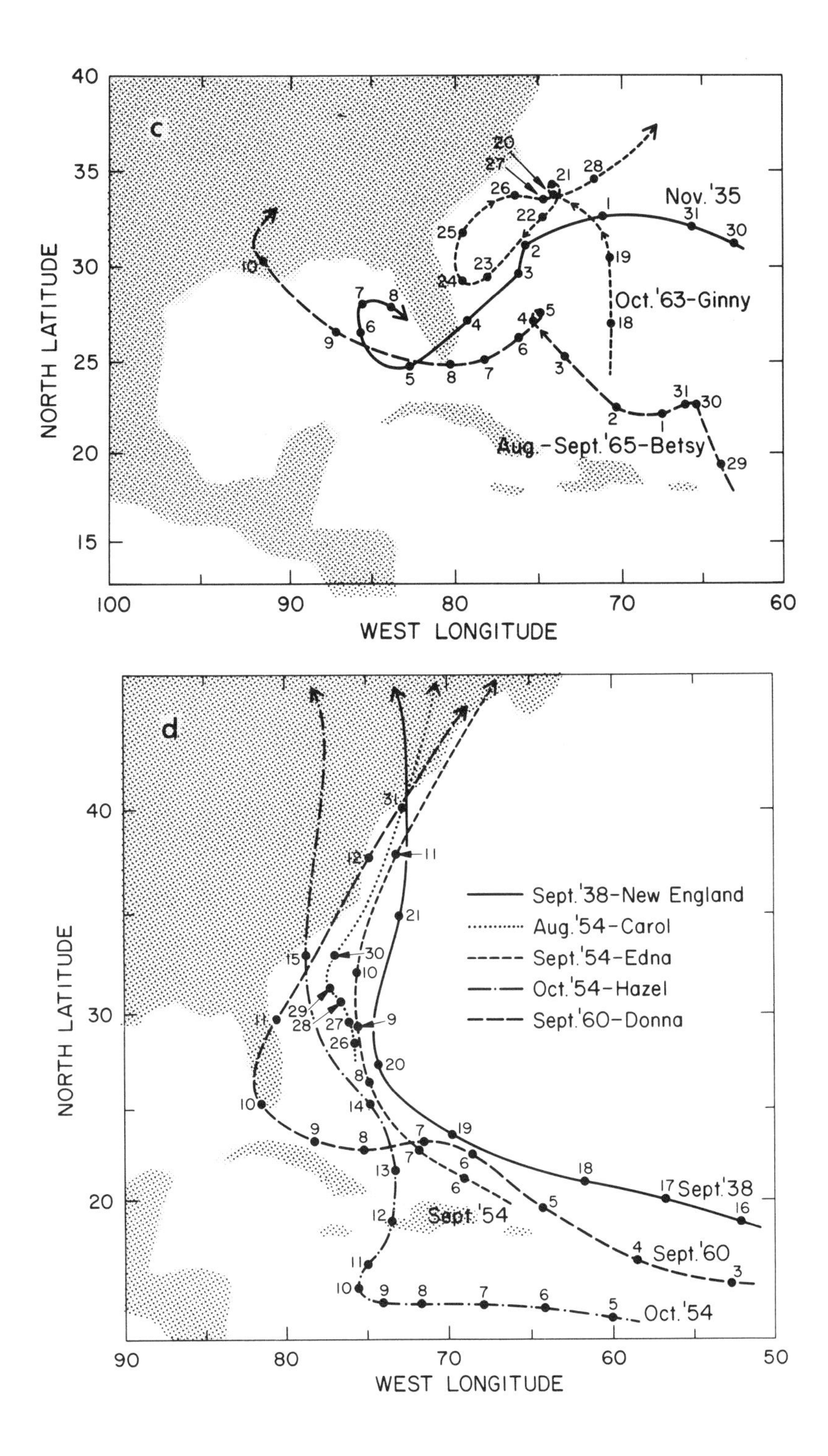
c
NORTH LATITUDE
WEST LONGITUDE
Nov. '35
Oct. '63-Ginny
Aug.-Sept. '65-Betsy
d
NORTH LATITUDE
WEST LONGITUDE
Sept. '38-New England
Aug. '54-Carol
Sept. '54-Edna
Oct. '54-Hazel
Sept. '60-Donna
Sept. '38
Sept. '54
Sept. '60
Oct. '54

Table 10. Central Pressures Measured or Estimated in Morning Hours for Hurricane Tracks of Figures 78a to 78d: Additional Readings Indicated

Hurricane	*Date*	*Pressure (mb)*	*Remarks*
Anita	*Aug–Sep 1977*		
	29	1,012	
	30	1,003	
	31	986	
	1	963	
	2, early	926	Large late deepening
	2, later	940	Slow filling at landfall
Camille	Aug 1969		See Figure 81
Beulah	*Sep 1967*		
	16	964	
	17	977	Slight filling only on Yucatan
	18	970	
	19	923	Strong deepening
	20	936	
	21	950	Slow filling at landfall
Betsy	*Aug–Sep 1965*		
	29	1,007	
	30	994	
	31	990	
	1	980	Deepening on turn west
	2	942	
	3	935	
	4	946	Weakens on cusp and loop (days 4–6)
	5	952	
	6	966	
	7	957	
	8	953	
	9	951	
	9, later	948	Landfall
Cleo	*Aug 1964*		
	23	950	
	24	950	
	25	n.a.	
	26	n.a.	On Cuba
	27	967	Miami low reading; strong recovery

Table 10—Continued

Hurricane	*Date*	*Pressure (mb)*	*Remarks*
Ginny	*Oct 1963*		
	21	983	Steady, minimal hurricane on large loop
	22	989	
	23	995	
	24	990	
	25	976	
	26	986	
	27	972	
	28	968–963	Deepening while accelerating northward to Newfoundland
	29	958–948	
Flora	*Oct 1963*		
	2	964	
	3	936	Caribbean
	4	970	After crossing Haiti Peninsula
	5	n.a.	On Cuba
	6	n.a.	On Cuba
	7	986	Hurricane center holds over shore
	8	983	Still holds
	9	975	Deepening on moving out of tropics
Donna	*Sep 1960*		
	3	947	
	4	952	
	5	958	Mid-Atlantic trough
	6	940	
	7	944	
	8	948	
	9	936	Very strong hurricane
	10	940	before landfall in Florida
	11	968	On Florida
	11, later	958	Back on Atlantic
Helene	*Sep 1958*		
	23	1,012	
	24	1,000	
	25	985	

Table 10—Continued

Hurricane	*Date*	*Pressure (mb)*	*Remarks*
	26	980–970	Deepening
	27	940–932	Sharp turn from coast
	28	942	
	29	942	
Audrey	*June 1957*		
	25	989	Formative, southern Gulf
	26	973	Deepening
	27	965–938(?)	Disastrous landfall; low pressure estimate
	28	995	Over southern United States
	29	974	Extratropical deepening
Janet	*Sept 1955*		
	23	996	
	24	995	
	25	988	
	26	n.a.	Extreme deepening; reconnaissance plane lost
	27	938	
	28	920	Landfall Yucatan
	29–30	n.a.	Second landfall Mexico
Carol	*Aug 1954*		
	26	1,011	
	27	995	
	28	975	
	29	975(?)	
	30	966	Great acceleration north
	31	960	Landfall
Edna	*Sept 1954*		
	6	1,006	
	7	1,002	
	8	991	
	9	978	
	10	980	Starts acceleration
	11	978	Touches Cape Cod; landfall Maine
	12	978	Toward Newfoundland

Table 10—Continued

Hurricane	*Date*	*Pressure (mb)*	*Remarks*
Hazel	*Oct 1954*		
	9	990	
	10	990	Turns north
	11	986	
	12	988	Crosses Haiti peninsula
	13	987	
	14	987	
	15	976	Deepening close to landfall; great acceleration north at night
	16	981	Southern Canada
	Oct 1947		
	10	1,003	
	11	990	Crosses western Cuba
	12	996	Crosses southern Florida
	13	n.a.	Sharp turn west, apparently intensifying
	14	n.a.	
	15	971	Deepening at landfall in Georgia

Nov 1935, Sept 1938, Sept 1947 not available.

Changes in Intensity

Some hurricanes undergo a second and major stage of intensification after traveling as minimal or even moderate storms for as much as a week. This event must be distinguished from the oscillations, up and down, of maximum wind and minimum pressure over a period of 1 to 3 days. A major and lasting intensification, not so rare, is most often, though by no means always, coupled with a turn of path toward a more westerly direction from northwest or north. Other hurricanes, in their initial deepening stages, continue growing and attain intensities far beyond what may normally be expected.

We return to the altogether extraordinary hurricane Camille of 1969 (the formative stage has been described in Chapter 4 and the late stage over land will be taken up in Chapter 10). It so happened that this hurricane moved on a path that took it through territory much better covered by upper-air observations than were other cyclones with similar development.

Following the initial vortex stage on 12–14 August 1969, Camille attained hurricane intensity while moving along the normal August hurricane path through the Yucatan Straits (Fig. 78a) toward northwest. At this time an upper cyclonic circulation of very unusual strength for the season was located near the central Gulf coast of the United States (Fig. 79). The illustration shows winds at 200 mb, the level of strongest wind, at which values as high as 75 kt were registered, and also the 500-mb temperature, which may represent the temperature of the troposphere. The cold core was very intense for the latitude and season.

At the first sighting one may not relate the hurricane to this distant upper cyclone, which had been in existence for several days, and anticipate Camille to travel as a moderate hurricane through the eastern Gulf. However, when Camille did enter the Gulf 24 hours later, an extraordinary development had taken place in the

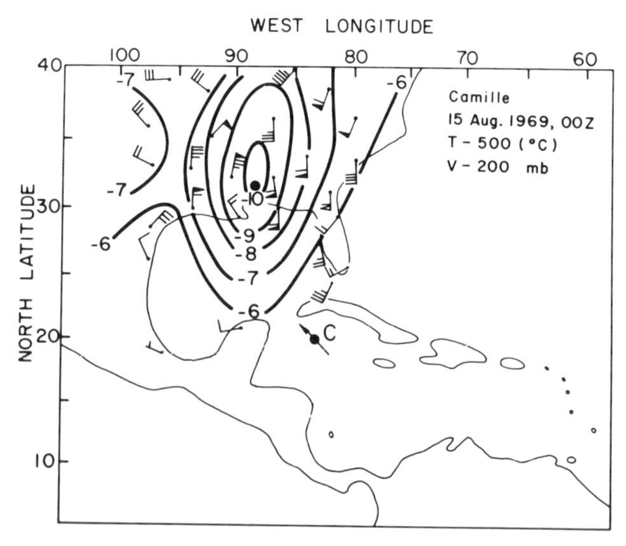

Fig. 79. Upper cyclonic vortex of great strength over the lower Mississippi basin on 15 August 1969 as seen by 200-mb winds and 500-mb temperatures representative of the troposphere.

southern United States. The upper cold air mass had completely subsided, and the cold center near the coast had disappeared (Fig. 80). At the same time, the cyclonic flow at 200 mb had degenerated to a residual line of sharp cyclonic shear, enclosed by the broken line in Figure 80. Subsequently, this shear line also vanished.

As Camille proceeded northwestward in the Gulf of Mexico on 16 August, reports of extreme low pressure near 900 mb at the center were received from reconnaissance aircraft in the later part of the day, leading to the barograph trace following the center, in Figure 81 (in Greenwich time, 6 hours ahead of local time in the Gulf). At the same time the satellite-depicted outline took on the ominous configuration of Figure 82, not unlike the photograph of cyclone Tracy on the day before destroying Darwin in northern Australia (Fig. 42). A very tight and hard central circular cloud mass is seen, with the eye open at the center, and is surrounded by a narrow ring of almost clear air with the cirrus outflow bands beyond. The central pressure is in the range of all-time lows and would sustain a maximum wind well above 120 kt, at least in the eastern sector to the right of the direction of displacement. The extreme intensity was confirmed; simultaneously, the upper out-

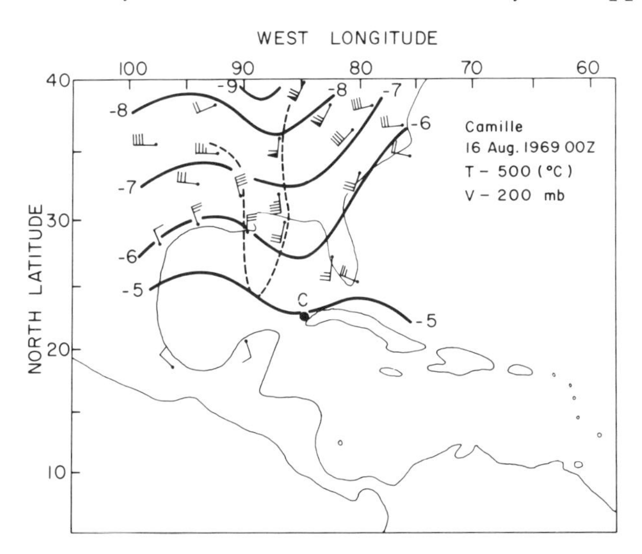

Fig. 80. Collapse of the upper vortex with sinking, 1 day later.

flow anticyclone swelled to the proportions illustrated in Figure 72 and sustained the northwestward direction of hurricane motion against other indicators pointing to recurvature.

How did intensification so extraordinary and unusual in any ocean area occur? The extreme event tends to show up the viable possibilities most clearly. Here, in the sense of release of potential energy through sinking of cold air, one can relate the pressure fall to the collapse of the deep cold center over the lower Mississippi basin. This would be the second upper cold core benefiting Camille within 3 days. The cold air intrusion east of the wave in Figure 37 provided the stimulus for intensification from wave to hurricane vortex. One can picture the second sinking of a cold mass in quite a different relative location as sustaining continuation of the hurricane growth to truly extreme proportions. Since the occurrence of such extreme events is still a matter of study, a final answer cannot be reported here. But it may be noted that the suggested sequence is a plausible one. Other hypotheses relating intensification, for instance, to a stimulus from release of latent heat

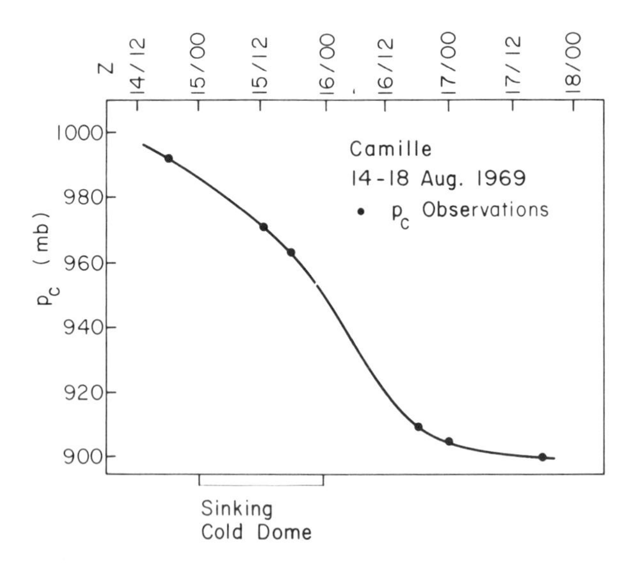

Fig. 81. Central pressures observed in hurricane Camille, 14–18 August 1969. Strongest deepening begins with collapse of the upper vortex in Figure 79 and is completed 12 hours after its disappearance.

Fig. 82. The great increase of Camille's intensity, 16 August 1969, seen by evolution of a very hard central core, with eye and width of about 500 km; surrounded by more filmy banded cloud spirals. ATS III photo.

of condensation suffer from the fact that the situation in this respect differed little, if at all, from many other cases in which hurricane development remained weak or moderate.

Camille was still well over a day away from landfall when the maximum event occurred. Two days later the center passed almost precisely through the position occupied by the upper cyclone in Figure 79. The need for an accurate concept of mechanisms of intensification becomes greatest when a center moves head on toward the coast and is a day or less away from landfall. Most often nothing dreadful happens; in fact, many centers begin to weaken on close approach when somewhat cooler, and especially drier, air from the continent may be drawn into the circulation. Then the warm core of the hurricane begins to weaken and therewith its central strength—a subject examined further in Chapter 10.

A rather different configuration was provided by hurricane Janet of 1955, which at first traveled steadily as a small center between 22 and 27 September 1955 through the southern and central Caribbean (Fig. 78a). This hurricane, moving from the east toward Jamaica, approached an upper cold air mass embedded in a pressure trough at 200 mb. Superficially, the situation is inverse to that of 12 August 1969 (Fig. 37) in which the trough was east of the wave that became Camille. But, the cold air sank in both cases; in fact, the whole trough disappeared on 26–27 September 1955 rather like the upper low of Figure 79 and the upper trough preceding Edith of 1971, prior to its attaining hurricane strength (Fig. 41). Thus it matters little on which side of a hurricane center the cold mass is found; important is whether it sinks and thereby makes new energy available to the storm. On 27 September 1955 the cold mass sank and, additionally, Janet made contact with a very large upper anticyclone centered over the northern Gulf of Mexico, left behind by a hurricane that had entered Mexico a week earlier. The result of this double event was the same as in Camille. Rapid deepening to about 915 mb ensued; the weather station on Swan Island in the western Caribbean was destroyed, as vividly described in Gordon E. Dunn and Banner I. Miller's *Atlantic Hurricanes*.[1] On landfall on the Yucatan peninsula a day later strongest winds were estimated from the damage to be near 200 kt.

We could go on at length with additional spectacular illustrations. But the cases described illustrate extraordinary intensification and the critical importance of recognizing the impending event. The reverse side of the coin also occurs, far less critical, to be sure, but nevertheless of much interest. As described in Chapter 5, hurricanes regularly lose intensity over the ocean when they move over a really cold ocean current such as the California current in the eastern Pacific. But there are other cases in which cold water does not furnish a ready explanation. The opener of the 1963 season was Arlene, which was found to be a full hurricane by reconnaissance aircraft on 2–3 August, east of the Lesser Antilles. Yet, as described in the report on the 1963 season, "There is no doubt that it (Arlene) was a well-developed hurricane and was so described by the [reconnaissance] plane's meteorologist, yet rapid deterioration of the eye structure as well as reduction of winds took place during the night of August 3–4 in an area where this rarely occurs, and by midday of the 4th Arlene was, at most, a tropical depression."[2] This decay, as well as its sequel, cannot be described very well, since the path of the storm was north of all upper-air stations the whole time. From low-level data no clue can be obtained. At high levels we only know that west to northwest winds were blowing along the Antilles—not the wind field usually associated with tropical hurricanes. There is a suggestion of upper cold air moving over the center, not alongside as in the earlier examples. Beyond this point, no evidence is available. Hurricane advisories were discontinued, but the cloud mass kept moving toward the northwest and regenerated to hurricane intensity near the point of track recurvature on 7 August. It is of interest that the second hurricane of the season intensified to hurricane strength 3 weeks later in just the same area where Arlene decayed.

Due to greatly increased upper-air observations, including winds calculated from geostationary satellite observations, far more definite conclusions can be drawn in later instances of extraordinary weakening of hurricanes over sea. At the end of September 1974 hurricane Gertrude approached the Lesser Antilles from the southeast while, by coincidence, a research vessel equipped with weather radar, paralleled its track about 200 km to the north and kept observing it. When reconnaissance aircraft reported 40–50 mps wind on 30 September, the hurricane was still in the immature stage but increasing. One day later, however, a cyclonic circulation with no more than 15 mps winds occupied the northern semicircle. On 2 October the circulation had entirely vanished, as did the cloud structure seen by the geostationary satellite and by the ship's radar.[3]

The 200-mb charts appeared to furnish a clue as to the weakening. In the early hurricane stage an upper trough preceded the center on its western side by about 500 km, not unlike hurricanes Edith and Janet, and others. But strengthening of the circulation occurred in this trough, especially north of 20° N; a large cyclonic center formed at 200 mb, highly unfavorable and just the opposite of the development needed for hurricane intensification. The upper center moved eastward and dragged along the lower portion of the trough, which assumed northeast-southwest orientation, over the hurricane core. Herewith, the cold air did not sink but overspread the central convective region and killed the warm core.

Changes of Intensity on Leaving Land. In general, hurricanes weaken upon landfall (Chapter 5). But upon leaving an island or peninsula, the center usually will recover hurricane strength and become a threat to the next coastline. Thus, hurricane Donna of 1960, a large and intense storm, regained hurricane strength no less than twice on its path (Fig. 78d). It first struck Florida but then regenerated to its northeast, passed over North Carolina, and regenerated once more to become the one hurricane to strike all parts of the Atlantic coast most exposed to hurricanes, causing damage from storm-generated waves well into New York harbor.

Hurricane Beulah of 1967 entered the southern Gulf of Mexico as a well-organized storm after crossing the Yucatan peninsula (Fig. 78a). As pointed out in the report for the season, forecasters were faced with the rather grim situation of this well-organized system coming out upon the expanse of the western Gulf of Mexico. Beulah fulfilled their expectations and entered southern Texas as a great hurricane bringing the rain catastrophe described in Chapter 11.

On 26 August 1964 hurricane Cleo crossed western Cuba from south to north on a path headed just about for southern Florida (Fig. 78b). Because the distance is very small, the question of regeneration and precise path became of immediate urgency. Indeed, the center regained central speed above 100 kt very quickly. The tendency toward recurvature was halted; Cleo turned somewhat west of north, similar to Camille on 16 August 1969, and passed directly through Miami from south to north during the ensuing night. This type of situation definitely warrants more study; the regeneration does not always occur.

During 2–3 September 1958 hurricane Ella moved westward over Cuba, was downgraded to tropical-storm intensity, and then entered the Gulf of Mexico, where immediate recovery to hurricane strength was expected in this well organized cyclone. However, it traveled the whole length of the Gulf with pressure near 1,005 mb, winds of gale force over a wide northern semicircle, and a precipitation shield in the north. But no deepening whatsoever took place in spite of superficially favorable circumstances over the warmest water of the Atlantic hurricane region. Ella (1958), discussed by Riehl, in *Climate and Weather in the Tropics*, continues to be an enigma.[4] Similarly, the rapid intensification of hurricane Audrey just before landfall in late June 1957, and of hurricane Celia in September 1970, both in the western Gulf of Mexico, as yet remains without satisfactory clarification. These kinds of situations must be understood and become predictable before any claim can be made that the problem of changes of hurricane intensity has been mastered.

Sudden Changes of Path

As a hurricane approaches land, the problem of the time and place of its landfall becomes highly critical. The severe weather sector usually is not very wide, and it is asymmetrically located to the right of the direction of motion in most instances. Thus, in a northward-moving cyclone it matters very much whether the core travels 50–100 km farther to the east or west. We have just encountered this problem in the case of Cleo, which went far enough west to strike the city of Miami instead of bypassing the coast and remaining over water—a more normal course.

Tendency to Remain Over Water. Since hurricane cores show an undoubted affinity for remaining over water, a hurricane forecaster's decision to lead a hurricane off the comfortable aqua path onto land at a very small angle, as on 26 August 1964, is difficult indeed. In the Caribbean it is well known that hurricanes seeking to exit northward to the Atlantic try to choose either the Mona Passage between Puerto Rico and Hispaniola or the Windward Passage between Hispaniola and Cuba. Because of the double-pronged, westward-protruding peninsulas of Hispaniola, at least the southeast one is likely to get hit during such passage. Farther west, hurricanes prefer the Yucatan Channel (Fig. 52), whereas those that pass the northern part of the Lesser Antilles gain the Atlantic with preference east of Puerto Rico via the smaller Virgin Islands.

From the tendency of hurricanes to choose overwater trajectories, readily understood from their need to retain a warm core (Chapter 6), we can sympathize with the problem Cleo caused at Miami in 1964. Even on a direct approach to land, hurricanes will often hesitate before crossing the shore. An outstanding example is Dora (1964), which approached northern Florida from east-southeast and slowed down virtually to a standstill for a day or more just off Jacksonville, Florida. Continuation of a westward path or northeastward recurvature hung in the balance; in the end the westward advance was resumed and the hurricane made landfall. A few years earlier hurricane Helene (1958) approached the Georgia-South Carolina coast while still deepening (Fig. 78b). Hurricane warnings were in full force. But the storm, marching straight onto the coast from southeast with 938-mb central pressure, halted 100 km offshore and made a 90° right turn northeast and remained over water. Only its weak side covered the coastal areas. To be sure, northward turning was foreseen from the influence of a trough advancing from the west, but the turning normally would be gradual, bringing the center inland. In contrast, hurricane Hazel (1954) entered without hesitation in the same area (Fig. 78d), and so have many other hurricanes. It is quite critical how fast the intercepting trough in the westerlies is advancing and how far east it has penetrated at the time of projected landfall.

The Limit of Predictability. When a hurricane attains the latitude of the subtropical anticyclone center, roughly near 25°, there may be no steering current or interaction with any nearby trough or cyclone, as took place in hurricane Dora on 8–9 September 1964. In extreme cases a hurricane may wander aimlessly for a few days and cause much damage from winds and heavy rain, as happened over eastern Cuba on 4–8 October 1963 (Fig. 83). We see a slow, winding path, even with a small loop, and no organized northward drift from internal forces. Clearly, the atmosphere was indeterminate in this situation, precluding any valid forecasting method other than persistence. East Cuba rainfall ranged from 100 to 200 cm on 3–8 October, and over 7,000 persons died in Cuba and Haiti, the greatest loss of life in many years.[5]

After some time, perhaps several days, an outside influence, usually from higher latitudes, will become dominant—especially the arrival of an upper trough from west—and then the hurricane will react by picking up speed toward the northeast, as Flora did.

Sometimes, the indeterminacy occurs under very trying circumstances, like those experienced east of Florida in September 1947. A hurricane of great intensity was approaching from the east-southeast on a quite normal track (Fig. 78a), until the question arose whether it would continue moving normally and recurve toward northeast. A trough over the United States was advancing eastward, so that chances looked good. But then the southern portions of the trough died out and were replaced by an anticyclone. In response, the hurricane, after slowing its movement toward northwest, turned gradually toward west-southwest, striking with great force the area between Miami and Palm Beach and then continuing across the eastern Gulf to hit New Orleans.

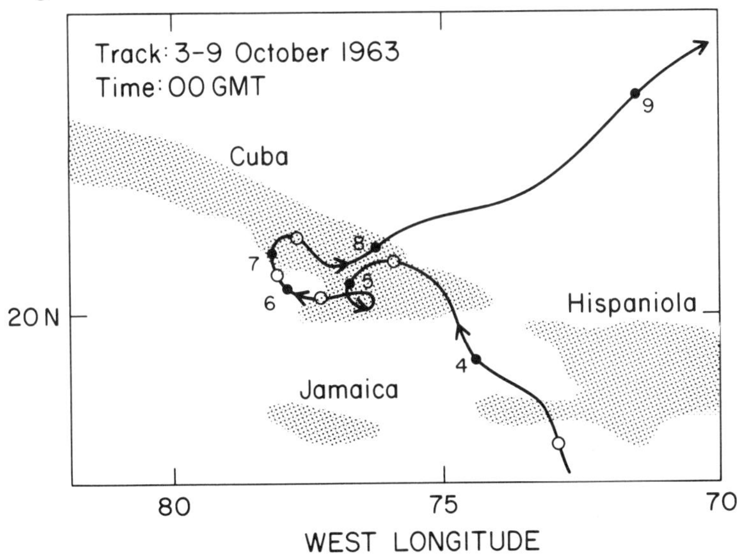

Fig. 83. Sojourn of hurricane Flora over Cuba, 3–8 October 1963. G. E. Dunn, "The Hurricane Season of 1963," *Monthly Weather Review*, XCII (1964), 128–38.

Strange Paths. Temporary indeterminacy of path, followed by a decisive turn, as in September 1947, is seen frequently on track charts like those in Chapter 5. At times the turns become extreme and very strange, as further discussed in Chapter 14. It appears that, except for supertyphoons and very large hurricanes, internal driving mechanisms in the subtropics are generally weak compared to external influences. Often hurricanes rather become a playball of several circulations of much greater size.

Just a month after the Florida event of September 1947 a hurricane of lesser intensity developed in the southwestern Caribbean and immediately moved north-northwest, then north-northeast, quite typical for the season (Fig. 78b). On 12 October it crossed Florida just north of Miami, and supposedly the last had been seen of this hurricane. But a major surprise was to happen. On 13 October the storm turned suddenly and sharply westward, and on 15 October it made landfall on the Georgia coast while still gaining intensity.

This odd track attained wide publicity. Similar configurations occasionally occur in all oceans. Yet it still is a very rare event for a hurricane to strike the United States a second time under such circumstances. Banking on the track northeastward away from the continent, this was the first hurricane subjected to experimentation with dry ice. A big question arose whether man's interference had caused the most unusual turn. It turned out that this position could not be maintained. Just as in the September case one month earlier, the upper trough guiding the hurricane northeastward weakened and was replaced by an upper anticyclone to the north. Under the influence of the reversed steering, now from east, the hurricane responded with its odd turn. It was also ascertained that the cloud seeding took place after the dramatic change in path, which therefore must be classified as an extraordinary natural event.

While the list of extraordinary paths in the several oceans is virtually inexhaustible, just a few more examples will serve to illustrate the vagaries of nature. One of the strangest paths on record is that of 30 October to 8 November 1935 (Fig. 78c). It did everything the wrong way. Very late in the season it started at a very high latitude—35° N—traveled westward and then southwestward to pass through Miami from the northeast! A further recurve back toward east followed in the eastern Gulf of Mexico. Rather similar to the October 1947 situation, an anticyclone overlay the western Atlantic at high levels in the atmosphere and protected an incipient center from the westerlies, so that it became a tropical rather than an extratropical low-pressure center. The upper anticyclone then strengthened over the eastern United States and guided the storm along its eastern flank southward. Again we see that in principle the situation is not hard to understand.

Hurricane Betsy of 1965, as well as another hurricane in the following year, presents an exaggerated version of the complex September 1947 case. After an early sharp turn westward at the end of August, recurvature started in the Bahamas. The history of September 1947 was repeated, but not before Betsy had almost escaped from the tropics. Thus we find a "cusp" point and a small loop, followed by westward passage just south of Florida, as in November 1935, and finally a landfall near the city of New Orleans (Fig. 78c).

Hurricane Ginny of 1963 produced tense days off the Atlantic seaboard (Fig. 78c). Approaching North Carolina on 20 October, the hurricane turned suddenly away from the coast toward northeast, but then completed a small clockwise loop to wander southwest for several days. It completed a second large clockwise loop off the coast and finally accelerated out of the subtropics on 28 October. It so happened that circumstances for strong deepening did not arise, and Ginny remained a minimal hurricane with 80–90 kt maximum speed. But the potential for landfall on many parts of the long coastline, plus intensification, certainly remained latent for almost a week.

Sudden Northward Acceleration. We come to a series of cases that have proved most dangerous for the higher latitudes of the United States and Canada as well as the Japanese islands. Beginning near New York the coastline juts out toward northeast all the way to Newfoundland. Thus hurricanes that curve northward out of the tropics can readily approach this coastline and make landfall. Because ocean temperatures decrease northward quickly, below 25°C beginning at latitude 40° or even farther south, one would think that such hurricanes will weaken rapidly and produce no more of an upheaval than the cyclones normally generated and experienced in the higher latitudes. Indeed, this often happens. However, some of the West Atlantic cylones attain great size and intensity given the right circumstances of formation. As noted in Chapter 5, a hurricane arriving at the limit of the tropics when a cyclone is about to form there will act as a powerful nucleus and may take on new life and great size and intensity. In such a setting hurricane Ella of September 1978 encountered the great liner Elisabeth II well to the east of Newfoundland in the central Atlantic, caused much destruction on the ship, and delayed its westward journey by 1 day.

The circumstances conducive for strong extratropical cyclone formation are well known, especially for the *Cape Hatteras secondary cyclones*, which bring on the severe east coast snow storms of winter. An upper trough of great intensity and with a very cold central portion approaching the east coast furnishes the prime setting, here illustrated for 15 October 1954 (Fig. 84) as hurricane Hazel made landfall in North Carolina (Fig. 78d). This 500-mb chart, depicting the extreme situation of record, has become a classic. It illustrates maintenance of hurricane intensity even over land coupled with extraordinary northward acceleration, which led Hazel from the Carolina coast to Canada in just 1 day.

The Hazel case will be explored in more detail in Chapter 10 as an event over land, but the setting is identically dangerous for hurricanes situated in the vicinity of latitudes 30°–35° offshore. Since the warm ocean water extends almost to the New England shoreline, chances of an explosive development are even greater there. Fortunately, the spatial and time coincidence required for an extreme event to occur is rare. Nevertheless, the problem arises and must be handled with great care in every instance when a hurricane moves north not too far east of Cape Hatteras on the North Carolina shore.

In recent history, the most remarkable events were those of 1938, 1944, and 1954; no less than three northward-accelerating cyclones with landfall occurred in 1954. The prototype is the New

England hurricane of 21 September 1938, which had been followed since 10 September, when a cyclonic weather system passed the Cape Verde Islands. As I. R. Tannehill wrote, "It was not definitely placed until the 16th near 20° N, 50° W. There was nothing extraordinary about it until after it began to turn to the northward late on the 20th. From that time it moved with increasing speed and on crossing Long Island late on the 21st was moving forward at the phenomenal rate of 56 mph."[6] From the scant upper-air data of 1938, analyzed by C. H. Pierce, we may assume that on 20 September the hurricane entered an environment similar to that in Figure 84, displaced to the east.[7] The great acceleration began in the morning hours of 21 September (Fig. 78d); the speed over 50 kt persisted until the center reached northern Vermont near 8 P.M. LT on that day. Upon crossing the coastline of Connecticut the center passed over the city of New Haven with central pressure near 950 mb, taken from the barograph there (Fig. 55). Damage from this hurricane in one of the most densely populated areas of

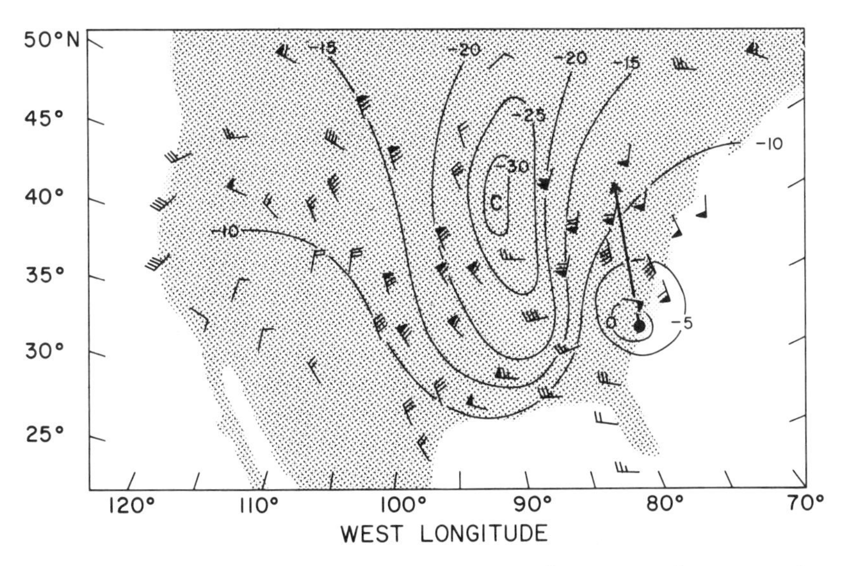

Fig. 84. Winds (kt) and temperatures (°C) at 500 mb on 15 October 1954 at the time of entry of hurricane Hazel into the southeastern United States. After E. Palmén, "Vertical Circulation and Release of Kinetic Energy During the Development of Hurricane Hazel into an Extratropical Storm," *Tellus*, X (1958), 1–23.

Fig. 85. Ship deposited on railroad track on Connecticut shore, 21 September 1938. In rear a wrecked railway coach. From Ivan Ray Tannehill, *Hurricanes: Their Nature and History*, 9th rev. ed. (copyright © 1956 by Princeton University Press), Fig. 17, p. 35; reprinted by permission of Princeton University Press.

the United States was widespread and extreme. It included long-lasting effects such as widespread destruction of forests in the White Mountains, still visible many years later. Among the immediate problems, a ship shown in Figure 85 has "landed" on top of a dock.

Lacking the satellite and aircraft reconnaissance facilities of later years, there is a question whether the hurricane intensified in the rapid phase of its northward track, in spite of motion over somewhat cooler water near landfall. In the atlas of hurricane tracks, the hurricane is indicated as transforming to an extratropical center beginning early on 21 September, but this hypothesis may be doubted.[8] Without question, a well-marked, nearly stationary front oriented north-south became established in New England

(Fig. 86); the hurricane center traveled almost due north on the east side of the front. But the front was pushed westward as the center approached, shown by Pierce from numerous thermograph traces.[9] Then it returned eastward with the westerly winds south of the center, cutting off any supply of tropical air from the south beginning shortly after noon. Understandably, ship reports are lacking; they could have supplied highly interesting details. In any event the impression is that of a warm-core center—though not quite as warm as Camille and other storms at much lower latitudes—surrounded by cold air that seeks to enter the center, yet is prevented from doing so by the very rapid motion. Whenever the cold air, whose entry into the core would destroy it, reached a cen-

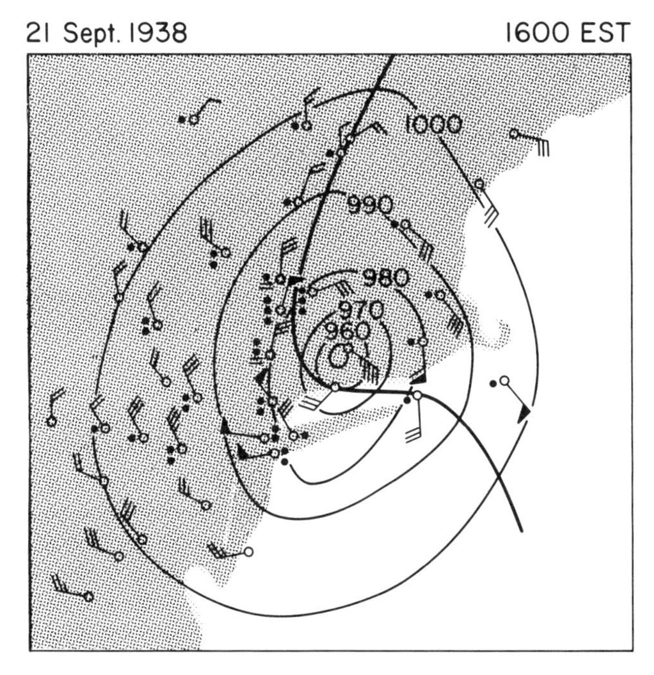

Fig. 86. Surface winds and isobars (mb) over the northeastern United States at 1600 EST, 21 September 1938, at the entrance of hurricane into New England from the south. *Heavy curve*, front between tropical and polar air. Rain is represented by: *single dot*, light rain; *double dot*, moderate; *triple dot*, heavy. After C. H. Pierce, "The Meteorological History of the New England Hurricane of September 21, 1938," *Monthly Weather Review*, LXVII (1939), 237–85.

ter position, the latter had already moved on by some 60 miles. Thus, through the very rapid displacement the warm core structure was maintained!

Because of this motion the air entering the center came mostly from the warm area over the northeastern coastal region, with surface temperatures above 22°C (Fig. 87) and high dew points. While over water, the local energy source undoubtedly remained active until close to the Long Island shore, where water temperature near 20°C is indicated on climatic charts. Over land, temperature decreases with decreasing pressure, as indicated in Figures 86 and 87. There is no surface energy source for the hurricane over land; heat energy remains constant along the trajectories, relative to the hurricane.

An extratropical mechanism for maintaining or strengthening the center must have come from fairly far west, as in Figure 84,

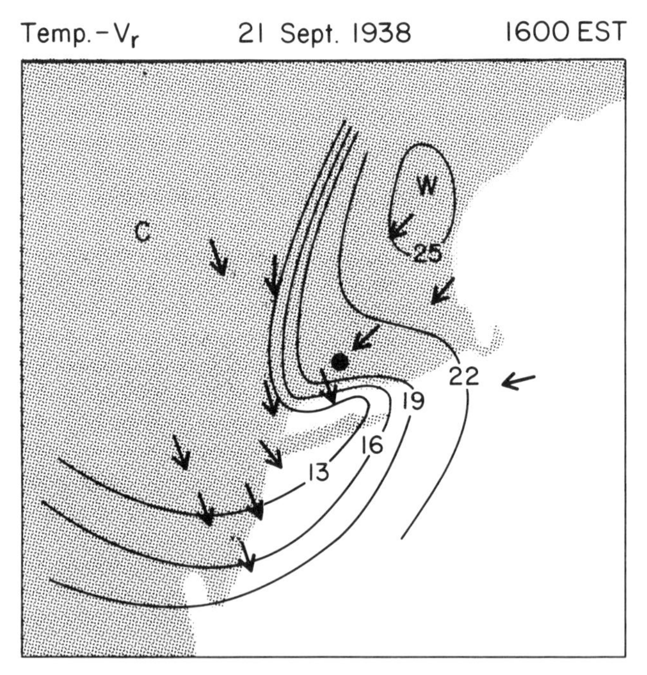

Fig. 87. Isotherms (°C) and wind directions relative to the moving center (*i.e.*, center motion is subtracted from actual winds) for the same time as Figure 85. *Black dot*, center.

where the cold upper air could sink. The low-level cold air, with a temperature of about 13°C near the front, obviously did not sink, since it was here that most of the moderate or heavy rainfall was reported. Thus the cold air itself converged near the ground to make the spectacular overrunning possible. While the cause of the acceleration of the hurricane is not obvious in this and other cases, the maintenance of the warm core through the acceleration should be regarded as the principal mechanism maintaining and perhaps strengthening the 100 mph+ (45 mps) winds near the center and the hurricane-force winds up to well over 100 miles (160 km) distant.

The hurricane that became known as the Great Atlantic Hurricane of 1944 followed a quite average parabolic track situated sufficiently far west to cover the whole of the United States and the Canadian coast as it moved parallel to the coast over a great distance northeastward. Forward acceleration occurred as in 1938 but only to a speed quoted as 40 mph (18 mps). Except for short stretches near Cape Hatteras, Rhode Island, and eastern Massachusetts, the strongest side of the hurricane remained averted from land until it crossed New Brunswick and then Newfoundland. Thus the extreme and damaging ocean surges of 1938 did not occur; total damage remained well below that of the predecessor. However, this hurricane apparently was of much greater intensity initially than the New England hurricane; highest wind speed of 134 mph (60 mps) was measured near Cape Hatteras close to the center on its western side. Due to lack of published upper-air charts in wartime, further discussion of this hurricane is not possible. Incidentally, this was the first major hurricane explored by reconnaissance aircraft—a daring undertaking at that time, for a small plane in a severe hurricane.

Hurricanes Carol and Edna of 1954 followed each other less than 2 weeks apart on very similar yet significantly different tracks (Fig. 78d). Carol developed slowly while drifting northward over 4 days close to the United States east coast. The upper-air circulation was indifferent; no obvious trough of monstrous proportions as in Figure 84 was marching eastward from west. Rather, the trough developed from a weak wave in the westerlies while traveling over the Great Lakes region; it extended itself southward rapidly on 30 August. On the afternoon of that day the situation began to resemble that in Figure 84, though a weaker cold core, as appropriate for August, was present; central temperatures near 500 mb were about −18°C. Correspondingly, terminal speed of Carol on approaching Long Island was only about 40 mph (18 mps), and the track maintained a component toward east, in contrast to September 1938. The bad part was that the acceleration, which nevertheless was large compared to the very slow earlier pace, commenced in late afternoon, carried the hurricane close to its landfall during the night hours, and arrived at the time of high tide. Its path and destructive effects over New England differed little from the New England hurricane of 1938.

Edna, following barely 10 days later, of course caught everyone in a great state of nervousness and awareness on 10 September. Here, also, the determining upper-flow pattern was not simple. An upper trough was moving rapidly eastward across the United States. Its position was near 110° W on 8 September, 9 A.M. EST; near 97° in the Middle West on 9 September; and at 85° on 10 September, intensifying though not nearly so deep as the trough of Figure 84. Nevertheless, the timing of the trough motion relative to that of Edna—never so slow as that of Carol—led to a highly dangerous and widely recognized flow pattern on the afternoon of 10 September, close to the initial position from which Carol had started its run. Acceleration indeed occurred. But the fast motion toward New England took place mainly during daytime on 11 September. All during that day widespread watch for the progress of the hurricane was in force. In angular measure the two paths seen in Figure 78d differed by only a few degrees, but the difference was large enough to keep Edna's center and the dangerous eastern side just off Cape Cod, leaving Boston in the weaker wind quadrants.

The variety of potentially explosive situations described, and the precision of prediction required in distinguishing between two paths such as those of Carol and Edna, make hurricane prediction a difficult task. Computer calculations may eventually be able to predict the major fraction of hurricane paths and their intensities. To what extent the extraordinary events can be so captured remains to be seen. The present state of the art and what people should and should not expect from hurricane prediction are described in Chapter 13. The road leading to the present skill has been long, slow, and arduous. While there is always hope for the future, one may well wonder how the evident problem of frequent indeterminacy of the atmosphere can be attacked with methods that deny the existence of indeterminacy.

A Very Long-lived Hurricane. Although numerous tropical cyclones maintain hurricane-force winds only for a day or two, September 1971 brought one of the longest-lived hurricanes of record. The storm was tracked for 31 days; hurricane intensity, mostly minimal, was maintained for 20 days. It first drifted eastward into the middle of the Atlantic from its starting position, and then it returned on very nearly the same course, until it finally turned toward northwest to make landfall in North Carolina (Fig. 88). A detailed description has been furnished by Robert Simpson and J. R. Hope.[10] Lowest pressure of 959 mb and strongest winds near 90 kt (45 m/sec) were attained at the eastern end when, under the influence of the east winds south of a developing upper high to the north, it was forced to reverse its track. This reversal was associated with a central pressure rise to 993 mb. After passing to the south of Bermuda, the hurricane once more turned toward southwest with intensification to a central pressure of 975 mb. On turn-

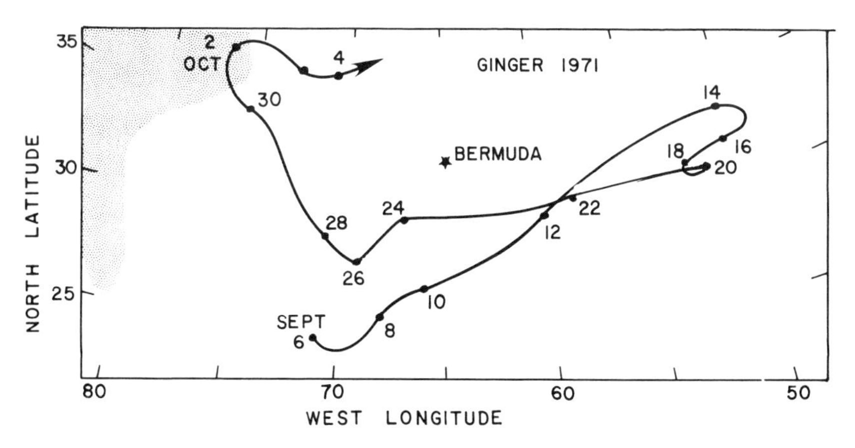

Fig. 88. Track of hurricane Ginger, 1971.

ing to the North Carolina coast, it made landfall with central pressure of 993 mb and strongest sustained winds of 65 kt (33 mps), barely hurricane force. Another recurvature brought it back to the Atlantic as a weak depression, which was finally absorbed by a cold front.

This hurricane is of special interest, because seeding missions were carried out in it on 2 days as it approached the coast. At that time its structure resembled a subtropical cyclone more than a hurricane. Clouds extending to the upper troposphere were few, but there remained the difference in temperature between the core and colder air surrounding the center in middle levels. The attempt to modify the storm by seeding failed as expected, since there were really no suitable clouds for modification. Damage on impact at the coast was slight. Yet Simpson and Hope note that rains of 25–33 cm fell in parts of eastern North Carolina, where there was severe damage to crops. One always must marvel at the rain-producing potential of old hurricanes, even in the case of Ginger, in which a vast array of observations offshore could not detect many cumulonimbus clouds.

The record also makes it quite clear that there is no equivalence between the intensity of a storm, measured in terms of pressure and wind, and the water yield from precipitation. For instance, in late July 1979 a wave traveling westward from the Caribbean intensified to minimal tropical-storm strength and slowly approached the east Texas coast from south, with strongest wind speed near 20 mps. This storm, called Claudette, hesitated for more than a day while crossing the shore, and then continued gradually north toward Oklahoma. No less than about 77 cm (30 inches) of rain fell in the area of Houston, while 50 cm (20 inches) was recorded in parts of Texas over 3 days' duration. This is a very large amount of devastating impact, which matches or exceeds total rain usually expected from fully developed hurricanes coming ashore.

REFERENCES

1. Gordon E. Dunn and Banner I. Miller, *Atlantic Hurricanes* (Baton Rouge: Louisiana State University Press, 1960), 1–7.

2. G. E. Dunn, "The Hurricane Season of 1963," *Monthly Weather Review*, XCII (1964), 128–38.
3. B. M. Lewis and D. P. Jorgensen, "Study of the Dissipation of Hurricane Gertrude (1974)," *Monthly Weather Review*, CVI (1978), 1288–1306.
4. H. Riehl, *Climate and Weather in the Tropics* (London: Academic Press, 1979), 475–78.
5. Dunn, "The Hurricane Season of 1963."
6. Ivan Ray Tannehill, *Hurricanes: Their Nature and History* (Princeton, N.J.: Princeton University Press, 1956), 220.
7. C. H. Pierce, "The Meteorological History of the New England Hurricane of September 21, 1938," *Monthly Weather Review*, LXVII (1939), 237–85.
8. C. J. Neumann *et al.*, *Tropical Cyclones of the North Atlantic Ocean, 1871–1977* (Washington, D.C.: NOAA, Government Printing Office, 1978), 161.
9. Pierce, "The Meteorological History of the New England Hurricane," 237–85.
10. R. H. Simpson and J. R. Hope, "Atlantic Hurricane Season of 1971," *Monthly Weather Review*, C (1972), 256–67.

PART IV

The Hurricane Impact

CHAPTER 8

The Impact of Hurricane Winds

In Chapters 4 to 7, we described the classical aspects of a hurricane and of the seedling disturbances from which it grows, the typical circulation dynamics, and the energetic properties applicable to their customary environment—the tropical ocean. In the next two chapters, we direct our attention to the impact of this destructive system on coastal structures as a hurricane reaches landfall and to the local changes that may be observed as it approaches the coastline or overtakes a vessel at sea.*

Without doubt, the most important overall impact of a hurricane is from the stresses its winds impose on a water surface. These stresses generate the spectrum of powerful coastal waves and longshore currents that initiate beach erosion. They also place enormous volumes of water into systematic motion, setting the stage for storm surges at ocean and bay shores—surges that may inundate the coastal plain and, at times, extend the destruction from the seacoast far inland (see, for example, Fig. 89a,b). In the 2 decades beginning with 1955, at least 90% of the deaths related to hurricanes in the United States were ascribed to drownings. And more than half the nearshore damage from most hurricanes can be attributed to the inundation and scour in battering action that occur as the sea intrudes upon the coastal plain. However, a discussion of the physical nature of these intrusions and their impact is reserved for Chapter 9.

* The term *local* as used here refers to numerical values recorded or changes occurring at a fixed observing position.

a

b

Fig. 89. Invasion of the coastal plain and destruction due to storm surge and wave action. *a*, Storm surge from the great Atlantic hurricane of 1944, which inundated coastal areas of North Carolina several miles inland; *b*, a shrimp-boat fleet swept from moorings, smashed together, and grounded in a harbor at Rockport, Texas, during hurricane Celia, 1970. *a*, Official U.S. Navy photo; *b*, photo by R. H. Simpson.

The Habitation Layer

This chapter will consider the structure of the hurricane wind system in the surface layer of importance to building design and construction and to agriculture, ornamentals, and forestry. The *habitation layer* is defined as the lowest 500 m of the atmosphere. The concern here will be primarily with *local* changes in wind velocities—speed and direction—and the specific forces at work as a hurricane approaches a threatened area along its track. Here we will be concerned most explicitly with what an observer on the beach would see as a hurricane approaches landfall and with understanding the changes in damage potential that occur as the severe wind circulation crosses the coastline. Within this context, we shall restate some of the physical principles developed in Chapters 4 to 7 with reference to those forces that develop and drive the hurricane in a large-scale ocean environment.

Horizontal Distributions of Wind Velocities

Within 200 km of the hurricane center, the pressure field and its isobars are very nearly circular and symmetric around the eye, as in Figure 90. Streamlines of the mean wind, and wind speeds in a habitation layer, however, are neither circular nor symmetrically distributed. Air parcels spiral inward toward the eye in response to the imbalances generated by frictional stresses at the ocean or land surface.

Figure 91 illustrates the local forces (observed from a coastal location) that govern the movement of air in the habitation layer as it enters the hurricane circulation. The sum of the centrifugal (Ce) and Coriolis (Co) forces is directed outward, while the sum of the pressure-gradient forces (P_g) and frictional stresses (F_s) is directed inward (normal to the streamline, S); this resultant supports an acceleration of the tangential component of wind as the air spirals inward toward the low-pressure center.

As a nearly circular storm system, the hurricane is unique in concentrating the greatest damage potential, highest winds, and dangerous seas within a few tens of kilometers of its center. This is in contrast to the highly asymmetric wave cyclone of higher latitudes, where the greatest damage potential may be located several hundred kilometers from the center, and the belt of severe weather may be much broader (Fig. 3). In a mature tropical cyclone (hurricane or typhoon), the increases in wind speed from the vortex periphery to the ring of maximum winds (RMW radius being generally less than 35 km) occur primarily in the tangential, or rotational, component of the wind:

Eq. 8.0 $$u \equiv V \cos \beta$$

where V is the total wind speed and β the crossing angle between streamlines and isobars. Increases in u result from a partial conservation of absolute angular momentum carried into the vortex from the environment. If we neglect losses due to surface friction, the angular momentum per unit mass, Ω, transported from the environment is

Eq. 8.1 $$\Omega = ur + \frac{fr^2}{2}$$

where the two terms on the right represent the angular momentum relative to the earth's surface and that due to the earth's rotation, respectively. Here f is the Coriolis parameter and r the radial distance from a local observing position to the center of rotation: the hurricane center as applied here. Rewriting Equation 8.1, we have

Eq. 8.2 $$u = \frac{\Omega}{r} - \frac{fr}{2}$$

As a basic principle of inviscid fluid motion, Ω is conserved. As air spiraling around the hurricane center transports angular momentum from the environment toward the center, the mean value of u at any radius, r, increases in accordance with Equation 8.2, friction not considered. The maximum value of u occurs when pressure-gradient forces just balance the centrifugal and Coriolis forces (*i.e.*, a gradient balance) or when—in the context of discussions in Chapter 4—cyclostrophic balance is achieved and the crossing angle, β, approaches zero. However, if Ω were not progressively reduced by the surface friction, the increase of u with decreasing r would quickly generate centrifugal forces that would

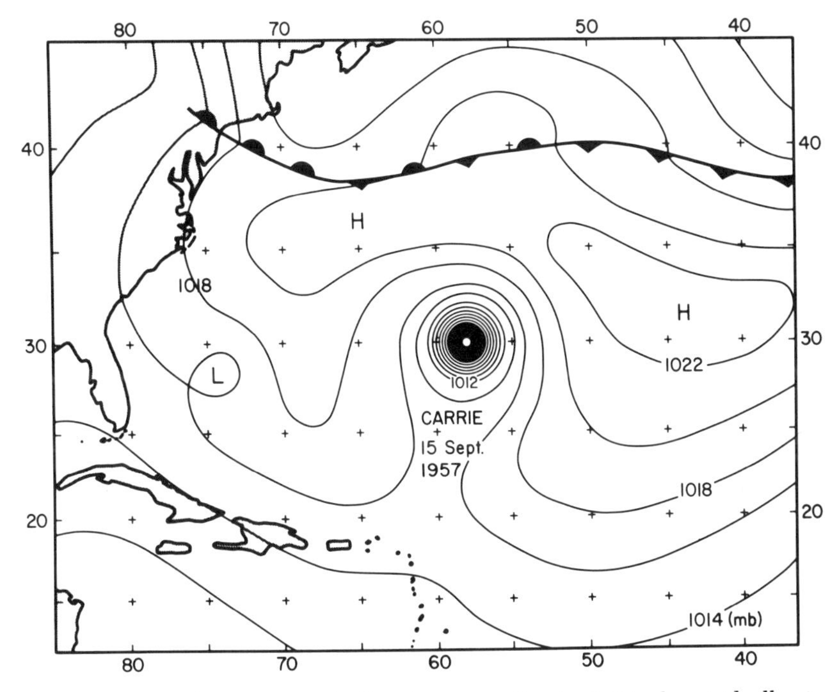

Fig. 90. The surface pressure field in hurricane Carrie, 1957, almost ideally circular on the synoptic scale.

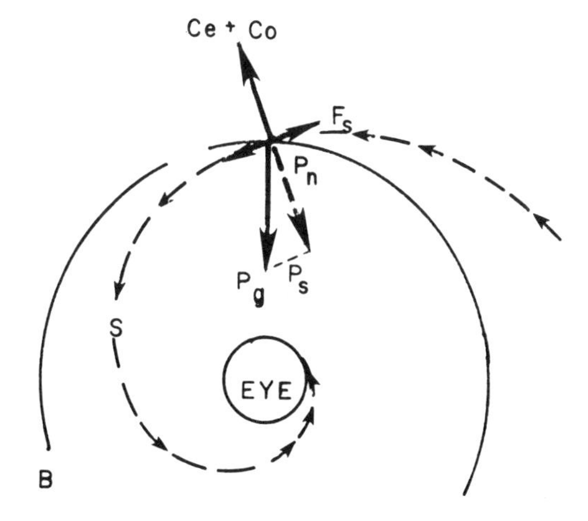

Fig. 91. Vector forces that determine sustained wind speeds in the habitation layer. *Ce*, centrifugal force; *Co*, Coriolis force; P_g, pressure-gradient force; P_n and P_s, the components normal and parallel to the wind; F_s, frictional force; *B*, an isobar; and *S*, a streamline.

overcome the greatest achievable pressure-gradient forces (Fig. 92), and the air would spiral outward toward higher pressure before it could generate and support hurricane-force winds.[1]

Thus surface friction plays multiple roles in generating and maintaining a hurricane system. First, it contributes to an initial crossing angle between streamlines and isobars so that the pressure forces can draw air from the environment toward the vortex center. Second, it progressively consumes a sufficient amount of the imported angular momentum so that the centrifugal forces associated with increasing u do not overbalance the pressure-gradient forces before an RMW can be established near the center. Here the imported air rises, creates a convective eye wall, and generates the central core of warm air that sustains pressure forces in the surface layer. Finally, friction generates turbulent eddies

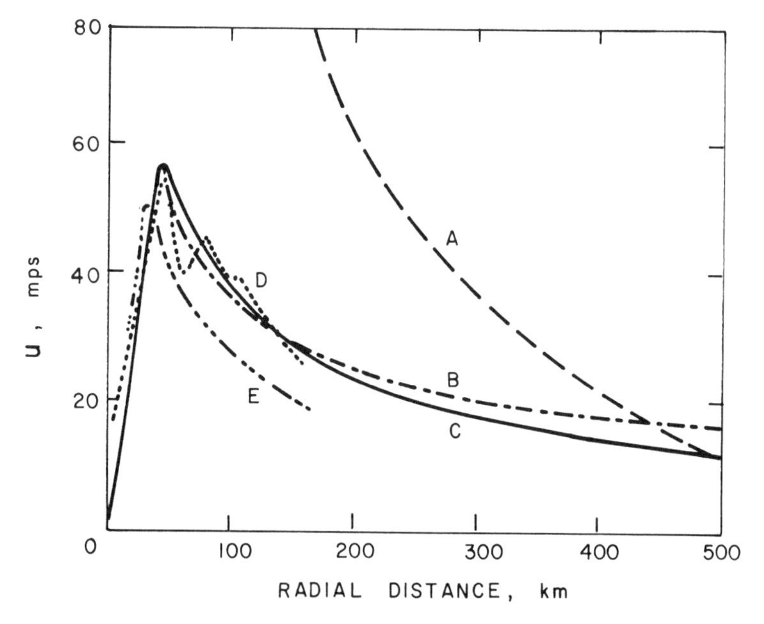

Fig. 92. Increase in rotational (tangential) wind speed, u, approaching a hurricane center. *Curve C* is from J. Malkus and H. Riehl, "On the Dynamics and Energy Transformation in Steady-State Hurricanes"; *B*, a modified Rankine vortex with x = 0.5 (see Eq. 9.3); *A*, a profile with constant absolute angular momentum; *D*, profile observed in hurricane Beulah, 1963, at 2,100-m altitude; *E*, in hurricane Hilda, 1964, at 1,000 m.

and the gusty character of hurricane winds, which are of critical importance to structural design.

From the RMW to the eye center, winds decrease almost linearly to zero, the point of zero wind speed ideally being displaced to the left of the track of lowest pressure at a distance proportional to the speed of vortex movement. The mean horizontal wind profile is approximated by the combined Rankine vortex, an ideal nondivergent rotating circulation in which the area inside the RMW is in solid rotation; that is, u/r is constant. In the annulus outward from the RMW, the wind speed diminishes approximately in accordance with the exponential relationship

Eq. 8.3 $$u r^{x} = \text{constant}$$

In the Rankine vortex, $x \equiv 1$, although in the hurricane, as we shall see, $0.4 < x < 0.8$.

Figure 92 compares a theoretical profile of realistic mean wind speeds in a hurricane to a computed profile of u that assumes conservation of Ω without losses to friction.[2] Also shown is a computed profile of wind speed for a (modified) Rankine vortex where $x = 0.5$, together with wind profiles from hurricanes Beulah (1963) and Hilda (1964) observed from reconnaissance aircraft.

As a corollary to this discussion, it is important to note that the role of friction in a hurricane is that of a catalyst, not a forcing function, in producing kinetic energy. Moreover, as a catalyst, frictional stresses have a critical effective maximum that, if exceeded, would erode the angular momentum, Ω, at such rates that extreme values of V in the vortex would be significantly reduced and a discrete eye wall could not be maintained. This occurs when a hurricane moves far inland. However, in such cases, the overall reduction in wind strength generally results more from a loss of an oceanic heat source than from increases in the surface friction.[3]

From this background, we may now examine the changes in wind speed and direction to be expected in a coastal zone as a hurricane approaches. The local crossing angle, β, at the vortex periphery is a function of frictional stresses, the momentum of air entering the vortex, and the speed and direction of vortex movement. For a stationary system, a line-integral value of β over water may vary from about 15° to 20° at the vortex periphery in a hurricane of moderate strength and from 20° to 40° in one of great strength. Air moving into a vortex from a land trajectory may exceed these values by 10° to 30° because of greater friction over land. However, (in the Northern Hemisphere) the momentum entering the right semicircle is appreciably greater than in the left semicircle; and in a moving system, β observed locally varies with the speed of hurricane approach, increasing in rear quadrants, decreasing in forward quadrants. Therefore, the changes in β and in wind speed as radial distance from the center increases differ from quadrant to quadrant, and from a westward-moving system—*e.g.*, Anita, 1977 (Fig. 78)—to a northward-moving system—*e.g.*, Eloise, 1975.

Variation of Wind Speed with Height

Above the first few hundred meters, the strongest winds of a hurricane vary little up to a height of 4 to 5 km. This is illustrated in Figure 93, which is based on electronically measured winds from a reconnaissance aircraft flying radial tracks across hurricane Hilda in 1964.[4]

Within the habitation layer, however, the variation of hurricane-force winds with height is a function of the decrease in frictional stresses with height. These are difficult to measure, and only since the late 1970s has there been a means of measuring or obtaining good approximations of wind variation with height in the lowest few hundred meters over a water surface. Theoretically, wind speeds increase on a relatively smooth surface from virtually zero at the surface to a nominal *surface wind speed* at 10 m above ground level—the international standard elevation for surface wind measurement.

The variation of winds with height in the so-called surface layer is a function of surface roughness or irregularities, which reduce wind speeds near the surface and create turbulent eddies. These eddies extend upward to some level where horizontal motions are free of surface friction. It should not, then, be surprising that turbulence generated by air moving over an ocean surface—whose irregularities, wave and swell, are themselves in motion—is quite different from that generated by fixed irregularities of land surfaces. A convincing demonstration of the differences imposed by friction on a hurricane as it passes from sea to land was made in 1978 by researchers at the Geophysical Fluid Dynamics Laboratory in Princeton using computer simulations of circulation changes.[5]

It is, perhaps, less obvious that, in large convective storms like

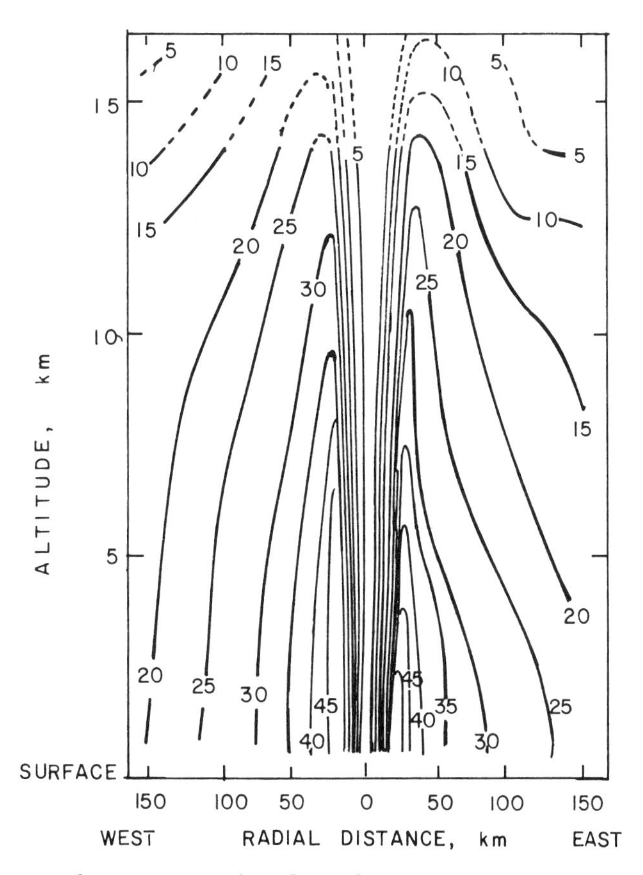

Fig. 93. Vertical cross section of wind speed (mps) relative to the moving eye in hurricane Hilda. Adapted from H. F. Hawkins and D. T. Rubsam, "Hurricane Hilda, 1964: II. Structure and Budgets of the Hurricane on October 1, 1964," *Monthly Weather Review*, XCV (1968), 617–36.

hurricanes, the influence of friction extends through a deeper layer than in extratropical storm systems. In the hurricane, the momentum in lower layers, together with its turbulent properties, is rapidly transported upward by strong convective currents, often beyond midtroposphere. The eddies created by surface friction are responsible for the gusty character of the wind, the mean value of which may vary by more than 60%, depending upon the time period over which wind speeds are averaged.

Table 11 is the result of a meticulous analysis of a long record of winds measured at the Royal Observatory in Hong Kong. Here the probable maximum wind for various time periods from 5 seconds to 10 minutes has been computed as a function of hourly mean wind speeds ranging in value from 30 to 100 kt.[6]

Table 11. Expected Values of $\bar{V}(t)$ in Knots

Hourly mean wind speed (kt)	*Time intervals (seconds)*					
	600	*300*	*60*	*30*	*10*	*5*
30	32	33	38	43	49	52
40	42	44	51	57	66	69
50	53	55	64	71	82	86
60	64	66	77	85	98	103
70	74	76	90	100	115	120
80	85	87	102	114	131	138
90	95	98	115	128	147	155
100	105	109	128	142	164	172

NOTE: Expected wind-speed averages (kt) for time intervals ranging from 5 to 600 seconds as a function of hourly mean wind speeds. This table considers only overwater trajectories from the southeast.

SOURCE: Courtesy of Gordon Bell, Royal Observatory, Hong Kong.

Since both the maximum speeds at anemometer level and the vertical wind shear in the habitation layer are so important to structural design, we shall now turn our attention to the influence of surface friction on wind-speed distributions and examine several numerical means for estimating the variation with height over water at the shoreline and inland. From turbulence theory, the variation of a representative mean wind speed with height in the surface (friction) layer is given by

Eq. 8.4

$$\bar{u} = \frac{u_*}{k} \left[\ell n \frac{z}{z_o} - \psi \right]$$

where $\bar{u}$ is the mean wind at a height, z; u_* is the *friction velocity*—a function of the surface stress; k is a scaling constant whose value is generally agreed to be approximately 0.35; and z_o is a measure of the surface roughness that depends upon the average height and spacing of surface irregularities.[7] ψ is a function of the static stability of the atmosphere. Its value is near zero in the hurricane over water, but over land the value is positive for stable and negative for unstable stratifications of the atmosphere. In practice, z_o may be experimentally determined by measuring wind speeds for various strong-wind cases at several levels on a wind tower. Over a rough but moving water surface, z_o may be considered to be much less than 1 cm. Experimentally, using aircraft wind measurements in two hurricanes, z_o was computed to be 0.15 cm.[8]

From Equation 8.4, a mean wind measured at the anemometer level (10 m) can be used to compute the friction velocity, u_*. Then $\bar{u}$ can be computed for various elevations in the friction layer. For example, if a sustained wind of 30 mps is measured 10 m above the water surface from an ocean buoy and if we use the value $z_o = 0.15$ cm, the friction velocity is computed to be 1.19 mps and the wind speed at 100 m is 38 mps.

In the friction layer, engineers commonly use Hellman's formula:

Eq. 8.5

$$V_z = V_{10} \left[\frac{z}{10} \right]^x$$

where V_z is the speed at any given altitude, z; V_{10}, the speed at standard anemometer height (10 m); and z, the height (m) at which V_z is observed. In building codes and standards, x is assigned a value that relates to the roughness of terrain. Table 12 lists some values of x that are commonly used. In the previous example, which used Equation 8.4, the computed increase in wind speed is equivalent to that obtained from Equation 8.5, if $x = 1/9.7$.

From Table 12 consider an isolated 10-story building near the shore under siege of a hurricane. If the fastest-mile surface wind speed is 50 mps, the speed at the top would be 59 mps, using the value x = 1/7. At the top of a 50-story building, the speed would be 74 mps.*

Table 12. Commonly Used Values of the Exponent x in Hellman's Formula

x	*Character of the terrain*
1/7	Open country near the coastline
1/4.5	Suburban and wooded areas
1/3	Cities or hilly, rough terrain

SOURCE: Values in table recommended in a letter from Herbert Saffir to Robert Simpson, November, 1978.

Herbert Thom used this formula to reduce the wind data from anemometers at nonstandard heights to values that would have been measured at 10 m, and then developed a climatology of extreme surface winds in the United States for various return periods.[9] This is shown in Figure 94 for a 100-year return period. These expected wind maxima are still widely used for engineering-design purposes.

However, for the design of structures located at the seashore or bay shore, these "probable maximum" values may be deceptively low, first, because the sample of wind records is small. Only a small percentage of anemometers survive their exposure to the maximum winds of powerful hurricanes. Second, many of the records that *are* available are from locations more than a kilometer from the open coast or in urban areas where frictional stresses are much greater than at the water's edge.

In Figure 95, vertical profiles of analyzed wind speeds measured at the surface and at the top of the habitation layer are compared. Values at the top of the layer are those measured by an aircraft equipped with a sensitive inertial navigation system coupled with other probes to measure ambient wind speeds; those at the surface are based on observations from a research buoy at the same locality

* This computation assumes a 6-m height for the first floor and 3-m for succeeding floors.

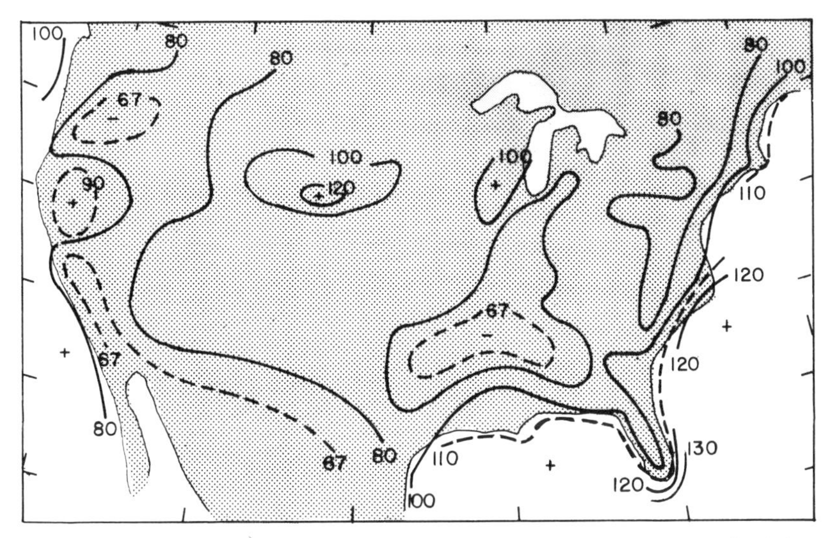

Fig. 94. Extreme winds in the United States, expressed as fastest mile (kt) at standard anemometer level with 100-year recurrence interval. Adapted from H. C. S. Thom, "New Distribution of Extreme Winds in the United States," *Proceedings of the American Society of Coastal Engineers, Structural Division*, LXXXVI (April, 1960), 11–16.

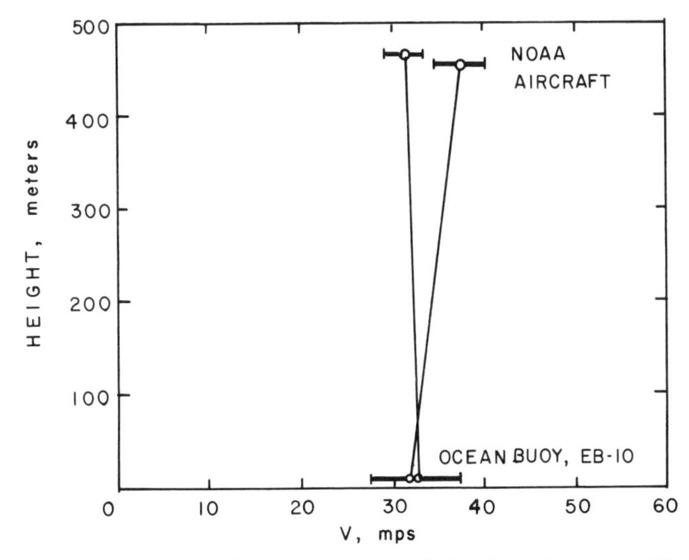

Fig. 95. Average wind speed and variations with height in hurricane Eloise, 20 September 1975, at distance 75 miles east-northeast of center, based upon data from ocean buoy EB-10 at 10-m height during two passes of a NOAA aircraft at 450 m and 460 m above the buoy at almost the same time. After John Bates, *Vertical Shear of the Horizontal Wind Speed in Tropical Cyclones*, Technical Memo ERL WMPO-39 (N.p.: Department of Commerce-NOAA, 1977), 4.

and time. Within the limits of observational accuracy, there is, at most, about a 20% increase with height through the habitation layer (over water). This represents the increase to be expected in only the first 36 m, using Hellman's power law with $x = 1/7$. For isolated tall structures at the water's edge, the vertical shear of the horizontal wind may be better approximated by the logarithmic relation in Equation 8.4. In any event, indications are that the vertical wind shear must increase rapidly with distance inland from an open beach; in fact, it increases significantly in the first half-kilometer inland. This has been demonstrated in the observations at Hong Kong from an instrumented building structure and four adjacent wind towers, where large gradients of vertical shear in typhoons occur in the first few hundred meters inland from a shoreline.

Sustained Winds Versus Gust Speeds

The international standard for *sustained winds* is the average speed for a 10-minute period. In the United States, the sustained wind is a 1-minute average. For extreme winds in severe storms in the United States, the unit of measure is the *fastest mile*—the highest speed at which 1 mile of wind passes the anemometer. Thus, in a hurricane, the maximum sustained wind is the fastest mile. The *peak gust* is the highest, instantaneous wind-speed value observed. This value, of course, varies with the inertial response of the anemometer and may account for variations of 30% or more when measuring gust speeds.

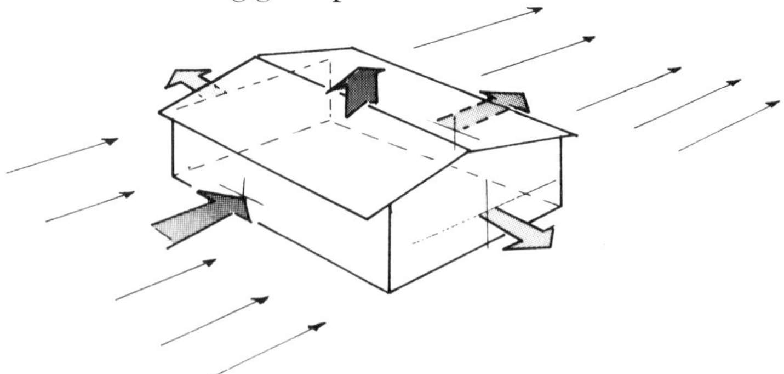

Fig. 96. Aerodynamic forces on a building.

G. N. Brekke concluded in 1959 that the so-called gust factor (the ratio of the maximum gust speed to the sustained wind speed) is 1.3, albeit an approximation with a large standard deviation.[10] It is questionable, however, whether either the sustained wind, as defined above, or the peak gust is a satisfactory unit for use in designing for wind loads on structures, for reasons discussed in the following section.

Wind Loads on Structures

The damage to building structures by the maximum sustained winds of a hurricane or from hurricane-generated tornadoes may be attributed to:

- aerodynamic forces,
- atmospheric pressure differentials between the interior and exterior of a structure, or
- the impact of airborne debris (missiles).[11]

Aerodynamic forces, commonly the most important of the three, are generated externally by the potential flow of air over and around a structure. As in Figure 96, aerodynamic forces are directed inward on the windward wall of a structure and outward on its roof, the leeward wall, and side walls.

Rapidly changing atmospheric pressure, as in the case of a fast-moving hurricane tornado, will provide a component of force acting outward on all exterior surfaces, the magnitude of which is proportional to the airtightness of the structure or the rate at which adjustments to pressure changes can be accommodated.

The pressure drop in a hurricane-generated tornado is unlikely to exceed 10 to 30 mb, and the rate of pressure change is unlikely to exceed about 5 mb sec^{-1}. The maximum rate of local pressure change due to the approach of an extreme hurricane is about 0.03 mb sec^{-1}. Even for the most "airtight" buildings, the explosive (outward-acting) forces due to atmospheric pressure differentials are small relative to the aerodynamic forces from sustained hurricane winds. This conclusion is less valid, however, for the massive tornadoes of the midwestern United States, especially those which represent the merger of several smaller vortices that, in combination, generate what Theodore Fujita calls "suction spots."[12] K. C. Mehta and his colleagues state that, from analyses of tornado-damage patterns, "it can be inferred that *conventional buildings are not damaged by atmospheric pressure-change effects but rather by forces due to aerodynamic effects of wind.*"[13]

A potentially more important contributor to the rupture of structures than natural atmospheric pressure changes is the effect of internal ram pressures that may be generated by the failure of exterior window or door openings on the windward side of the structure. Herbert Saffir estimated that ram pressures of this type can provide interior forces 1.5 times those in a fully enclosed building, acting outwardly on all faces of a structure and augmenting explosive effects from other sources.[14] The impact of airborne debris is of importance primarily to glass openings and to unreinforced concrete-block construction.

One means of expressing the damage potential of winds on a structure is in terms of the *uniform wind pressure.*[15] This may be expressed as

Eq. 8.6
$$p = 0.00256 V^2 C_d$$

where p is the wind pressure in pounds per square feet, V is wind speed in miles per hour, and C_d is a pressure coefficient or shape factor.

The constant, 0.00256, is based on the density of air at United States standard temperature and pressure. C_d values refer to external wind forces acting on an enclosed structure. If a building has openings or leakages that allow the wind to gain entry and develop internal pressures, the coefficient, C_d, *must* be replaced by

Eq. 8.7
$$C_f = C_p - C_{pi}$$

which is the difference between external and internal forces where the sign convention is positive for inward-directed forces. With knowledge about the construction design of a building, engineers use such expressions as Equations 8.6 and 8.7 to calculate maximum wind speeds that cause such damage.

We now return to the definition of a sustained wind in a damaging wind storm. If the wind damage can be considered to be due mainly to aerodynamic forces, then a *sustained wind in severe weather events should be defined as one that persists for a sufficient period to develop optimal aerodynamic forces on a complex-shaped building structure.* Wind-tunnel tests on simple building models show that *steady dynamical forces are established in a period of about 3 seconds.* For buildings with more complex shape factors, the time to achieve a steady state may be considered to be less than twice this period, varying, however, with wind speed. Although more experimentation and modeling are needed for numerical validation, the existing experience may be extended to conclude conservatively that *a nominal wind speed that persists for as long as 6 seconds is sufficient to establish the maximal pressure-force loading* on nearly all structures. A climatology of expected maximum winds sustained for 6 seconds would unquestionably indicate a need for upgrading the structural-design criteria for protection from hurricanes.

Uniqueness of the Hurricane Gust

In addition to the damage from sustained winds discussed in the last section, two additional sources of damage are recognized as attributable to a fluid acceleration, or what is termed *gust sensitivity.* A wind gust, as defined in the first section above, represents deviations from the mean wind caused by turbulent eddies. Turbulence theory assumes that those deviations have a Gaussian distribution. In the hurricane, however, observational evidence suggests a uniquely skewed distribution. These eddies, which are a function of the roughness of the surface and have dimensions proportional to the mixing length of the fluid, appear locally as a rapid acceleration of wind speeds to peak values that, in the absence of irregular surface features, may exceed the sustained wind speed by an average of about 30%. At a coastline, where the terrain sometimes rises abruptly or building structures may impose irregular mechanical turbulence, gusts may occasionally exceed sustained wind speeds by more than 50%. Figure 97 is a typical example of hurricane gustiness recorded at a coastal location.

Structural responses to hurricane gusts are of two types: 1) the pressure force and 2) the fluid-acceleration effect. The pressure-force term is analogous to that in Equation 8.6 for sustained wind

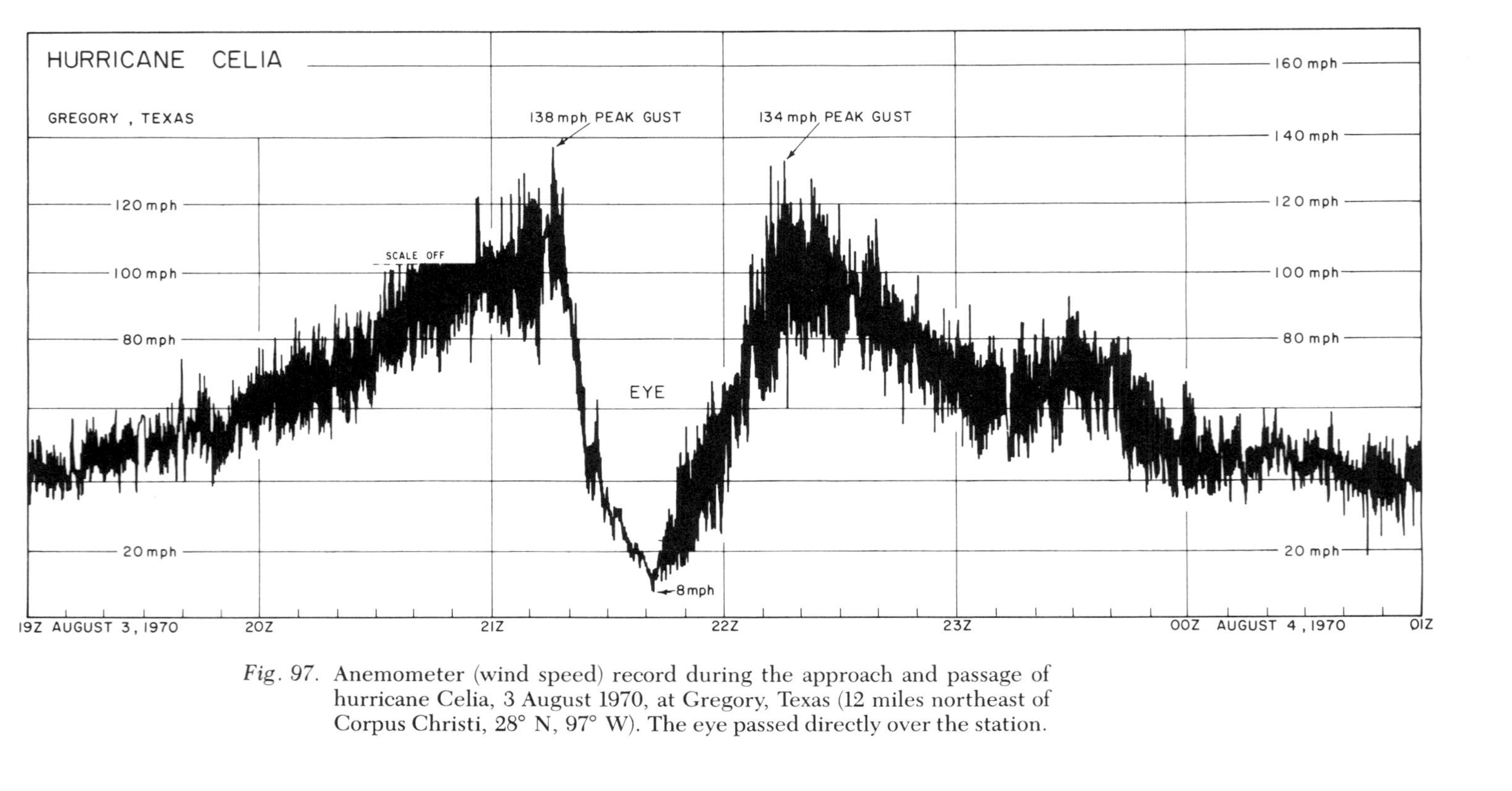

Fig. 97. Anemometer (wind speed) record during the approach and passage of hurricane Celia, 3 August 1970, at Gregory, Texas (12 miles northeast of Corpus Christi, 28° N, 97° W). The eye passed directly over the station.

speeds but with a different value for C_d. The pressure effect due to gustiness is necessarily more complex because it involves both an extreme pressure force and a time-derivative term that varies nonlinearly with the local duration and strength of a turbulent eddy. Experimentalists do not always agree as to the value of pressure coefficients that should be used. The most common usage in the United States is that recommended by the American National Standard Institute (ANSI).[16]

The second type of gust response—fluid acceleration—which some investigators believe to be more important than the pressure-force term, relates mainly to the natural frequency of oscillation of some structures.[17] Turbulent eddies in hurricanes are of unique shape and duration and are associated with a minimal variation in the wind direction. For this reason, they may contribute significantly to the amplitude of oscillation in structures having natural periods that are harmonics of the hurricane-eddy period. Most low-rise conventional structures and their components, roof and wall sections, are not sensitive to this type of acceleration. However, the catenaries of electric power lines *are* susceptible, and they, in turn, impose large stresses and torques on the steel towers that support them. Some types of high-rise structures may also be susceptible to this source of damage.

Few observations of gusts during hurricane-force winds have been recorded with sufficient resolution to describe the rate of wind accelerations. However, some experienced hurricane observers describe these gusts as a *gradual increase* in the wind speed to peak values followed by a sudden reduction, or relaxation, to speeds well below the mean value. Only a few reliable measurements have been made in hurricane winds from aircraft using sensitive gust probes or hot-film anemometer systems. Figure 98 is one example of such data obtained from hurricane Eloise (1975).[18] This sample was obtained with a gust probe at a flight elevation of 362 m along an upwind flight track nearly parallel to a streamline. The mean wind speed, 21 mps, and linear changes in the mean along the 13-km flight track were removed from the record to reveal the eddy wind-speed configuration. The eddy cycle, in this case, has a duration of about 90 seconds of flight time at an air speed of 103 mps, a wind-path length of 9.3 km. At this elevation, the gust factor, or ratio of expected peak gusts to sustained wind speeds, was 1.14, less than half the gust factor for anemometer levels. Additional data obtained from the same locale at elevations of 150, 540, 660, 900, and 1,210 m within about 20 minutes of the time of the sample in Figure 98 showed similar gust configurations and dimensions. Frictional dissipation of kinetic energy at each level was almost uniform, consistent with a conservation of gust factor with height in this layer. However, the dissipation rate for the same locale, computed from aircraft data at 85 m and from ocean-buoy data from Eloise at a later time, indicated a rapid change of gust factor in the lower 100 m.

Figure 99, a vertical profile of mean winds obtained from the seven sampling levels in Eloise on 17 September 1975, shows very little variation of mean wind speed with height in the layer 100 to 600 m. With substantially larger frictional dissipation rates in the lower 100-m layer, the mean winds should increase steadily from the standard anemometer level to about 100 m and stay nearly constant through the remaining habitation layer, 100 to 500 m.

In 1977 John Bates, using wind data obtained by aircraft and ane-

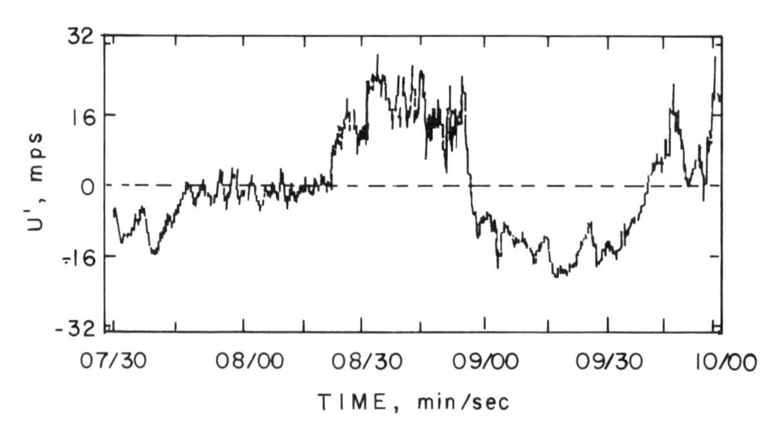

Fig. 98. Gust profile from hurricane Eloise, 1975, measured with a sensitive gust probe on a NOAA research aircraft over water, altitude 362 m. The mean wind speed, 21 mps, has been subtracted from this record. The aircraft was moving into the wind along a streamline traveling 103 mps. After M. S. Moss, *Low-Layer Features of Two Limited-Area Hurricane Regimes*, Technical Report ERL 394/NHEML 1 (Springfield, Va.: Department of Commerce-NOAA, March, 1978), 23.

mometer, normalized the wind-speed values with reference to the 150-m level and derived an empirical model for vertical wind shear in the lower 10 km. This profile is reproduced in Figure 100.[19] In this model, the wind shear in the lower 100 m is given for open water (or immediate shoreline locations) and for overland locations more than half a kilometer inland. The latter is based upon a modification of the wind-shear power law by R. H. Sherlock.[20]

The recently acquired gust-probe data from the hurricane lower-boundary layer, exemplified in Figure 98, provide significant new insight to the physical character of the hurricane gust and confirm qualitative accounts of the hurricane gust and its character.

In summation of research experience, including the limited samples using aircraft gust probes, we may propose the following description: Hurricane gusts, generated and sustained almost en-

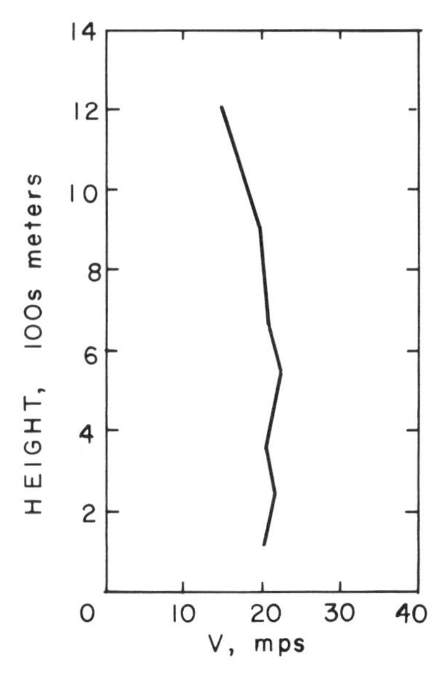

Fig. 99. Vertical profile of sustained wind speeds in hurricane Eloise, 17 September 1975, measured by a NOAA research aircraft with a coupled inertial navigation system. The measurements were made along streamlines of the horizontal wind at a radial distance about 110 km north of the center. After John Bates, *Vertical Shear of the Horizontal Wind Speed in Tropical Cyclones*, Technical Memo ERL WMPO-39 (N.p.: Department of Commerce-NOAA, 1977), 5.

tirely by wind shear in the 100-m layer nearest the surface, recur at intervals of several minutes. They involve a gradual local acceleration of wind speeds for a period of 5 to 10 seconds followed by an abrupt decrease to values much less than the mean within a second or two. The gust factor diminishes upward from about 1.3 to 1.15 in the lowest 100 to 150 m; otherwise the size and configuration of the eddies are conserved throughout the habitation layer.

Conclusions Concerning Hurricane Winds in the Habitation Layer

At a coastal location where a hurricane is moving directly shoreward at a nominal speed of 12 to 15 kt, winds can be expected to reach hurricane force (33 mps, 64 kt) near or just off the open shoreline 3 to 6 hours before landfall. In a hurricane or typhoon of

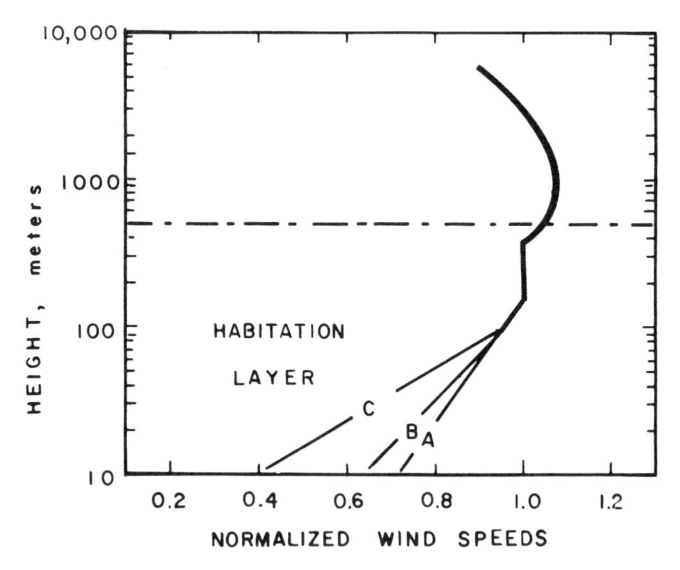

Fig. 100. Model of wind-speed variations with height in tropical cyclones; *A* is over water and *C*, over land. *B* indicates the profile for 1/7 power law. Speed factors are normalized with reference to those measured by aircraft in the layer 150 m to 350 m where the least change-with-height was observed. After John Bates, *Vertical Shear of the Horizontal Wind Speed in Tropical Cyclones*, Technical Memo ERL WMPO-39 (N.p.: Department of Commerce-NOAA, 1977), 6.

moderate strength, winds will increase exponentially with time as the hurricane center approaches, reaching maximum sustained speeds of 45 mps at anemometer level (10 m), with peak gusts of about 60 mps at the time the center moves onshore. In an *extreme* hurricane or typhoon, sustained winds may reach 75 mps or more, with peak gusts of about 97 mps.

Using the Bates model, Figure 100, the maximum winds in the habitation layer and the wind shear acting upon a shoreline structure during a *moderate or extreme* tropical cyclone can be computed. These values appear in Table 13 for each of three shoreline locations: position B, situated directly on the track of the eye center; and positions A and C, abreast of the center position at landfall, a radial distance equal to the radius of maximum winds. For example, the center is assumed to approach at a speed of 7 mps and move inland at right angles to the shoreline. These positions are shown in Figure 101. The difference in the maximum sustained wind to either side of center reflects not only the hurricane movement (7 mps, which adds to winds abreast of the center on the right and subtracts on the left) but also the fact that the trajectory approaching the point of maximum for position A is entirely over water, whereas that at C is over land. Gust factors at anemometer level are considered to be 1.3, and at 100 to 500 m, 1.15.

Table 13. Maximum Wind Speeds Affecting Shoreline Structures in the Habitation Layer

Elevation (m)		*Maximum wind speeds (mps)* Position A	B	C	Position A	B	C
		Moderate hurricane			*Extreme hurricane*		
10	*Sustained*	*45*	*39*	*31*	*75*	*69*	*57*
	Gust	*58*	*51*	*40*	*98*	*90*	*74*
100	*Sustained*	*62*	*55*	*48*	*104*	*96*	*89*
	Gust	*71*	*63*	*55*	*120*	*110*	*102*
500	*Sustained*	*62*	*55*	*48*	*104*	*96*	*89*
	Gust	*71*	*63*	*55*	*120*	*110*	*102*

NOTE: Positions A, B, and C relative to the track of the center are those identified in Figure 101. Hurricane tracks are assumed to be perpendicular to the coastline, the approach speed 7 mps and RMW 25 km.

Wind-Speed Distribution Relative to Landfall Positions

From Figure 101, the horizontal distribution of wind-damage potential considers that the frictional dissipation at the surface for a distance inland of 10 km or more is characteristic of an overland trajectory but that no change in dissipation occurs above 100 m either at the shore or inland. The damage potential from pressure forces, proportional to the square of the surface wind speed, has been normalized with respect to the value at position A. A drop of more than 50% in damage potential occurs in the first 10 km inland. However, a narrowing zone of major damage, 30%–35% of the shoreline maximum, extends much further inland.

This model appears reasonably consistent with the wind damage experienced from severe hurricanes moving directly inland. However, there are notable exceptions, especially when a hurricane does not approach at right angles to the coastline or when it encounters a new source of energy as the track parallels a cold front and maintains a great strength far inland. An exceptional example, described in Chapter 7, is the case of Hazel in 1954.[21] A slow-moving hurricane with a very large vortex, such as Carla in 1961 at the Texas coast, may not exhibit as much asymmetry of damage potential as implied in this model; and a few may show irregular distributions of wind maxima at the coastline, such as Celia in 1970 at the Texas coast and Frederic in 1979 near Mobile, where maximum winds occurred in the left semicircle. Despite the anomalous cases, the distribution of wind speeds and of damage potentials in Figure 101 and Table 13 should be useful and realistic guides in estimating the wind-damage potential along a hurricane-prone coast.

Measurement of Extreme Winds

The climatology of hurricane surface-layer winds in the United States has been handicapped by a lack of well-exposed anemometer systems in the coastal zone. Several tall towers, most notably the one at Brookhaven Laboratory on Long Island, New York, have been carefully instrumented with recording anemometers at several levels. Nevertheless, few are located directly at a shoreline,

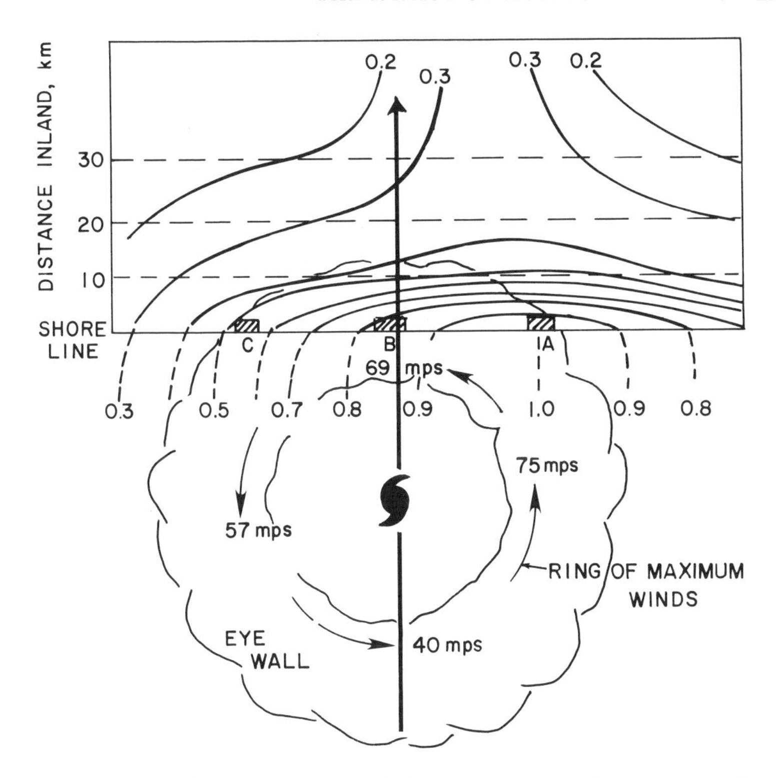

Fig. 101. Maximum surface-layer winds from an extreme hurricane. Relative wind-damage potentials for shoreline structures along the track and abreast of an extreme hurricane. Damage potentials are normalized with respect to the maximum at position *A*.

and complete records of a hurricane passage have rarely been acquired. Operational anemometers at weather stations are not often found at shoreline locations; and, because of instrument or power failures, it has been difficult to capitalize on opportunities for data acquisition when hurricanes *have* affected such locations.

Interesting engineering wind-tunnel tests and numerical modeling of building responses, conducted during the 1970s, supplied new knowledge of the impact of sustained winds on large building structures.[22] However, the United States has not yet conducted experiments in the real atmosphere to determine the impact of hurricane gusts on structures. This will require either the shoreline installation of many new and superior anemometer systems or the elaborate instrumentation of high-rise buildings in coastal locations to record wind stresses and the resultant strains imposed on the structures.

A program to accomplish this has been implemented in Hong Kong, where a 10-story building frame and a cluster of four instrumented towers were erected explicitly to measure the impact of typhoons. In London, a multistory commercial building has been instrumented for similar purposes. At the Hong Kong structure, records have been acquired from a number of typhoons, and work by Gordon Bell and his colleagues at the Royal Observatory is underway to interpret these data.

The sampling of hurricane winds by ocean buoys and aircraft sensors will undoubtedly continue to provide useful information about turbulent eddies and the structure of hurricane gusts over water. This kind of data acquisition is too hazardous to extend across, or even near, a coastline under hurricane conditions. The search for an alternate means of acquiring such data deserves high priority.

Finally, it should be remarked that the system for recording and archiving extreme-wind data should permit climatic summarizations to be made for time intervals other than a sustained-wind and peak-gust period as presently defined. Proper archiving of extreme-wind data should allow spectral analyses of wind-speed fluctuations to be made.

From the preceding discussions, it is clear that a *significant extreme wind* for wind storms should be defined as being considerably shorter in duration than the fastest mile. Moreover, the definition of peak (or maximum) gust speed needs to be reexamined with a view toward specifying a unit that has both dynamical significance as to impact and realism with regard to the ability to measure and record it.

Hurricane Tornadoes

The impact of hurricane-generated tornadoes has not been considered in earlier discussions and will receive only cursory attention here for two reasons: First, the probability of this event affecting any given structure is quite small; second, the damage potential from such events is generally less than that of the sustained winds and gusts of the mature hurricane.

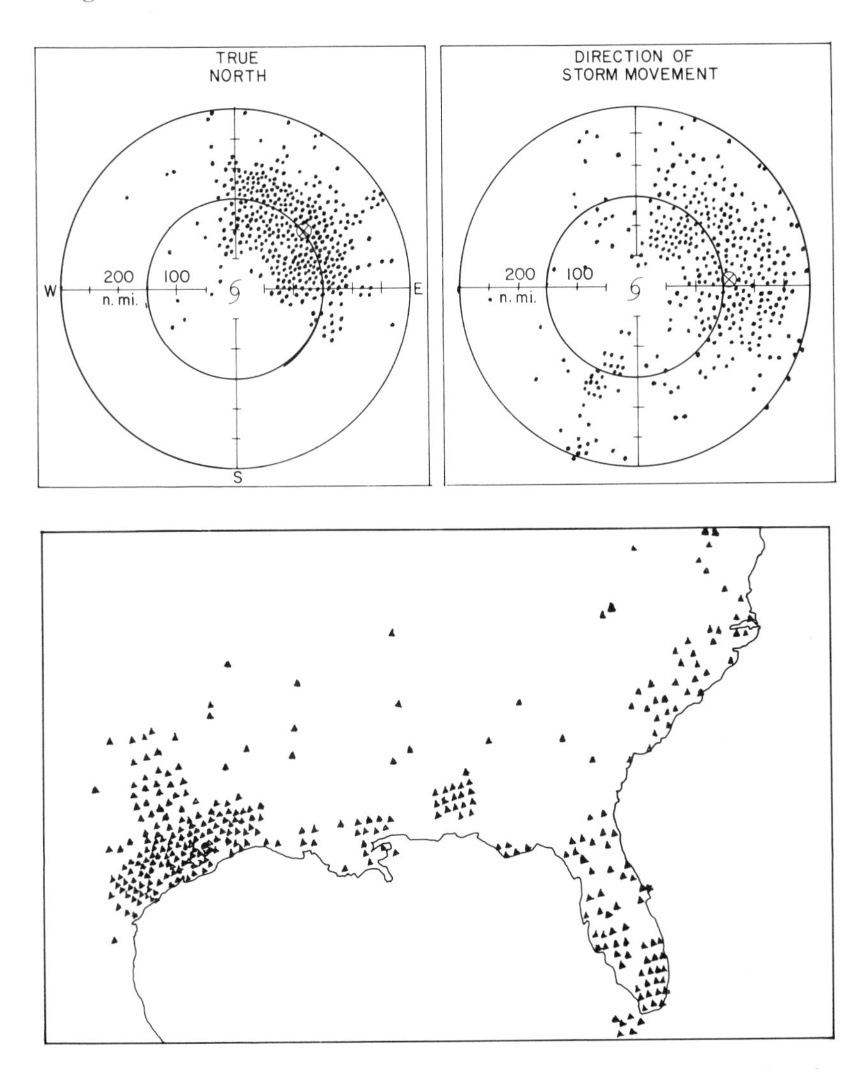

Fig. 102. Distribution of hurricane tornadoes in the hurricane vortex and in the coastal zones of the United States, with respect to the hurricane vortex. After W. M. Gray and D. J. Novlan, "Hurricane-Spawned Tornadoes," *Monthly Weather Review*, CII (1974), 476–88.

Hurricane tornadoes develop in the spiral rainbands, mostly in the right-front quadrant outside the areas of sustained hurricane- or gale-force winds. Figure 102 shows the centroid and distribution of hurricane tornadoes. Although some hurricanes produce families of tornadoes, the individual event is a small, rope-type vortex similar to a waterspout. It has a short path length, maximum wind speeds are usually less than 50 mps, and pressure drops in the funnels are believed to be no more than about 20 mb. For further information on these small-scale systems, the reader is referred to the works of Allen Pearson and Alfred Sadowski, William Gray, D. Novlan, and Joe Golden.[23]

The Saffir/Simpson Damage-Potential Scale

In 1972, the United States National Weather Service inaugurated a program to relate the known meteorological characteristics of hurricanes at sea to the patterns of damage that such systems might produce as they make landfall. The damage potential from wind and from storm surge is expressed on a scale of one to five. The damage-potential-scale concept originated with Herbert Saffir, who drew upon broad experience in surveying hurricane-wind damage to structures, street signs, and ornamentals in order to assign damage expectancies from sustained wind speeds in terms of an increasing scale of zero to five.[24] The patterns of storm-surge damage potential relate less to structural integrity than to inundation levels and the requirements for public evacuation. This part of the scale is based upon surveys of surge damage by Robert Simpson and his computations of peak storm surges, which correspond to those ranked by Saffir with respect to wind damage.

The program, initially used as guidance for disaster-relief agencies, was first introduced in public advisories in 1975. The primary purpose was to enable coastal residents to compare the relative severity of an approaching hurricane with what had occurred in the same coastal zone during previous hurricanes. A description of the damage-potential scale is contained in Appendix C together with the damage-potential ratings for historic hurricanes.

REFERENCES

1. J. Malkus and H. Riehl, "On the Dynamics and Energy Transformations in Steady-State Hurricanes," *Tellus*, XII (1960), 1–20.
2. *Ibid.*
3. B. I. Miller, "Characteristics of Hurricanes," *Science*, CLVII (September, 1967), 184–95.
4. Harry F. Hawkins and D. T. Rubsam, "Hurricane Hilda, 1964: II. Structure and Budgets of the Hurricane on October 1, 1964," *Monthly Weather Review*, XCV (1968), 617–36.
5. R. E. Tuleya and Y. Kurihara, "A Numerical Simulation of the Landfall of Tropical Cyclones," *Journal of Atmospheric Science*, XXXV (1978), 242–57.
6. G. J. Bell, "Surface Winds in Hong Kong Typhoons," *Proceedings of the U.S./Asian Weather Symposium* (1961).
7. R. A. Pielke and Y. Mahrer, "Representation of the Heated Planetary Boundary Layer in Mesoscale Models with Coarse Vertical Resolution," *Journal of Atmospheric Science*, XXXII (December, 1975), 2288–308.
8. M. S. Moss, *Low-Layer Features of Two Limited-Area Hurricane Regimes*, Technical Report ERL 394/NHEML 1 (Springfield, Va.: Department of Commerce-NOAA, March, 1978), 46.
9. H. C. S. Thom, "New Distribution of Extreme Winds in the United States," *Proceedings of the American Society of Coastal Engineers, Structural Division*, LXXXVI (April, 1960), 11–16.
10. G. N. Brekke, *Wind Pressures in Various Areas of the United States*, Building Materials and Structures Report 152 (Washington, D.C.: Government Printing Office, April 24, 1959).
11. K. C. Mehta, "Windspeed Estimates: Engineering Analyses," in R. E. Peterson (ed.), *Proceedings of the Symposium on Tornadoes: Assessment of Knowledge and Implications for Man* (Lubbock: Institute of Disaster Research, Texas Tech University, 1976), 89–103.
12. T. Fujita, "The Lubbock Tornadoes: A Study of Suction Spots," *Weatherwise*, XXIII (1970), 160–73.
13. Mehta, "Windspeed Estimates: Engineering Analyses"; E. W. Kiesling, K. C. Mehta, and J. E. Minor, "Protection of Property and Occupants in Windstorms," *Proceedings of the American Society of Civil Engineers, National Convention* (April 29, 1977).
14. H. S. Saffir, "Designs and Construction Requirements for Hurricane-Resistant Construction," *Proceedings of the American Society of Civil Engineers, National Convention* (April 29, 1977), 161–66.
15. Mehta, "Windspeed Estimates: Engineering Analyses."
16. American Society of Civil Engineers, Committee on Wind Forces, "Wind Forces on Structures," *Transactions from the American Society of Civil Engineers, Part II*, (1961), 1124–98.
17. A. G. Davenport, "Gust Loading Factors," *Proceedings of the American Society of Civil Engineers, Structural Division*, XCIII (June, 1967), 11–34.
18. Moss, *Low-Layer Features of Two Limited-Area Hurricane Regimes.*
19. John Bates, *Vertical Shear of the Horizontal Wind Speed in Tropical Cyclones*, Technical Memo ERL WMPO-39 (N.p.: Department of Commerce-NOAA, 1977).
20. R. H. Sherlock, "Variation of Wind Velocity and Gusts with Height," *Proceedings of the American Society of Coastal Engineers*, XVIII (1952).
21. E. Palmén, "Vertical Circulation and Release of Kinetic Energy During the Development of Hurricane Hazel into an Extratropical Storm," *Tellus*, X (1958), 1–23.
22. K. Suda, S. Seiji, and K. Takeuchi, "Wind Tunnel Experiments for Studying a Local

Wind," *Proceedings of the Sixth Conference of the U.S./Japan Panel on Wind and Seismic Effects*, (May 15, 1974), 1–11; J. A. Shanahan, "Evaluation of and Design for Extreme Tornado Phenomena," in R. E. Peterson (ed.), *Proceedings of the Symposium on Tornadoes: Assessment of Knowledge and Implications for Man* (Lubbock: Institute of Disaster Research, Texas Tech University, 1976), 251–82.

23. A. D. Pearson and A. F. Sadowski, "Hurricane-Induced Tornadoes and Their Distribution," *Monthly Weather Review*, XCIII (1965), 461–64; W. M. Gray and D. J. Novlan, "Hurricane-Spawned Tornadoes," *Monthly Weather Review*, CII (1974), 476–88; J. H. Golden, "Waterspouts and Tornadoes over South Florida," *Monthly Weather Review*, XCIX (1970), 146–54.

24. Developed by H. Saffir, consulting engineer, Dade County, Florida, and R. H. Simpson. See R. H. Simpson, "The Hurricane Disaster Potential Scale," *Weatherwise*, XXVII (1974), 169, 186.

Fig. 103. Gale-force winds lash a coastal village and harbor—only the beginning of the destructive onslaught ahead—as a hurricane approaches.

CHAPTER 9

Waves and Tides

The tropical cyclone can be the most dramatic manifestation of large-scale fury in nature. Yet it is not the wind fury but the rage of the sea—its impact at the coast and across the coastal plain—that is the main source of the drama of destruction and the struggle to remain alive (Figure 103).

While the primary concern is for protection of lives and property in a coastal zone, the rapid increase in the ocean traffic, especially in the number of privately owned yachts and small boats, emphasizes the need to understand the hurricane impact on vessels operating in coastal and offshore waters. In the greater Miami area alone, more than ten thousand small boats must seek some form of safe shelter when a hurricane threatens. A rapidly developing hurricane near shore can overtake hundreds of small boats in offshore waters before they can make safe harbor. Therefore, in this chapter we examine the consequences of hurricane wind stresses on the sea—in deep water, across the continental shelf, and in shoal water—so that we may understand the source and nature of the threat at sea and at the coast, and why the threat may vary from one coastal site to another and with the size and movement of the hurricane.

The Growth and Propagation of Ocean Waves

A complete description of ocean waves, their generation, size, and modes of propagation is a complex exercise in nonlinear fluid mechanics, beyond the scope of this discussion. However, a reasonable approximation of wave sizes and of wave movement generated by hurricane wind stresses on the sea can be obtained by simple applications of progressive-wave theory. To understand the hurricane impact at sea, we need to write several fundamental statements that govern size and movement of waves in deep and in shallow water.

From linear wave theory and the model in Figure 104, three categories of waves may be defined as functions of water depth:

Deep-water waves: $d/L > 0.5$;
Shallow-water waves: $d/L < 0.04$;
Transitional waves: $0.04 < d/L \leq 0.5$;

where d is water depth and L is wavelength.[1]

The propagation of a nonbreaking wave theoretically involves a local displacement of water particles, as shown in Figure 104, without a net transport or accumulation of water mass. These displacements describe circular orbits in deep water and elliptical orbits in shallow water. The sinusoidal profile used in linear wave theory underestimates the curvature and height of a crest (above still water) and overestimates the curvature and depth of a trough. The actual shapes of troughs and crests are better described by the second-order Stokes profile, which will not be discussed here.[2]

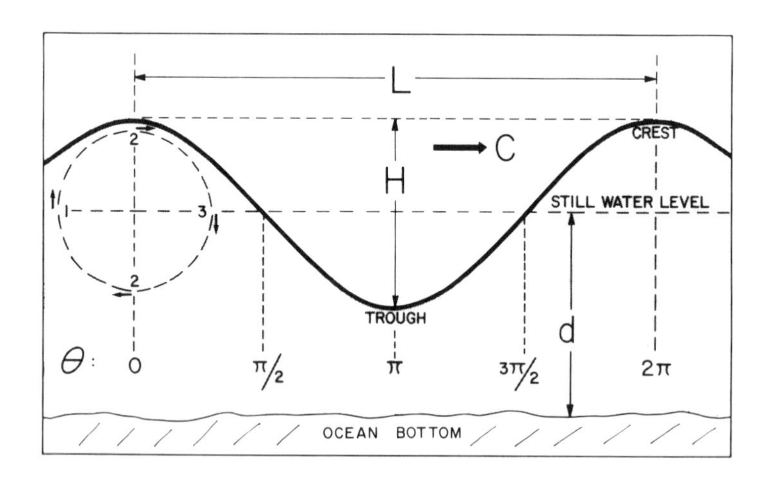

Fig. 104. Model of a linear sinusoidal water wave, including a definition of its properties: *L*, wavelength; *H*, wave height; *C*, celerity; *d*, mean water depth.

The *celerity*, or speed of wave propagation, C, depends upon the water depth and wavelength. This may be expressed

Eq. 9.1
$$C = \sqrt{\frac{gL}{2\pi} \tanh \left[\frac{2\pi d}{L}\right]}$$

where g is acceleration of gravity, and other symbols are those defined in Figure 104.

In deep water, $d/L > 0.5$ so that $\tanh(2\pi d/L) \rightarrow 1$ and the celerity for deep-water waves may be written approximately

Eq. 9.2
$$C_{DW} = \sqrt{\frac{gL}{2\pi}}$$

Since the wavelength, T, can be expressed $L = CT$, where T is the wave period or time between the arrival of two successive crests, Equation 9.2 may be written in the more convenient form

Eq. 9.3
$$C_{DW} = \frac{gT}{2\pi}$$

T being easier to measure than L.

In shoal water, $d/L < 0.04$ so that $\tanh(2\pi d/L) \rightarrow (2\pi d/L)$ and, with good approximation,

Eq. 9.4
$$C_{SW} = \sqrt{gd}$$

celerity depending upon water depth alone.

Breaking Waves

The above relationships relate to wind-driven waves in a *fully arisen sea*, one in which the generating wind source has achieved a quasi-steady-state, or balance between the stresses applied at the water surface and the spectrum of wave-form responses. To understand the impact of hurricane winds on the sea state in deep water and at the coast, it is necessary to examine the destabilizing processes that cause waves to break and, in the process, to transport a mass of water in the direction of wave motion. In the open sea, the breaking of large waves is the principal threat to vessels of all sizes. Near the coast, it is the means of generating a spectrum of smaller waves that may cause beach erosion, damage beach structures, and

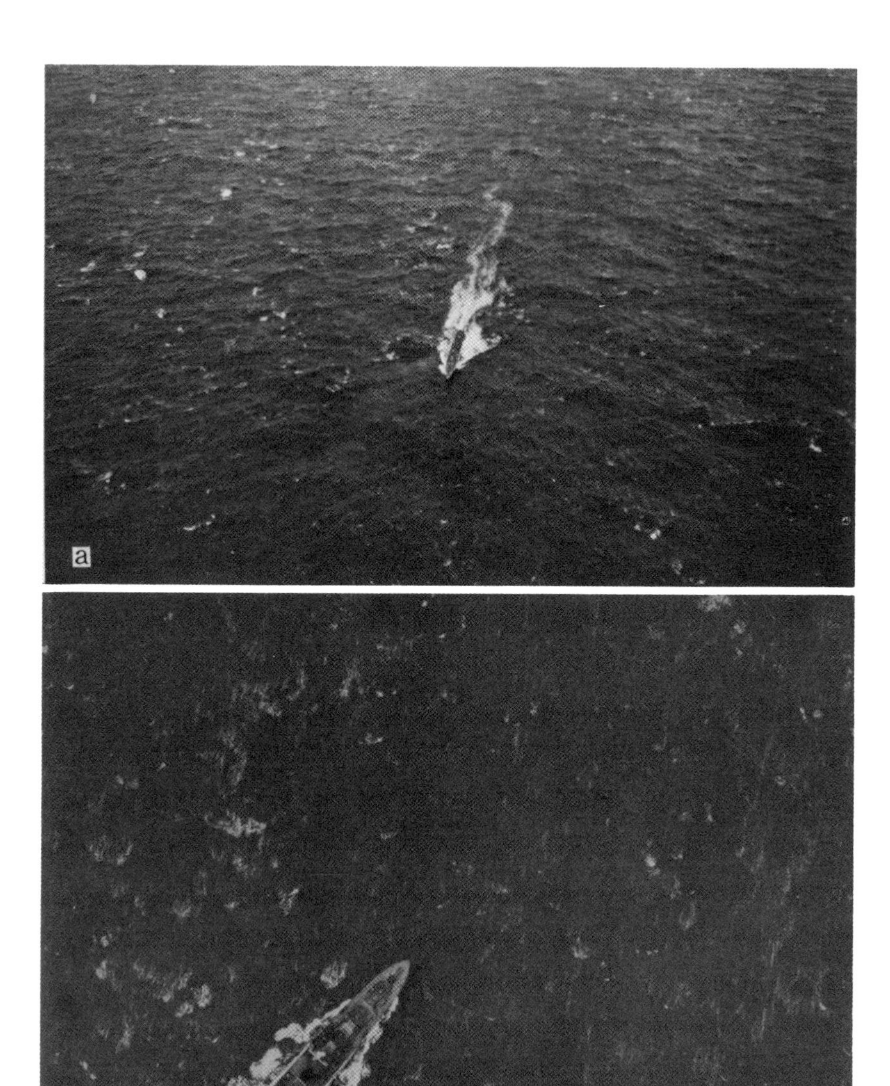

Fig. 105. The ocean surface viewed from a low-flying reconnaissance aircraft. *a*, Surface wind speed, 25 kt (strong wind); numerous whitecaps; some waves over 3 m. *b*, Surface wind speed, 35 kt (just gale force); waves longer; streaks of foam visibly windborne; some waves over 9 m. *c*, Surface wind speed, 65 kt (just hurricane force); long rolling waves; exceptionally high sea, nearly covered with streaks of airborne foam. *d*, Wind, approximately 100 kt (severe hurricane); very dangerous, high seas; water surface barely visible through airborne streaks of foam. Official U.S. Navy photos.

c

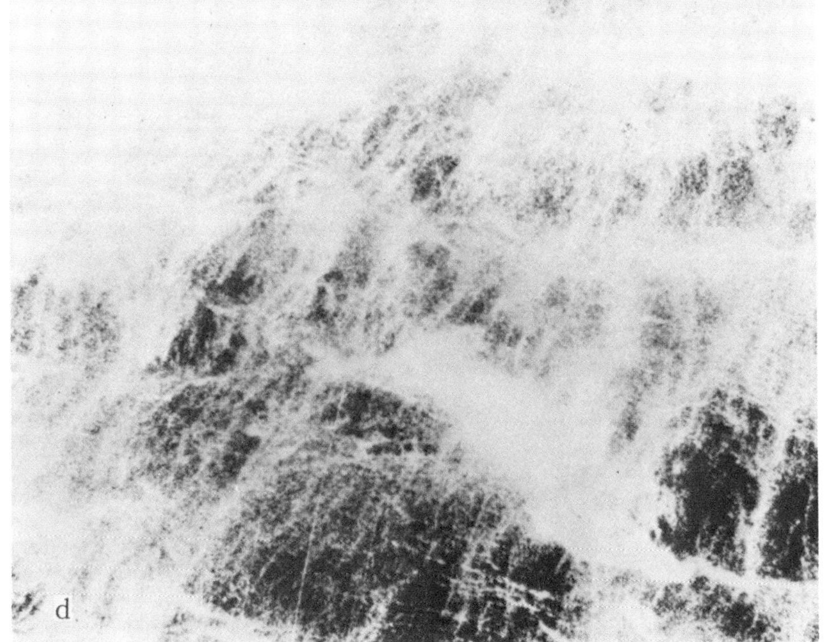
d

cause the shoreward transport and accumulation of water, increasing the inundation potential from storm surges.

Waves of any size become unstable and break when local water-particle speeds at the wave crest exceed the celerity of the wave. Ordinarily, no organized breaking of waves is observed for sustained wind speeds less than about 12 kt, irrespective of water depth. A few whitecaps appear at 13 kt, with numerous small breaking waves at 15 kt. Figure 105 shows the changes in deep-water-wave appearance from a low-flying airplane as strong winds increase to gale and hurricane force.

In deep water, the *significant wave*, or dominant wave form (Table 14), tends to have superimposed upon it a spectrum of shorter waves that break at lower wind speeds.* The larger, organized, progressive waves grow in height and length with increasing wind speed and duration of the stresses. These waves sometimes reach heights in excess of 20 m, although it is rare for severe conditions to persist long enough in space and time to generate an organized field of significant waves of this height. In hurricane Eloise (1975), an example of a moderate-strength system, instrumented sea buoys recorded maximum significant waves of 8.8 m under sustained winds of 37 mps. These waves had a celerity of 12.6 mps and a wavelength of 102 m.

Table 14. Terminologies Used and Relative Heights in a Spectrum of Deep-Water Waves Generated by Hurricane Winds

	Wave terminology	*Relative height*
H_s :	Significant wave	1.00 m
H :	Average	0.63 m
H_{rms} :	Root-mean-square (RMS)	0.71 m
H_{10} :	Highest 10%	1.27 m
H_1 :	Highest 1%	1.67 m
H_x :	Probable maximum	~1.76 m

* Significant wave height, as defined by Munk, is the average of the highest one-third of the waves measured. It relates numerically to other wave-height designations as indicated in Table 14. W. H. Munk, "The Solitary Wave Theory and Its Application to Surf Problems," *Annals of the New York Academy of Sciences*, LI (1949), 376–462.

The limiting factor in growth of deep-water waves is the wave slope. Theoretically, the maximum slope for a nonbreaking wave in deep water is

Eq. 9.5 $$H/L = 0.142$$

a ratio of one to seven.[3]

In shallower water, the breaking of waves is a function of the water depth, bottom friction being the primary contributor to instability. The *breaking-water depth* for waves in shoal water, based upon a theoretical relationship, is given by

Eq. 9.6 $$d_b = 1.28\,H$$

where d and H are as defined in Figure 104.[4]

Calculating the Significant Wave Height

The calculation of the significant wave height as a function of the persistent wind speed is a complex task, made less arduous and more precise in recent years with the help of high-speed computers. If the generating winds have straight trajectories for great distances, as in the warm sectors of wave cyclones, the computation of the wave height can be simplified by employing the concept of *fetch*, or distance over which a wind of uniform strength applies its stresses at right angles to the line of wave crests.

A nomogram much-used for this purpose is reproduced in Appendix F, Figure 148.[5] This diagram provides a reasonable estimate of significant wave height in extratropical storms where very long fetch winds occur in the warm sector, and a fair approximation in hurricanes where longest fetches occur in the right-rear quadrant. However, because of the large curvature of wind trajectories in most of the hurricane vortex, the concept of fetch is more uncertain and difficult to apply. Figure 106 is a reasonable approximation of the expected wave-height distribution, normalized with respect to the maximum wave height.[6]

A more precise calculation of significant wave-height distributions can be generated with numerical models using iterative procedures for applying instantaneous stresses to waves moving through a changing field of wind motion.[7] Figure 107 is an example of a hurricane wave-height pattern computer-produced in an operational time frame using a parametric model.[8]

In severe tropical cyclones, the maximum significant wave heights are probably not as great as those in some severe extratropical storms because of the limited fetch over which extreme wind speeds persist. However, in hurricanes, the more challenging and dangerous element in deep-sea navigation is the interference pattern between two or more wave trains whose generating fetches have different orientations. This is a critical problem in the right-front quadrant (Figure 108).

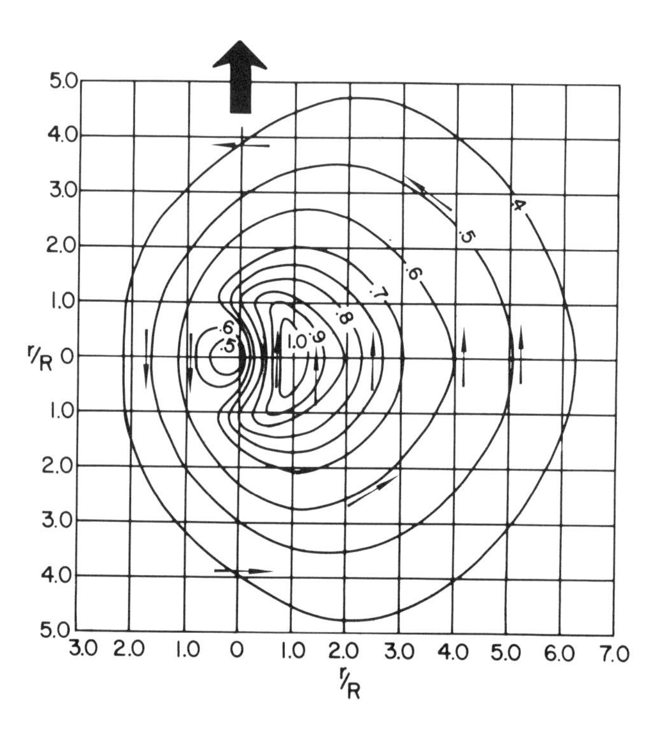

Fig. 106. Isolines of significant wave height for a slow-moving hurricane scaled with respect to the maximum significant wave. *Small arrows*, approximate wave directions; *large arrow*, motion of hurricane; R, radial distance to maximum significant wave; r, radial distance to point of interest. Adapted from Department of the Army, Corps of Engineers, Coastal Engineering Research Center, *Shore Protection Manual* (Washington, D.C.: Government Printing Office, 1973), Vol. I, p. 3.59.

The upper limit for significant wave heights in extreme hurricanes is probably about 15–18 m. This means that an average height for the one-third of wave population with greatest mean heights would be on the order of 15 m. Operationally the greatest interest will center around the question, What is the greatest wave height that a vessel could expect to experience in a nominal period of exposure to such seas? Using Rayleigh-distribution theory, the probability of experiencing a wave higher than the significant wave can be computed. A nomogram for this purpose is reproduced in Figure 149, in which the theoretical probability of seeing waves with heights greater than the significant wave is listed (line b). The broken straight line (a) is based upon observations of several thousand waves at a single site during one weather condition. For ex-

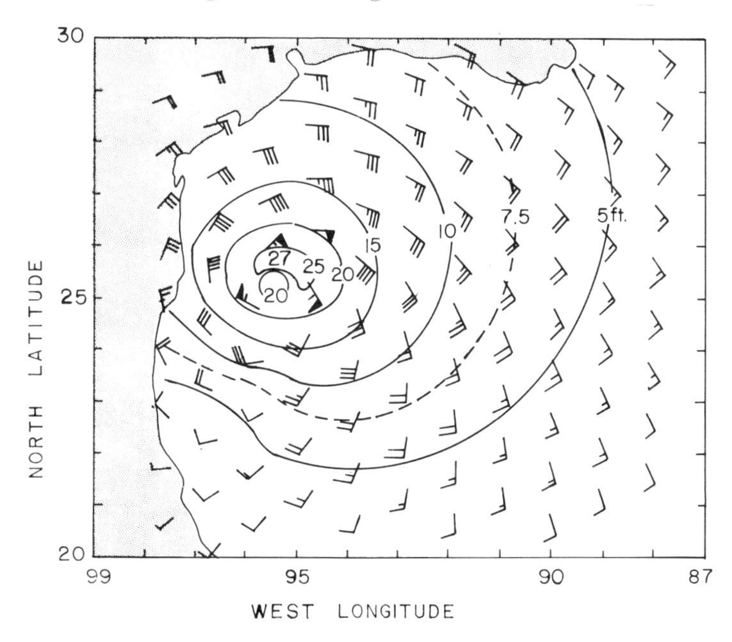

Fig. 107. Computation of significant wave heights in hurricane Anita, 1977, using a parametric model. Contours are for height in feet. Grid of wind vectors shows computed values of generating winds in knots (*half barb*, 5 kt; *full barb*, 10 kt, as in earlier figures). Adapted from Vincent J. Cardone, Duncan B. Ross, and Merlin R. Ahrens, "An Experiment in Forecasting Hurricane-Generated Sea States," *Proceedings of the Eleventh Technical Conference on Hurricanes and Tropical Meteorology* (Boston: American Meteorological Society, 1977), 693.

ample, let us assume a significant wave height of 15 m, a wave period of 10 seconds, and a local exposure to the extremes for 30 minutes. From Figure 149, there would be a theoretical expectancy of encountering one wave with a height of 26 m. If, however, the probability is computed in terms of the empirical relationship (based, however, upon observations in winds less than hurricane force), the expected extreme wave would be 24 m.

For a less-extreme hurricane such as Eloise, in which a significant wave height of 12.2 m and a period of 7.5 seconds were recorded at a sea buoy in an exposure of 1 hour, the theoretical extreme wave would be 23 m, and using the empirical curve, only 21 m. Either would pose serious operational problems for vessels of all sizes. The extreme waves in both cases, however, may be unrealistically high because of the tendency for the larger hurricane waves to break prematurely as a result of interference between wave trains from more than one generating source.

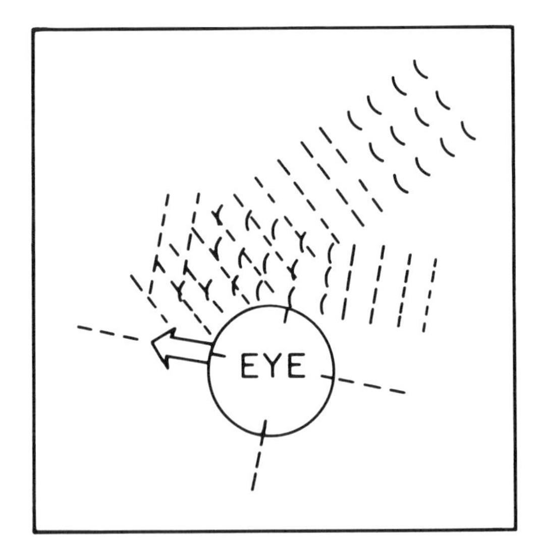

Fig. 108. Interference wave patterns from two separate generating sources that pose difficult problems of navigation and survival for vessels of all sizes. The problem tends to be maximized in the right-front quadrant. Here the developing wave train is parallel to the hurricane track; the decaying train provides interference.

The Hurricane Swell

When trains of waves outrun their generating source, they decay but continue to propagate in the direction of the generating fetch in a form known as *swell.* While the characteristic wavelength is conserved, the amplitude diminishes. Long trains of swell running ahead of the hurricane sometimes initiate beach erosion long before the arrival of strong winds.

For the concerned observer at the beach who does not have access to aircraft or satellite tracking information, changes in swell pattern, direction, and period may provide a rough measure of changes in the relative strength and track of the hurricane. The dominant swell with the longest period nearly always emanates from the right-front quadrant in the direction of hurricane movement, having been generated in the area of longest fetch: the right-rear quadrant. Appendix F contains a nomogram (Figure 150) that shows the rate of decay as waves outrun their generating sources but retain their identity after traveling several thousand kilometers. Their wavelength and height, propagating as swell, are an index to the distance from and strength of their generating source.

Cold Water in the Wake of a Hurricane

Oceanographic observations during the 1950s revealed that air-sea interactions in a hurricane cause important changes in the thermocline structure, often leaving a wake of colder sea-surface temperatures.[9] For years, this cooling was considered to be mainly an upwelling phenomenon resulting from the surface-layer water divergence generated by wind stresses near a hurricane center. In the early 1970s, however, the theoretical work of Jack Geisler and others, verified experimentally by Peter Black, described an intricate system of internal waves extending through the thermocline and conducting colder water to the surface by an eddy-mixing process.[10] This mixing can cool surface waters as much as 5°C and is most pronounced when the hurricane moves at speeds less than the internal-gravity waves (4 to 8 kt).

The consequences of this cooling, aside from marine-biological influences, is to present a sea surface sufficiently cool to sap the strength or inhibit the development of any subsequent hurricane that crosses the cooler surface within several weeks. One of the more dramatic examples of such an effect occurred in August 1955 when hurricane Connie, of moderate strength, crossed the North Carolina coast at a speed of 6 kt carrying full hurricane-force winds. Four days later, on 16 August hurricane Diane, a more severe hurricane at sea, crossed the track of Connie a short distance off the United States mid-Atlantic coast and rapidly lost strength. It moved inland a day later with winds of only moderate gale force (albeit with very heavy rains that caused record flooding in several states).

The Transition from Deep to Shoal Water

As a hurricane moves out of deep water, crosses the continental shelf, and approaches a coastline, the enormous energy imparted to the sea by wind stresses begins to be deployed more complexly, reordering the threat and the potential impact both in open water and at the shore. Massive waves, which in deep water break because of their excessive slope, break in shoal areas as the water, set in rotary motion by wave passage, begins to scrape bottom. This causes the formation of successively shorter, lower waves, each transporting large amounts of water. Near shore these transports accumulate and cause an upslope of the water surface shoreward. This in turn may generate powerful longshore currents, the first of a series of destructive, erosive actions that may eventually affect not only the beach and dune structures, piers, seawalls, and bulkheads, but also shoreline dwellings and buildings! The erosion at beaches from some hurricanes that never approached close enough to provide wind damage has often been impressive. The other important shoal-water process generated by hurricanes is *storm surge*, a superelevation of sea level due to a combination of direct wind-driven water and an uplift induced by the pressure drop; together they reach maximum heights as the hurricane center arrives at an ocean or bay shore.

Through these processes, the character of the hurricane threat is altered as the center approaches shore, sea levels rise, inundation begins and is incremented by the astronomical, secular, and seasonal tides, by freshwater runoff from heavy rains and the outfalls from rivers and streams. Together all these contributions constitute what is known as the *hurricane tide* (sometimes incorrectly called a tidal wave). This tide projects a still-water platform inland upon which the scour and battering of wind-driven waves can have access to structures that otherwise would remain high and dry.

First, let us consider the modification of wave structure with diminishing water depth and its contribution to the hurricane tide. The transport and accumulation of water due to the breaking of waves is known as *wave setup*. The term *setup* refers to a process that causes still-water surfaces to incline upward locally to elevations above the prevailing sea level. Conversely a process that depresses sea level is known as *set down*.

Computing Wave Setup

The computation of water accumulations and the elevation of sea level due to wave setup is complex; and the numerical results, except in more detailed individual case studies, can generally be considered, at best, approximations. The contribution of wave setup is sometimes incorporated implicitly into computations of storm surge by "tuning" the model results for real cases to agree with observed peak-surge heights. However, under some unusual configurations of the ocean bottom, especially where the water depth increases dramatically from the shoreline seaward, wave setup may constitute a significant fraction of the total hurricane tide; therefore, some means of estimating these exceptional contributions is useful.

One relationship, based on the work of M. S. Longuet-Higgins and R. W. Stewart, is

Eq. 9.7 $$S = 0.19\,H_b\,[1-2.82(H_b/gT^2)^{1/2}]$$

where S is the superelevation due to wave setup, and H_b is the "breaking height" where H is measured as in Figure 104; both S and H_b are expressed in feet.[11] One important instance in which wave setup contributed significantly to the hurricane tide occurred in hurricane Eloise (1975), when a tidal maximum of 4.9 m was observed. Postanalyses were unable to account for more than a 2.8-m rise due to storm surge, plus 0.7 m attributable to longer-term anomalies in sea level. The remaining 1.4 m is considered to have resulted from an unusual contribution from wave setup due to the peculiar *bathymetry*, or contours of water depth, nearshore. Here, water depths average less than 3 m nearshore and then drop rapidly to depths of more than 15 m in less than 1 km. The evidence is that significant waves of about 10.5 m approached within several kilometers of shore before breaking and cascading massive amounts of water shoreward as deep-water waves "climbed the nearshore cliff." Applying Equation 9.7 and assuming a wave period of 8 seconds,

Eq. 9.8 $$S = 4.2 \text{ ft} = 1.3 \text{ m}$$

This value is a good approximation of the extraordinary wave setup in Eloise. In most instances, the slope of the bottom is more gradual, and deep-water or transitional waves cannot approach as close to shore before breaking occurs, leading to the generation of a packet of shorter, lower waves. In such cases, wave setup is not so large a factor and can reasonably be accounted for, as mentioned earlier, by "tuning" or other adjustments of model output values to account for systematic deviations from observed values. When breaking occurs in deeper water, local accumulations of water and inclinations of still-water surfaces are continually restored to normal by lateral mass divergence—a process that, in the Ekman (friction) layer, becomes restricted as water depths diminish and bottom friction retards the restorations.

Storm Surge at the Open Coast

The most important component of the hurricane or storm tide is storm surge. For purposes of this discussion, storm surge is defined as the superelevation of the still-water surface that results from the transport and circulation of water induced by wind stresses and pressure gradients in an atmospheric storm.

Storm surge, in some more general contexts, has been considered the sum of all meteorologically generated components contributing to the superelevation of sea level. For quantitative evaluation and for our purposes here, however, storm surge is considered the separate and unique component of the hurricane-produced tide, as defined above.

In the wave cyclone of middle latitudes, storm surge is the result of the combination of rises caused by pressure gradients and rises due to wind setup from long-fetch winds—east or northeast winds in the *cold sector* of wave cyclones off the United States Atlantic coast, and west to southwest in the *warm sector* off the Pacific coast.

In tropical revolving storms—hurricanes or typhoons—the superelevation due to storm surge is more complex dynamically. In this case, there are large rotational components of the wind stress that generate a surface-layer convergence of water mass. In deep water, a compensating divergence prevents any appreciable water accumulation. However, this circulation of mass generates absolute vorticity that tends to be conserved through the Ekman layer (from the surface to a depth of about 90 m). In addition, the steep downward slope of the atmospheric pressure surfaces in the hurricane eye wall generates a hydrostatic rise of sea level known as the *inverse barometer*, which, as a static component, may account for an uplift of about 1 cm for each millibar of pressure drop from the environment to the hurricane center (1 m for a 100-mb drop). In deep water, this is the only contribution to changes in still-water levels.

As the hurricane system moves into shallower water, the conservation of absolute vorticity acquired in deep water, reduced only by bottom friction, results in a superelevation of the still-water surface. Moreover, in moving over shallow water, the inverse barometer is augmented by a dynamic component that substantially increases the amount of uplift from this source. The combination of rises resulting from conservation of vorticity and the dynamically enhanced inverse-barometer component creates a mound of water, the crest of which lies abreast and to the right of the hurricane center near the position of maximum winds. These components are illustrated in Figure 109. This mounding has been described by Jelesnianski as having the essential shape and properties of half a long gravity wave, the wavelength of which is the diameter of the annulus of maximum winds.[12] While this long gravity wave, like the tsunami wave, falls into the classification of a shallow-water wave ($d/L < 0.04$), it is a captive of, and travels with, the hurricane wind system. This long-wave embryo of hurricane storm surge grows in amplitude with diminishing depths, reaching a maximum at about the time of landfall, for approaches nearly perpendicular—an 80° angle—to the coastline (see Figure 110).

From these comments, we may summarize: The hurricane storm-surge height, H_s, is a function of three sources of setup:

Eq. 9.9 $$H_s = f(S_B, S_W, S_V)$$

where S_B is the setup due to the dynamic inverse-barometer effect and is concentric with the low-pressure center; S_W is the setup due to stresses from the irrotational component of the wind; and S_V is the setup due to the stresses from the rotational component, dependent upon the conservation of absolute vorticity as water depths diminish. S_W is the primary component of surge in small

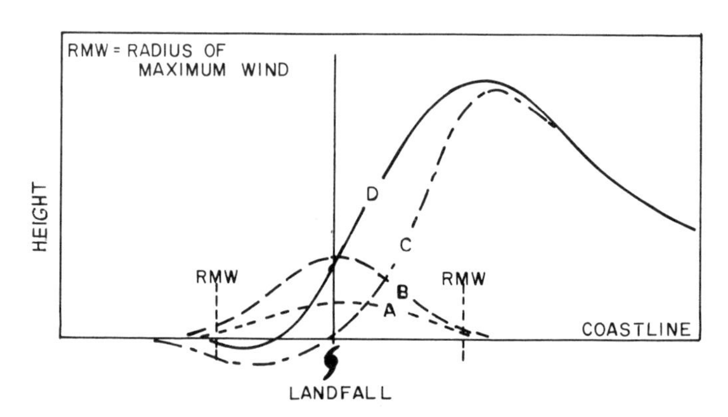

Fig. 109. Components of storm surge at a point of hurricane landfall. The profiles are relative to the coastline as viewed from the sea: *line A*, static inverse barometric effect; *B*, dynamic inverse-barometer effect as a hurricane moves over shoal water; *C*, component resulting from wind stresses; *D*, observed surge profile (exclusive of contributions from wave setup).

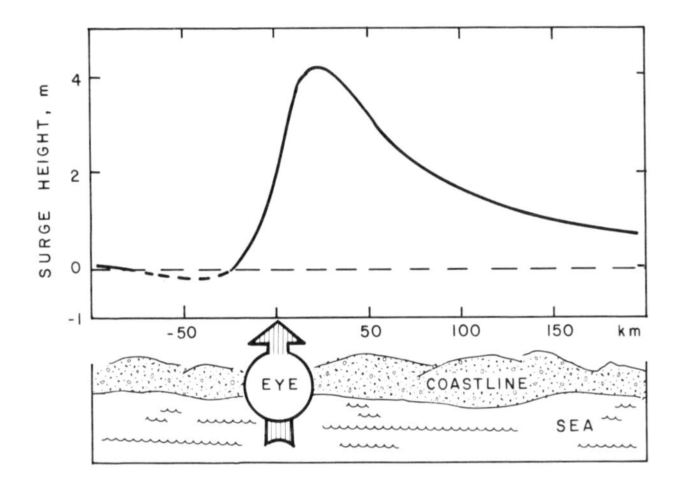

Fig. 110. Typical vertical section of storm surge heights as the center crosses a coastline. Viewed from the sea toward land.

enclosed basins, bays, or estuaries; in extratropical wave cyclones, it combines with the rise due to pressure gradients to provide the total surge. S_V is the major component of storm surge at the open coast in hurricanes and typhoons, especially those with small radii of maximum wind.

Whereas S_W is essentially the accumulation of water that is driven shoreward by the wind stresses, S_V is the uplift of still-water surfaces required to conserve the absolute vorticity or spin generated in deep and transitional water. As water depths decrease and the Ekman friction layer contracts vertically, the conservation of absolute vorticity requires a lateral expansion or divergence of the vorticity-laden column of water. While the drag of bottom friction acts to reduce the total vorticity of the water system, it also reduces the divergence in lower layers; and, as water depths diminish, it becomes difficult for the divergence in lower layers to keep pace with the convergence in surface layers as well as to satisfy the requirements imposed for conservation of vorticity. The net result is an uplift of the water surface. Thus a hurricane that moves smartly toward its landfall will ordinarily create a larger peak surge than if it approaches slowly.* Also, a coastal area in

* There is, of course, an upper limit of approach speeds beyond which the physical reasoning concerning the uplift caused by storm surge is no longer valid. This is rarely reached even by recurving hurricanes.

which shoal water extends a considerable distance seaward will experience a greater peak surge than one in which water depths increase rapidly seaward. In part, this is due to the increase in the dynamical component of S_B in shallow water.

Experience with numerical models of storm surge indicates that at an open coast, peak surge heights from a landfalling hurricane increase

- with lower central pressures;
- with the increase of the radius of maximum winds (RMW) to (but not beyond) about 50 km;
- with some increases in the speed of approach to a coast;
- with the decreasing slope of the bottom surface from the beach seaward for a distance equal to the diameter of the ring of maximum winds.[13]

For simplicity of computation, it is sometimes useful to define the storm-surge potential for various specified coastal basins in terms of a *shoaling factor*, or coefficient applied to the bathymetry of a standard basin. In Appendix F, Figure 152 shows the center points of basins on the eastern and southern shores of the United States and their respective shoaling factors. These are used in numerical predictions of storm-surge profiles for the National Hurricane Center in Miami. The surge heights expected from any given hurricane are directly proportional to the shoaling factor, which, in this case, is normalized with respect to a standard basin near Jacksonville, Florida. Figure 151 contains the nomogram for computing peak storm surges for this standard basin as a function of the radius of maximum wind and pressure drop, Δp (from environment to vortex center). The supplemental nomogram adjusts the peak surge for the velocity (direction and speed) of a hurricane approach at the landfall position. The corrected value can then be finally adjusted by applying the shoaling factor (Figure 152) for the landfall basin.

Dynamic models for computing storm surge obtain estimates of surge heights at the open coast by solving the basic hydrodynamic equations of motion. For purposes of warning and evacuation, the storm-surge hydrograph, or marigram, showing the rate of water rise at the open coast is an important by-product of most storm-surge prediction models. Figure 111 is a typical marigram showing the variations of sea level at various positions along a north-south coastline as a westward-moving hurricane makes its landfall and

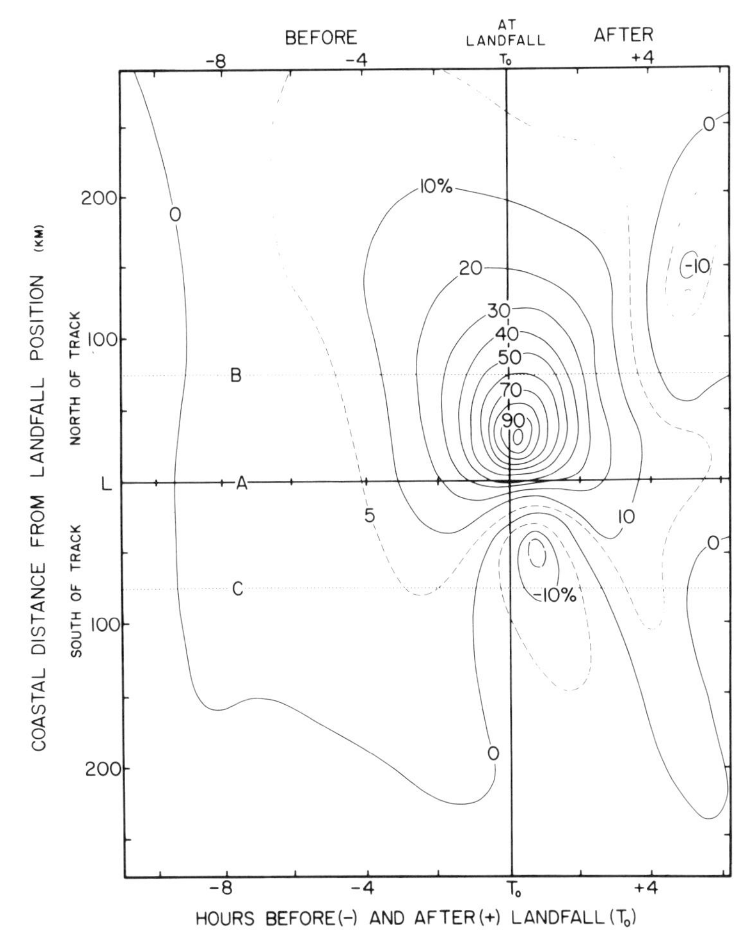

Fig. 111. Changes in sea level due to storm surge along a north-south coastline during a 16-hour period as a hurricane center moving westward at 7 mps (*line A*) approaches the coast and, after 10 hours, reaches its landfall, *L*, and passes inland. For this computation, the RMW is 25 km (see text). Data from C. P. Jelesnianski, *SPLASH (Special Program to List Amplitudes of Surges from Hurricanes): II. General Track and Variant Storm Conditions*, Technical Memo NWS TDL-52 (Springfield, Va: Department of Commerce-NOAA, March, 1974).

passes inland. The variations here are expressed as percentages of the computed peak surge height. As in Figures 109 and 110, the peak surge occurs abreast of the landfall position to the right of the track (north of center, in this case), while a depression of sea level occurs to the left (south) of landfall just after the center passes inland. At 75 km north of the track (line B), the rises begin about 30 minutes earlier than at the position of ultimate landfall, but they crest at heights only 60% of the peak surge observed at the RMW (25 km north of landfall). However, 75 km south of landfall (line C), the surge crests 2 hours before landfall but at a height only 5% of the peak surge value.

The preceding discussion has been directed toward storm surges from landfalling storms. However, surprising and dangerous surges sometimes occur from hurricanes that approach the coast but remain offshore. Hurricane movement along, or nearly parallel to, a shoreline is generally classified as "up the coast" when the coastline is to the left of the track, and "down the coast" when to the right of the track. The peak surge at the open coast precedes the arrival of the eye center (abeam of a coastal position) for up-the-coast movement but follows arrival of the center for down-the-coast movement. When the center position lies within a distance 2R of the coastline, where R is the radius of maximum wind speed, changes in surge height may occur very rapidly. For example, Figure 112 shows a predicted marigram for a severe hurricane moving rapidly northward in the Gulf of Mexico nearshore and abeam of Tampa Bay. This typifies not only the difficult meteorological problem of interpretation but also that of public response to warnings. In this case, the sea level is depressed at the open coast, and water rushes out of Tampa Bay as the hurricane center approaches; however, after the center has passed abeam of Clearwater (28° N, 82.5° W), water levels rise more than 6 m in 30 minutes!

Alongshore-moving hurricanes may generate a resonance or seiche-type phenomenon that tends to create a series of resurgences, or fluctuations in surge height, at 4- to 8-hour intervals.[14] In general this is a function of the bathymetry, or bottom contours, of storm size, and velocity of hurricane movement. It is a complex phenomenon that can be anticipated only by the careful application and interpretation of numerical simulation models; yet, as is obvious from Figure 112, it is of critical importance to public warnings and evacuations.

For alongshore-moving hurricanes, the following summary of characteristics can be stated:

- Peak surge heights and resurgences are generally greater for a down-the-coast movement than for an up-the-coast movement.
- Peak surges precede the center for an up-the-coast movement and follow it for a down-the-coast movement.
- Resonance, or seiche action, is restricted to fast-moving hurricanes (generally > 15 kt) and may occur both at an open coast and in enclosed basins.
- Maximum surges for a down-the-coast movement occur with relatively fast movement along a track that is nearshore (within a distance, R); for an up-the-coast movement, surges are at a

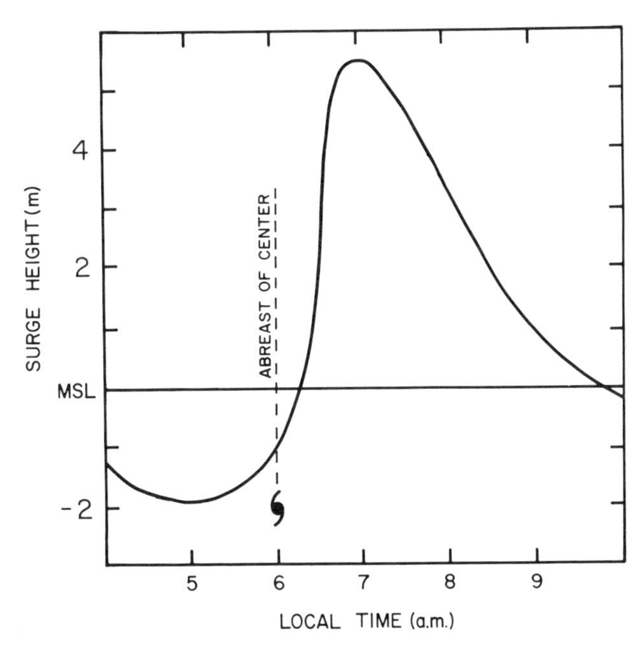

Fig. 112. Variations in sea level caused by storm surge as a hurricane moves down the coast nearshore (see text). This computation for the mouth of Tampa Bay, Florida, is for hurricane Donna after transposing its track to provide landfall at Clearwater, Florida (28° N, 82.5° W) rather than south of Fort Myers (26.4° N, 91.5° W).

maximum for a relatively slow movement along a track that is just onshore.

- Storm size is more critical in determining peak surges for an alongshore-moving hurricane than for a landfalling hurricane.
- For a recurving hurricane (on parabolic, up-the-coast track), maximum surge heights precede the center, and peak surges occur beyond the point of closest approach (Figure 113).[15]

Storm Surge in Bays and Estuaries

A superelevation of the sea level in bays and estuaries may exceed those at an open coast by a factor of 50% or more for slow-moving storms. One example is given in Figure 114 where an open-coast surge from hurricane Carla (1961) reached maximum heights of about 4 m; far inland in Lavaca and Galveston bays, water levels rose more than 6 m. Such rises inland represent an effect of the wind setup, S_W, in shallow-water basins rather than the simple accumulations of water transported from the open coastline. Here a

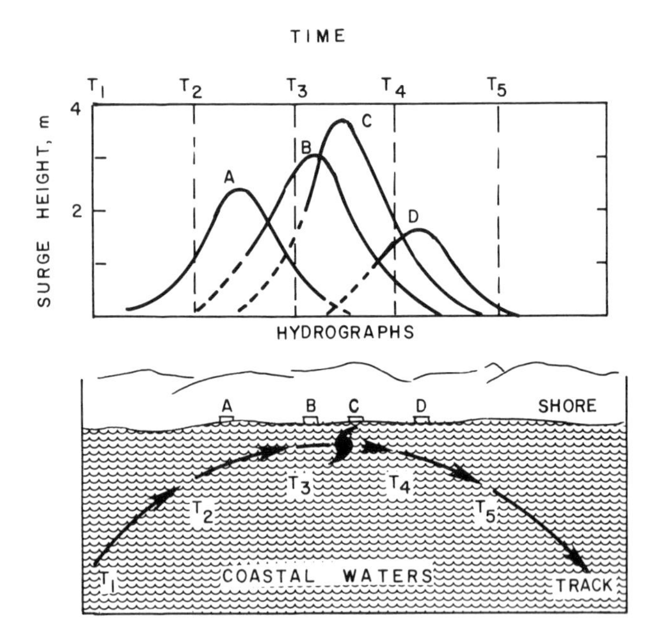

Fig. 113. Marigrams for several coastal locations as an alongshore hurricane moves up the coast and passes abeam.

still-water level acquires a slope that is the result of forces due to wind stresses driving the water shoreward and forces of gravity acting to reduce the slope. For basins exposed to the right semicircle of a landfalling hurricane, highest surges occur on the inland shores of the basin. Those mainly exposed to the left semicircle experience peak surges abreast and to the left of the center.

When a small, intense hurricane moves over an inland basin having a diameter greater than 2R, the setup due to the dynamical component of inverse barometer, S_B, may influence peak surges and augment the resonance and resurgences that tend to occur as

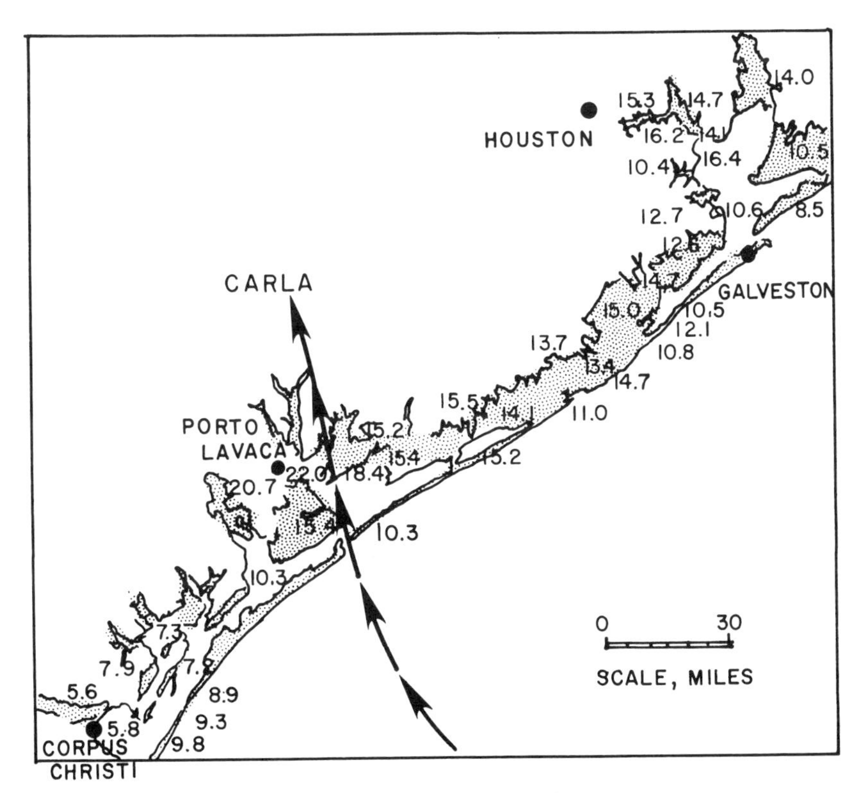

Fig. 114. Storm tides (feet) at the coast and inland during hurricane Carla (1961), whose center moved ashore along the track shown by the *arrows*. Highest tides occurred far inland, where they were 50% greater than those at the coast. *Shading* indicates area of flooding. Adapted from Department of the Army, Corps of Engineers, Coastal Engineering Research Center, *Shore Protection Manual* (Washington, D.C.: Government Printing Office, 1973), Vol. I, p. 3.88.

the eye center passes further inland and winds shift nearly 180°. In most cases, however, the bottom friction in most shallow basins acts effectively to dampen resurgences.

The worst erosion and channeling across a barrier island that are caused by a hurricane generally occur when the backwash from inland basins attacks the landward shore of the island as the wind reverses direction and couples with gravity to accelerate the water accumulated on inland shores in a seaward direction. In 1967, Padre Island on the south Texas coast was cut by hurricane Beulah at thirty-one points; in each instance, the cutting action was either initiated or completed by the backwash from Laguna Madre as the wind stresses on the inland water surface reversed direction.

The Hurricane Tide

The net superelevation of sea level, or hurricane tide, is an accumulation of water due to 1) wave setup, 2) storm surge, and 3) hydrologic contributions of the river and stream outfall and the precipitation runoff. Collectively, these are modulated by the rise and fall of astronomical tides.

When referring to tides of any kind, one must remember that the day-to-day sea level is variable and, like the weather-climate relationship, rarely coincides with mean sea level from surveyed datum planes.* In addition to seasonal anomalies, there are short-term variations, the source of which is often indiscernible, that may cause significant variations in the expected sea level. These anomalies constitute what is known as *preliminary rise* and may be positive or negative relative to the epochal mean sea level. For nearly 72 hours before hurricane Eloise (1975) made its landfall in Florida, there was a preliminary rise of 0.6 m along the coastal sector—a significant contributor to the destructive hurricane tide that peaked at 4.9 m.

Wave setup may cause early tidal rises well in advance of the rises that come from wind stresses and pressure gradients. Together with preliminary rises, this coastal inundation can begin

* Tidal levels of deviations from *mean sea level* are commonly computed with reference to *local mean sea level*, defined as the average height of the surface of the sea, observed at hourly intervals during a 19-year epoch as base.

early and limit the time available and the options open for public evacuation. For landfalling hurricanes, storm surge is the coup de grâce, augmenting—often quite swiftly—the other components of hurricane tide that peak at the time of landfall. The impact of wave setup is minimal for down-the-coast movement, and the hurricane tide peaks after the passage of the center. The range of astronomical tides during hurricane season is listed for various coastal locations in Table 35.

Although tidal ranges are quite small for the coastal sectors most frequently affected by hurricanes, there are some locations—particularly the Georgia and Carolina coasts, and in New England—where the tidal phase at landfall may make the difference between catastrophe and inconvenience. For example, at Charleston, South Carolina (33° N, 50° W), several hurricanes in this century, notably Gracie (1959), have carried the meteorological potential for enormous destruction. Yet, because each reached the coast at a time of low astronomical tide, the net hurricane tide did not produce major inundation, and losses (though significant) were not catastrophic. Had Gracie made landfall at high tide, Charleston would have faced major inundation and destruction.

Categories of Damage at the Coast

In a coastal zone, damage resulting from a hurricane tide can usually be attributed to wave scour, battering from waves and flotsam, and simple flooding. Of these three, the most important is wave scour. The importance of storm surge is not just the simple flooding it brings, but rather the elevation of still-water surfaces upon which waves may extend their cascade of energy, erosive action, and battering for hundreds of meters—sometimes kilometers—inland from ocean and bay shores.

The first victim of wave scour is the beach itself. As the sea level near a shore rises due to wave setup, waves may advance up the beach and across the berm. With further storm-surge rises, waves begin to attack the dune structure. In some hurricanes, vast amounts of sand are removed by scour and transported away by strong longshore currents. In hurricane Eloise, a remarkable example of beach erosion, nearly all of which occurred in less than 4 hours, an average of 45 cubic yards for every yard of oceanfront (31.5 m^3 per m) was removed over a distance of more than 15 km.

Breaking waves may quickly remove loose dune sand from around or beneath the foundations of a beach structure; and, where foundations have not included pilings or footings that penetrate well below the dune sand, the structure risks collapse. However, even a small, exposed frame structure may survive if constructed on adequate pilings and in accordance with a good code. Bulkheads and seawalls may provide some limited protection from erosive waves in a hurricane. However, unless properly designed and constructed, a bulkhead may fall prey to the scour of coastal waves as rapidly as dune sand.

Sand dunes, if left in their natural state, provide the most effective protection from an invasion of the sea during a hurricane or other coastal storm. However, when they are disturbed by the imposition of building structures, developed for public use in a way that inhibits the natural accretion and exchange of sand, *or worse*, if sections of a foredune are removed to enable beach access, the protection they afford, and even their own preservation, may be seriously jeopardized by the onslaught of one or two storms.

Barrier islands, subject to hurricanes, often suffer one or more breaches during a hurricane as water accumulations on the landward side of the bay, lagoon, or estuary rush seaward when the eye center passes inland. These breaches, or washovers, tend to repair themselves rapidly in most instances. However, once a breach has been made, it remains a weak link, subject to reopening by each succeeding hurricane and is, therefore, unsuited as a site for building construction.

Inland from an open coast or a bay shore, waves become diffracted and their energies dispersed by urban development or by forested areas. Consequently, wave erosion diminishes rapidly. Nevertheless, flotsam, including derelict boats, structures floated away from their foundations, telephone poles, and heavy wood members surging with the hurricane tide and accelerated by hurricane winds, may become battering rams, sometimes rupturing the exterior walls of well-constructed dwellings and destroying other installations as far inland as the tide provides flotation.

Aside from wave erosion and battering forces, simple flooding of the coastal plain with seawater, often bearing a surface coating of petroleum products, may not only cause extensive damage to the contents of dwellings and the loss of livestock but also may render agricultural lands useless for most crops until the sea salt and petroleum products are leached from the soil. Table 36 (Appendix F) shows a chronology of major hurricane tides in the United States and the relevant meteorological characteristics.

REFERENCES

1. Department of the Army, Corps of Engineers, Coastal Engineering Research Center, *Shore Protection Manual* (Washington, D.C.: Government Printing Office, 1973), Vol. I, pp. 2.1–2.137.
2. G. C. Stokes, "On the Theory of Oscillatory Waves," *Mathematical and Physical Papers, I* (Cambridge, England: Cambridge University Press, 1880).
3. J. H. Mitchell, "On the Highest Waves in Water," *Philosophical Magazine*, XXXVI, No. 222 (1893), 430–37.
4. W. H. Munk, "The Solitary Wave Theory and Its Application to Surf Problems," *Annals of the New York Academy of Sciences*, LI (1949), 376–462.
5. Department of the Army, *Shore Protection Manual*, Vol. I, pp. 3.1–3.49.
6. *Ibid.*
7. Samson Brand, J. W. Blelloch, and D. C. Schentz, "State of the Sea Around Tropical Cyclones in the Western North Pacific," *Journal of Applied Meteorology*, XIV (1975), 25–30.
8. D. B. Ross and V. J. Cardone, "A Comparison of Parametric and Spectral Hurricane Wave Prediction Products," *Proceedings of the NATO Symposium on Turbulent Fluxes Through the Sea Surface, Wave Dynamics and Prediction, Marseilles, France* (New York: Plenum, September, 1977), 1–18.
9. E. L. Fisher, "Hurricanes and the Sea-Surface Temperature Field," *Journal of Atmospheric Science*, XV (June, 1958), 328–33; D. F. Leipper, "Observed Ocean Conditions and Hurricane Hilda," *Journal of Applied Meteorology*, XXIV (1964), 182–96.
10. J. E. Geisler, "Layer Theory of the Response of a Two-Layer Ocean to a Moving Hurricane," *Journal of Geophysical Fluid Dynamics*, I, No. 1–2 (1970), 249–72; P. G. Black and W. P. Mallinger, "Means of Acquisition and Communication of Ocean Data," *World Meteorology Organization Report on Marine Science Affairs #7*, II (1972), 290–314.
11. M. S. Longuet-Higgins and R. W. Stewart, "A Note on Wave Setup," *Journal of Marine Research*, XXI, No. 1 (1963), 4–10.
12. C. P. Jelesnianski and A. D. Taylor, *A Preliminary View of Storm Surges Before and After Storm Modifications*, Technical Memo ERL WMPO-3/NHRL-102 (Springfield, Va.: Department of Commerce-NOAA, May, 1973).
13. C. P. Jelesnianski, *SPLASH (Special Program to List Amplitudes of Surges from Hurricanes): I. Landfall Storms*, Technical Memo II NWS TDL-46 (Springfield, Va.: Department of Commerce-NOAA, April, 1972).
14. C. P. Jelesnianski, *SPLASH (Special Program to List Amplitudes of Surges from Hurricanes): II. General Track and Variant Storm Conditions*, Technical Memo NWS TDL-52 (Springfield, Va.: Department of Commerce-NOAA, March, 1974); Jelesnianski, *SPLASH I*.
15. Jelesnianski, *SPLASH II*.

CHAPTER 10

The Hurricane over Land

The Normal Event

On crossing a coastline, a hurricane always loses some of its high speed, and central pressure rises. The supply of energy from the warm underlying water that makes and maintains hurricanes is suddenly cut off, so that quick loss of intensity in the inner core ensues. Moreover, the center crosses from the area of mobile ocean and its spray to a land surface that is rigid except for flying debris and has varying degrees of "roughness" (Chapters 8, 9). Wind distribution with height in the lowest 100 to 300 m changes quickly as the hurricane center passes the first 10–20 km inland. At 300 m height, winds as yet may remain almost unaffected, while lower down they are retarded. A wind profile typical of a boundary layer over land develops and replaces that typifying wind structure over the open ocean.

Within 12 hours after landfall all warnings at sea except for small-craft warnings are usually discontinued. But the subsequent behavior of a storm is not always uniform; rather it depends on the conditions at landfall. Important is the geography of the land itself. In mountainous areas and heavily built-up city areas, hurricane velocities may be expected to decrease most rapidly; a minimal storm will lose the core containing hurricane speeds completely. In the case of a large flat coastal area with many lakes and channels, all subject to flooding, the decrease of energy supply to the hurricane will be slower, so that hurricane speeds, though weakened, may still be found 24–36 hours after entry for a slow moving storm. Further, it matters greatly whether a hurricane approaches the coast from southeast or southwest (in the Northern Hemisphere). Coming from southeast to south, highest intensity may well be maintained on the shore itself, for instance, at the entry of hurricane Camille near Biloxi (30° N, 89° W) on the Gulf coast on 18 August 1969. Pressure onshore was still 900 mb; devastation from wind and the rise of water were widespread. A hurricane arriving from southwest may well have started to lose intensity over the coastal water when cooler and drier air—especially in the late hurricane season—entered it from the continental side. Then it would be difficult to find any coastal location at all that experienced a full hurricane-force wind.

Filling of Hurricane Donna (1960) over Florida. A very thorough study of the weakening of hurricane Donna (Fig. 78d) was made by Banner Miller.[1] An unusual number of observations from upper-air and surface stations as well as from research aircraft reconnaissance was available to demonstrate the physical processes leading to the decline of the storm as it passed from the Florida Strait to the west coast of Florida and then crossed the whole peninsula on a northeastward course (Fig. 115). Donna was an intense hurricane with several ups and downs; surface pressure touched a value just below 930 mb for the third time on 10 September. The hurricane gradually "slid" onto land in the flat Everglades region of Florida while still moving west of north. Thus, all of southern Florida was filled with undilute maritime tropical air advancing north with the high-speed winds to the right of the direction of motion. Minimum pressure was maintained when the center was crossing the Florida Keys; decay did not begin in earnest until the center made definite landfall. Then surface pressure rose rapidly to 965 mb and even 970 mb near Orlando, in the center of Florida, only 18 hours later.

Surface tangential wind speed measured by lighthouses offshore and later by anemometers inland was nearly constant at 36 to 38 mps, when averaged around a circle with a radius of 40 nmi. (74 km) on 9 and 10 September. Highest speeds were above 55 mps. Then they rapidly decreased to 24 mps, gale strength only, in northeastern Florida. Comparing radial-pressure profiles on 10 and 11 September (Fig. 115), it is interesting to see that while pressure in the interior rose sharply, it actually fell slightly (following the storm)

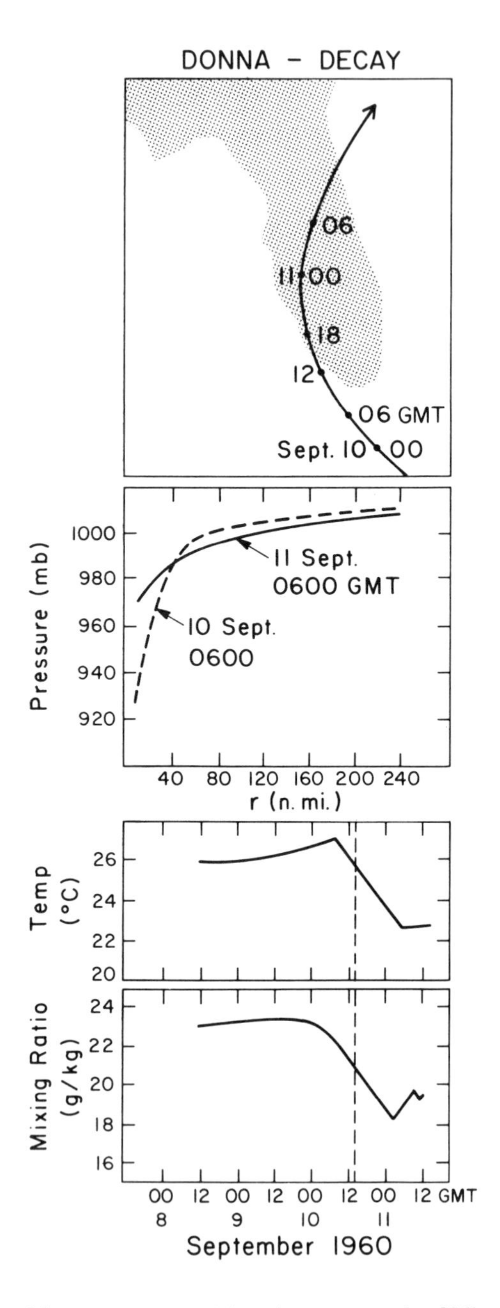

Fig. 115. Decay of hurricane Donna (1960) core upon landfall in Florida. *Top*, path; *below*, pressure, temperature, and mixing ratio at the center after entering land. B. I. Miller, "A Study of the Filling of Hurricane Donna (1960) Over Land," *Monthly Weather Review*, XCII (1964), 389–406.

beyond the 40-mile radius, indicative of a radial shift in mass distribution inward.

Most spectacular are the thermodynamic changes in Figure 115. Surface temperature measured just outside the eye was 26°C on 9 September, rising to 27°C on 10 September with the falling pressure, and then dropping abruptly to 23°C with sharply rising pressure on 11 September. Simultaneously, mixing ratio dropped from 23 to 19 g/kg. This clearly indicates that the oceanic energy source ended suddenly and that the Everglade swamps produced no compensation! The consequence of the decrease in low-level thermodynamic energy is seen in Table 15, in which upper-air temperatures before and after landfall are compared. While surface temperature decreased by 3°C, the drop at 200 mb was fully 11°C, a very large amount, which tended to destroy the inner warm core and therewith the ability of the hurricane to maintain its inner strength. Still, temperatures remained appreciably warmer than the mean tropical atmosphere, given in the right column of Table 15, and thus enabled the hurricane to hold together and maintain its identity until it regained the Atlantic waters east of Florida 1 day later.

The rapid cooling of the inner core on landfall is the main and

Table 15. Upper-Air Temperatures in the Central Parts of Hurricane Donna (1960) Before and After Landfall

Pressure (mb)	*Temperature (°C)*			
	Before	*After*	*Difference*	*Mean tropical*
200	−40	−51	−11	−55
300	−18	−27	−9	−33
400	−5	−12	−7	−18
500	+4	−2	−6	−7
600	+11	+6	−5	+1
700	+16	+11	−5	+9
800	+21	+16	−5	+15
900	+26	+21	−5	+20
Surface	+27	+24	−3	+26

SOURCE: B. I. Miller, "A Study of the Filling of Hurricane Donna (1960) Over Land," *Monthly Weather Review*, XCII (1964), 389–406.

overwhelming reason for the immediate decay of hurricanes reaching the shore. Whereas subsequent tests following Miller's analysis have shown the same development, and whereas at this late stage even numerical calculations are able to reproduce the events correctly, the credit for the most conclusive demonstration rests with Miller. In earlier days an increase of friction at the ground was proposed as the cause of a hurricane's collapse. Except for some decrease of surface wind due to touching rigid land, this explanation fails. As Miller shows, the decrease of surface pressure gradient in the core region eliminates the pressure force that provides large acceleration to the surface air over sea. In a budget of kinetic energy for the lowest kilometer of the atmosphere, he is actually able to show that the frictional dissipation of energy decreases as the storm moves over land; the frictional drag of the agitated ocean and masses of spray thrown up into the surface air is greater than that which the flat Everglades with bush vegetation can offer.

Herewith, it is suggested, a definitive answer on the initial filling of hurricanes over land has been provided. The reverse process may occur when an old hurricane regains the sea, and this happened in the case of Donna, not just once but twice (Fig. 78d)—after leaving Florida and again after passing over North Carolina. In New York harbor its waves did substantial damage to ships and docks, causing protracted law suits.

Tracks After Landfall

After as well as before landfall, the hurricane is subject to steering by the weather systems surrounding it, even more so after entry, since the amount of control the storm of average intensity has over its own destiny offshore is diminished through weakening of the central warm core and therefore of the upper anticyclone, which is the internal steering mechanism. Given a vigorous flow around a high-pressure area over the ocean, a clockwise turning path may continue undiminished and is often encountered when storms cross the Carolina coasts of the United States, Japan, the island of Madagascar in the South Indian Ocean, and eastern Australia. When such control is lacking, the storm may recurve very slowly inland; sometimes it regains the sea and once again acquires hurricane intensity, as did Donna and Cleo (Fig. 78b). When a hurricane enters a coast with east-west orientation, all depends on whether there is a trough of low pressure extending into the tropics from high latitudes. Such a trough will draw the storm or its remnants northward, as in hurricane Camille (Fig. 78a), or southward, as in case of cyclone Tracy (Fig. 42), which crossed the whole Australian desert toward southeast after destroying Darwin (Chapter 4). Given an inland anticyclone, a storm will tend to drift slowly northward to northwestward and even westward at times, with heavy rains attending it for days.

Thus we see that a general statement can only be made for the initial hours after landfall, and even then the decrease in strength takes place at variable rates. Beyond these initial hours, the forecaster's judgment or computation on the future course comes heavily into play. We now explore several alternatives in terms of rainfall and flooding, though rare cases in which hurricane winds are carried far inland will occur at times. The reader may recall Table 7 and the surprising conclusion that 1 day's rain from a moderate hurricane inside the 200-km radius produces rainfall equivalent to the total runoff of the Colorado River in an average year. The major fraction of the hurricane rainfall is likely to run off, especially in mountainous terrain but also in the plains, unless dry soil from antecedent drought is capable of absorbing a substantial fraction of the rapidly falling water. This would be a rare event in view of the normally slow percolation of water into the deeper ground levels.

Types of Flooding Situations

The old, even decadent, hurricane has the remarkable property that it frequently, almost normally, is capable of producing very heavy rain long after the hurricane characteristics have disappeared from the surface; at times one finds the rain-producing remnant only at upper levels of the atmosphere. The precise reason for this feature of old hurricanes, days after landfall, is not always obvious, but one must never lose sight of it. Nor are these rains always unwelcome. For instance, hurricane rains have more than once saved the cotton crop of the southeastern United States from prolonged drought. Large Texas ranches have been rescued from becoming complete deserts. Along the Mexican west coast, in the isolated city of Hong Kong, and in Southeast Asia, to name but a few examples, hurricanes have been the savior in times of failing water supply. These, however, are generally regarded as unspectacular events from the news viewpoint; the emphasis has too often been reserved for the cases in which the old hurricane rains brought flooding disasters.

Slowly Recurving Hurricane. An intense rain catastrophe was brought on by hurricane Beulah (1967), which crossed from the Gulf of Mexico into southwestern Texas near Brownsville near sunrise on 20 September 1967. Moving toward north-northwest at a speed of about 10 kt, it then slowed down for some hours and reversed course toward southwest (Fig. 78a). Over sea, central pressure was as low as 923 mb, almost a record for the western Gulf; the path was uncomplicated and warnings were timely. Lowest pressure at Brownsville was 951 mb; the hurricane slowly weakened over land on 20 September before it passed southwestward into Mexico, but it still retained hurricane force throughout the day.

In Figure 116 maximum gusts observed or estimated have been plotted in relation to the storm center. Gusts of 100 kt (185 km/h) were observed up to 65 nmi (120 km) from the center, and gusts of 70 kt (130 km/h) up to 120 nmi (200 km) in the northeast, indeed a powerful circulation over land.

The really destructive feature of Beulah was the precipitation, which, over 2 or 3 days, amounted to 50% to 100% of mean annual precipitation and brought on widespread flooding and severe destruction in the whole area. Total storm precipitation has been entered at all stations in Figure 116, although, of course, because of hurricane propagation, only part of the rain occurred in the indicated positions. However, precipitation on the heaviest day was half or more of the total rain, so that the pattern furnishes some measure of rainfall concentration and gradients. Totals ranged far above that calculated earlier for the average oceanic typhoon with 11 inches (28 cm) in 48 hours. The slow motion of Beulah plus its sharp recurvature no doubt contributed to the excessive rain event. However, the track was again curved in such a direction during the whole period that in the lowest kilometers maximum moisture was imported from the Gulf of Mexico and beyond.

As brought out vividly in the annual hurricane report: "Torrential rains fell in southern Texas and northeastern Mexico and produced major floods. Every river and stream in southern Texas south of San Antonio flooded. Storm rainfall ranged from 10 to 20 inches over much of southern Texas, and totals exceeded 30 inches in some areas. Previous flood records were erased as the rain water collected in the lower Texas rivers."[2]

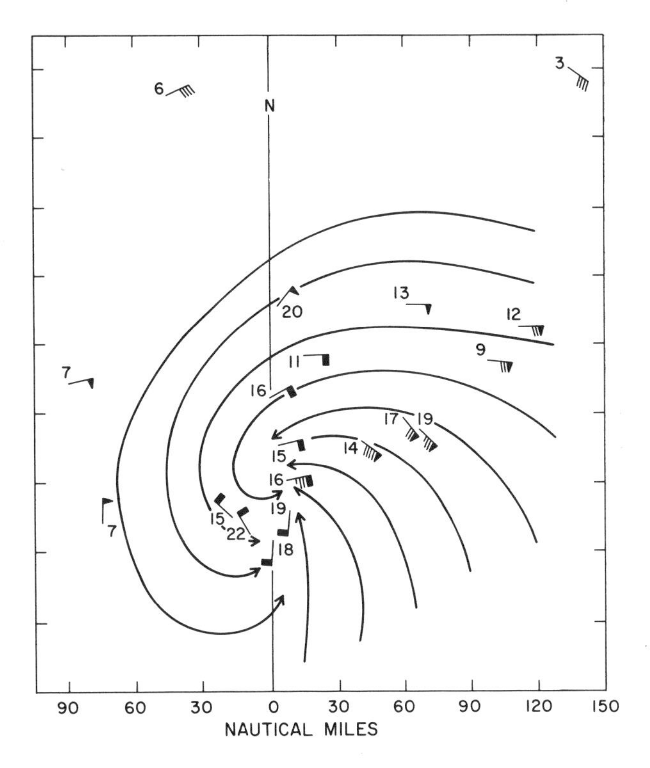

Fig. 116. Maximum surface gusts and total precipitation (inches) experienced at Texas stations during slow passage of hurricane Beulah, 20–22 September 1967 (Fig. 78a). *Short barb*, 5-kt winds; *long barb*, 10 kt; *heavy triangular barb*, 50 kt; *heavy rectangular barb*, 100 kt. All stations composited with respect to moving center.

Hurricanes and Mountain Ranges. When hurricanes cross or even pass alongside mountain ranges, the forced uplift on windward sides will greatly add to the hurricane precipitation there, while diminishing it on the leeward side. Thus, the rain pattern left behind by a hurricane will be grossly distorted from the normal precipitation pattern over the ocean.

A spectacular illustration is offered by the island of Madagascar for March 1959. The island, 1,600 km in north-south extent off

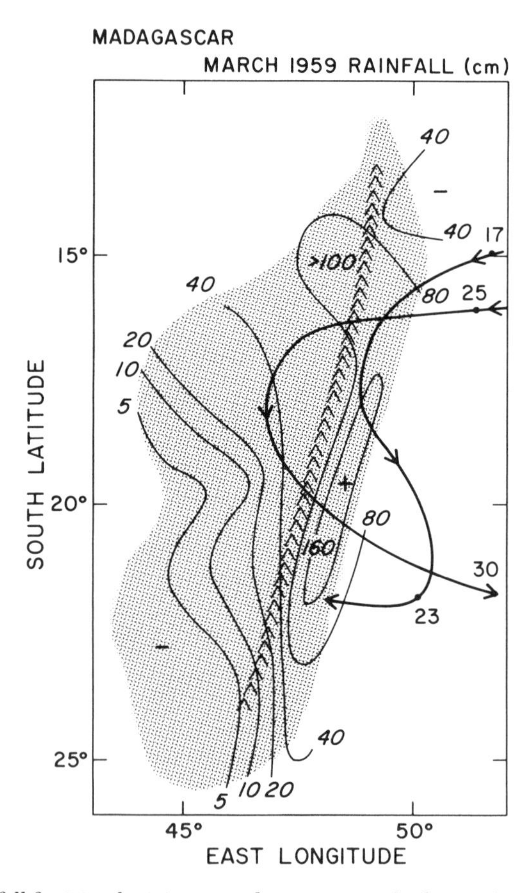

Fig. 117. Rainfall for March 1959 on Madagascar, mainly due to the two hurricane tracks (*heavy lines*). Precipitation in centimeters; logarithmic spacing of isohyets. Top of mountain range along center of island indicated. After A. Chaussard and L. LaPlace, "Les Perturbations dans le Sud-ouest de l'Ocean Indien," *La Meteorologie*, IV (1959), 323–66.

southeastern Africa in the Indian Ocean, has a central mountain backbone with average elevation near 1,200 m and some peaks well over 2,000-m height (Fig. 117). Two hurricanes occurred in March 1959, each affecting the island for about 6 days. The first, 17–24 March, took an erratic course, meandering southward along the eastern shore. It was closely followed by the second, and very strong, hurricane of 25–30 March, which effected a simple recurvature.

Precipitation for the month, mostly due to the two hurricanes, is depicted in Figure 117 with logarithmic spacing of isohyets, which must always be employed when very strong rainfall gradients are present. Along the eastern shore the highest isohyet is 160 cm, about six times the normal monthly rainfall, and amounts well over 200 cm (80 inches) were reported. In contrast, the area west of the mountain ridge averaged far less than 100% of average, generally less than 50%; and the southwestern coast was entirely rainless. The first hurricane delivered rain only east of the divide; the west was completely dry. It is only the recurvature of the second hurricane, well west of the mountain crest, that produced rain there. Nevertheless, the gross distortion of the rainfall pattern due to the topography remains evident.

Extreme rain may not always be associated with the immediate vicinity of a hurricane. Long trails of convergent air seen as cloud bands on satellite photos often extend far east to southeast of a hurricane moving in the trades. Such a band may cross an island that may have been missed entirely by the center itself and where inhabitants may think with a feeling of relief that all danger is past.

An event of this kind occurred when hurricane Hazel, in 1954 (Fig. 78d), moved northward out of the Caribbean through the Antilles into the western Atlantic. The island of Puerto Rico was 700 km to the east of this track. But a long convergence band extended from Hazel down the eastern Antilles, where heavy rain had just occurred from a preceding wave in the easterlies, 9–10 October. Herewith the stage was set for a flood disaster. The eastward band from Hazel intensified 11–12 October (Fig. 118) when it moved north of Puerto Rico and continued a slow northward drift with Hazel until the latter accelerated on its path toward the southeastern United States on 14 October.

Heavy rains set in on Puerto Rico on 11 October, reached their peak on 12 and 13 October, and then gradually decreased until 16 October. Over the 6 days, more than 20 inches (50 cm) were measured at numerous stations, and flashflooding occurred. We see the total rainfall for 6 days in Figure 118 (lower part). The main top-

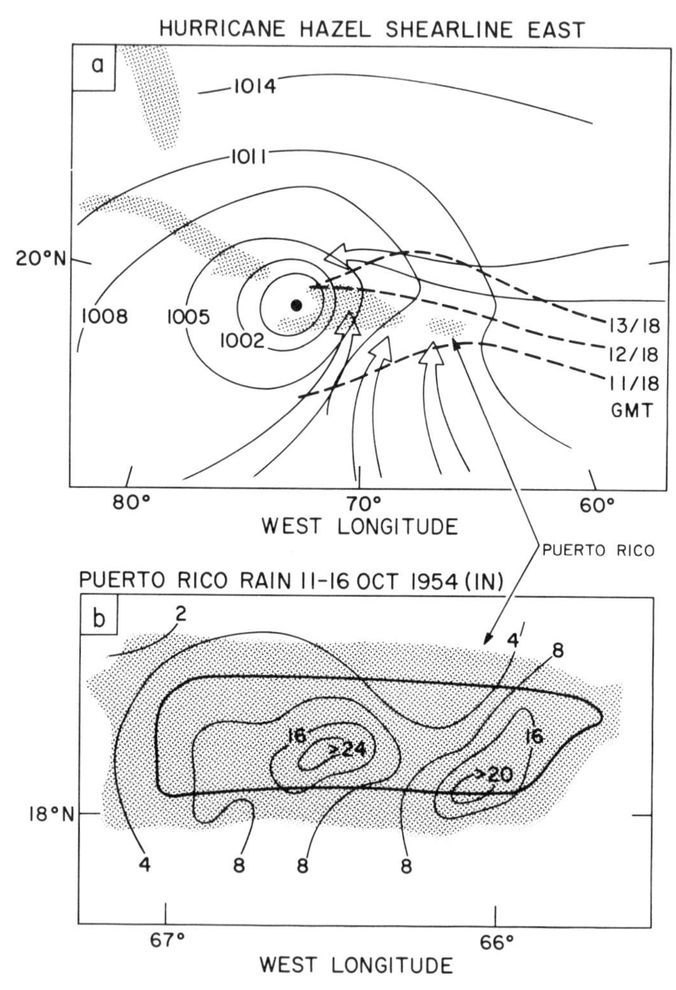

Fig. 118. *b*, Heavy rainfall on Puerto Rico, 11–16 October 1954, associated with *a*, the shear line extending far east of hurricane Hazel. Ralph Higgs, "Severe Floods of October 12–15, 1954, in Puerto Rico," *Monthly Weather Review*, LX (1935), 45–49.

ographic feature is a mountain range with average height of 700–1,000 m running east-west the whole length of the island. In view of persistent winds from south, the concentrated precipitation of 6–10 inches (15–25 cm) lies along the south coast, including two centers with well over 20 inches (50 cm), the western one near the highest elevations. In contrast, the northern side was relatively well protected, with 2–4 inches (5–10 cm)—not unusually large. There, highest 24-hour rain exceeded 10 inches (25 cm) at a few places only. The disaster, with flooding of as much as one-quarter of the island's area, resulted from the unrelenting accumulation. Some flooding even covered the north coast and was evidently outflow from the central area in northward-directed streams.

In view of the slow evolution of the situation, warnings appear to have been excellent, and major loss of life was avoided. Damage to property of course was extensive. Other instances of such flooding have been reported from the Caribbean and elsewhere, a larger-scale by-product of rather distant hurricanes and accentuated by mountainous relief. While rated as the extreme on record for Puerto Rico, the magnitude of the rain episode falls short, for instance, of the Manila rains of 17–21 July 1972, when nearly 40 inches (1 m) fell in 4 days, without there being any recognizable weather system, either a typhoon or, at least, a tropical storm.

Hurricanes That Remain Vigorous

Events Ahead of a Deep Trough in the Westerlies. Hurricanes will remain vigorous during motion out of the tropics, or they may even gain intensity, when their arrival in middle latitudes coincides with a situation there in which development of an extratropical cyclone could be expected from the middle latitude flow patterns present. Unless hurricane motion is very fast, the storm is "transformed" into an extratropical cyclone and develops frontal systems. It may retain considerable intensity, resulting now in large part from sinking of cold air in an upper trough to the west with respect to the warm air in the center. The oceanic heat source, cut off, no longer is required as a means to maintain the cyclone's temperature gradient; it is replaced by the middle latitude air-mass contrast between tropical and polar air, even though the cyclones continue to be marked by exceptionally heavy precipitation on many occasions.

As Chapter 7 showed, the hurricane will retain its central warm core and heavy rain pattern, if it is greatly accelerated northward on the eastern side of a large-amplitude upper-air trough in the westerlies. In case of the 1938 New England hurricane, the warm core was present at landfall and persisted until it reached the border between the United States and Canada some hours later. Our upper-air chart, lacking sufficient coverage in 1938, was borrowed from the case Hazel of 1954; as emphasized, this was an extreme upper flow and temperature display (Fig. 84). Hurricane Hazel had been gaining intensity while moving out from the Caribbean into the Bahamas (Fig. 78d) and arrived on the Carolina coast as a very powerful cyclone.

The entry of this strong hurricane into the extreme middle latitude situation of Figure 84 led to sustenance of hurricane-force winds along the long path northward that it took over land. Wind records were broken in many places, especially in the city of Washington. Forward acceleration had the magnitude of the 1938 case, and much of the same scenario was repeated. In Hazel it appears that the warm and moist air moved north with the center for a long distance, and heavy rains continued; the cold air was left behind as in Figure 86. The whole history of transformation unfolded with such rapidity that one can hardly tell whether it was just the old hurricane that was racing northward, or whether its role was taken over by another low-pressure center newly formed some distance ahead (to the north) along the extratropical front.

This somewhat academic argument continues to beset researchers. The whole entity, which we may well continue to call Hazel, covered the distance from landfall in the Carolinas well into southern Canada in less than a day. Forward speed was 50 kt (26 mps) during a wild night, in which precipitation fell at an enormous rate (Fig. 119) all along the path, and wind-speed records were being toppled. Compared to earlier precipitation charts in this chapter the amounts in Figure 119 do not appear excessive. But it must be recalled that duration of rain lasted from 18 hours up to several days to arrive at the earlier amounts! In contrast, the rain band of Hazel was laid down by the racing center in perhaps no more than 1 to 3 hours during its passage over one station; the heaviest flooding catastrophe actually occurred in the vicinity of Toronto on the northern shore of Lake Erie in Canada, where over 15 cm (6 inches) were recorded.

Events Associated With the Rear of Jet Streams. The trough model of Figure 84 illustrates one favorite type of case in which waves in the westerlies and the jet stream (core of strongest westerlies) act together to produce strong cyclone development. Even at 500 mb

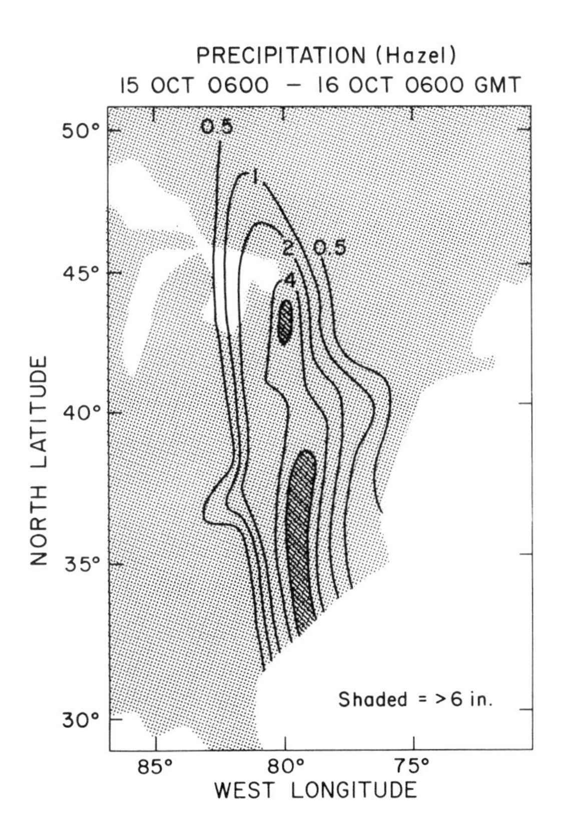

Fig. 119. The precipitation streak during rapid northward advance of hurricane Hazel in eastern North America (Fig. 78d). E. Palmén, "Vertical Circulation and Release of Kinetic Energy During the Development of Hurricane Hazel into an Extratropical Storm," *Tellus*, X (1958), 1–23.

we see a band of northwest wind with speeds of 70 kt (35 mps) on the rear side of the trough, but the fast wind dies out where the flow curves sharply cyclonically at its lowest latitude. Farther northeast, ahead of the trough, another jet stream forms and strengthens toward north from the outflow of the cyclone. A second type of heavy-rain producing, high-tropospheric situation is portrayed schematically in Figure 120. Here the jet stream center is located in the ridge of the wave pattern, accelerating upstream and decelerating downstream from there. The sector marked IV is particularly favorable for development of high-level divergence.[3] Air arriving underneath in the low levels will converge and ascend there; thus, sector IV is a well-known rain sector. The pattern often moves very slowly; it may even retrograde. Flooding events from protracted rains under this flow configuration, even without any surface cyclone, have occurred, mostly in winter but at times also in the warmer part of the year. Suppose now that an old hurricane over land makes contact with this right rear sector IV by moving into it from the south. As long as the supply of warm, moist air from the tropics has not yet been cut off (which could happen), an explosive situation becomes imminent.

Once more we turn to hurricane Camille of 1969, which has supplied so many excellent illustrations through this text. Figure 121

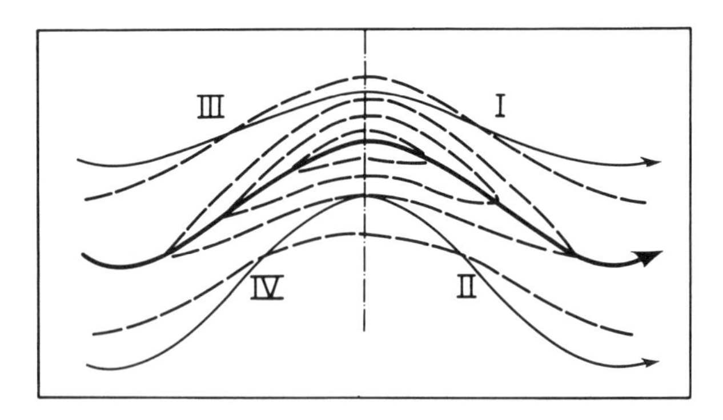

Fig. 120. Model of four sectors around jet stream maximum (indicated by broken lines) situated in long wave ridge, showing sectors favorable and unfavorable for precipitation through ascent from vorticity modeling. H. Riehl, *Jet Streams of the Atmosphere*, Technical Report No. 32 (Fort Collins: Department of Atmospheric Science, Colorado State University, 1962), 57.

provides another satellite photo 36 hours after landfall, near noon of 19 August 1969, when Camille was beginning to curve eastward in the middle of the eastern plains of the United States (Fig. 123, top). The scale is about the same as that in Figure 65, supported further by gridded pictures, which, however, do not reveal the fine structure quite so well as our illustration. Width of the rainy core is about 500 km, as at landfall; it has remained surprisingly solid after more than a day over land, and the eye is clearly visible. A fine band structure extending toward south and southwest is also unchanged from early on 18 August, and the more or less east-west oriented bands to the north have strengthened. Without the solid core at the western edge, one could readily interpret the picture as one of a cyclone about to start occluding.

Fig. 121. ATS III photograph of hurricane Camille at southern tip of Illinois near noon (local time) of 19 August 1969. Width of solid cirrus cloud about 500 km.

In spite of the solid appearance on the satellite photograph, winds and rain associated with Camille had been declining, and the history of this great hurricane appeared to be about terminated. Yet, as the center turned eastward, an enormous resurgence of rainfall took place in a belt less than 100 km wide, which paralleled the path of the cyclone and had a rain center about 50 km to its north (Fig. 122). In the illustration the lines of equal precipitation have been greatly simplified. Up to 30 inches (75 cm) fell in some areas during this night—it always seems to be at night—causing what has been termed "Unprecedented Rainfall in Virginia."[4] Amounts well over 20 inches (50 cm) occurred at numerous stations. As the old hurricane propagated eastward quite rapidly, the whole event took only 12 hours or less and, therefore, was far more intense in terms of rainfall and flooding than, say, the Puerto Rican episode of Figure 118. Mass curves of precipitation from recording rain gauges show the principal deluge confined to an interval of no more than 6 hours.

The upper-air trough, which drew Camille northward from the Gulf coast, at first was very well marked in the middle troposphere (Fig. 123, top). But it then weakened and transformed to a shear line with northeast-southwest orientation through the storm center at the time when the heavy rains were about to begin (Fig. 123,

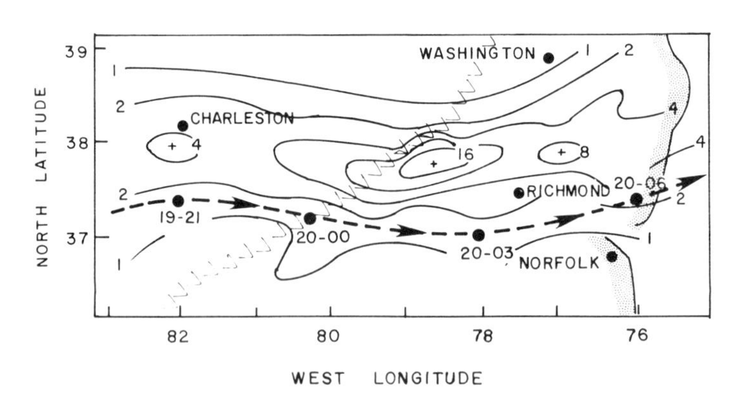

Fig. 122. Track of hurricane Camille through the east-central United States during the night of 19–20 August 1969; three hourly positions indicated. Total precipitation (inches, patterns simplified) derived during storm passage. F. K. Schwarz, "Unprecedented Rains in Virginia Associated with Remnants of Hurricane Camille," *Monthly Weather Review*, XCVIII (1970), 851–59.

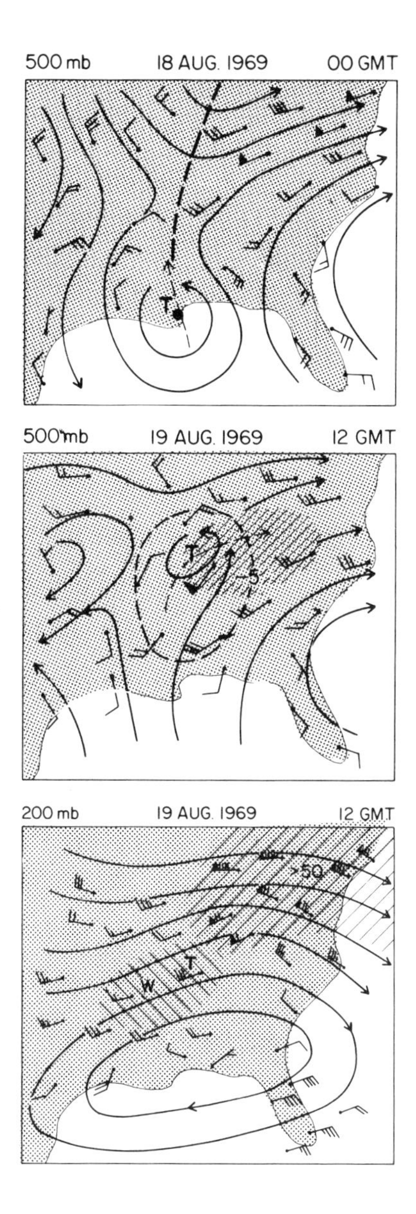

Fig. 123. Inland penetration of hurricane Camille at upper levels. *Top*: *solid heavy line*, 500-mb trough influencing Camille shortly after landfall; *broken line*, track. *Middle*: 500-mb chart 36 hours later; *broken line*, track; *hatching*, beginning rain area; *broken oval*, region with 500-mb temperatures warmer than −5°C. *Bottom*: 200-mb chart for the same period; *hatching*, *upper right corner*, region with winds greater than 50 kt; *hatching*, *opposite slant*, 200-mb warm area (*W*); *T*, position of tropical storm.

middle). A large area with temperatures higher than −5°C (broken ellipse) surrounded the center, and the beginning rain area extended eastward from there (shaded). One cannot invoke here the sinking of a cold dome, as in hurricane Hazel, because the chart indicates no such cold dome.

To be sure, warm and moist air obviously had traveled northward with the center. Further, some of the heavy rain was concentrated on the southeastern slope of the Appalachians, where the wind was blowing directly uphill in advance of the center. However, the 16-inch isohyet in Figure 122 lies east of the Blue Ridge Mountains; evidently, the whole maximum event of "unprecedented" proportions and the orientation of the precipitation band extending clear out into the Atlantic ocean cannot be understood in simple terms of upslope motion.

Inspection of Figure 123 (bottom) now reveals that the hurricane had made contact with sector IV of a jet stream at 200 mb that, from previous maps, had been stationary in the area for 2 days. It is suggested that the superposition of this quadrant of the jet stream on the well-maintained hurricane center brought on the precipitation disaster. Herewith the kind of general weather pattern related to heavy outbreaks of precipitation from old hurricanes after landfall is shown to be quite variable, but of an easily recognizable type that may serve as a useful tool in warnings. It is not without interest that Camille regained minimal hurricane strength after moving offshore into the Atlantic Ocean from Virginia on 20 August 1969.

Anomalous East Pacific Tracks

The majority of tropical cyclone formations on the Pacific side of Central America is due to impulses crossing the narrow corridor of land from the Atlantic. Most tracks then lead along and away from the northwestward-sloping coastline into the open ocean, where the cyclones encounter lower ocean temperatures, especially when attaining latitudes above 20° N. Less frequently one finds recurving tracks toward Mexico, especially near Mazatlan (17° N, 95° W), toward lower California or even the coastline near 30° N and southern California. Normally, the hurricanes are reduced to tropical storm intensity during the travel over the waters with unfavorable decrease of ocean temperature in the direction of propagation. But the capability to produce heavy precipitation remains intact, much as in case of old tropical storms elsewhere. Heavy flooding has occurred in many areas of the mountainous Mexican Pacific coast.

Actual tracks into the United States are rare and very intermittent. For years in succession the general wind circulation of the eastern Pacific is such that old hurricanes do not even come into the vicinity of southern California. But then, with the vicissitude of the upper winds, amply demonstrated earlier in this chapter, situations arise when the upper flow guides centers halfway between the recurving tracks and those directed toward west. For instance, this has happened in each of the years 1976 to 1979 once or twice, a rare sequence that has brought up to 10 cm of rain to the arid parts of California, Nevada, and Arizona. Since these incursions take place in August and September, in the middle or later part of the normal dry season, especially in California, one would think that the unusual augmentation of the water supply in the very dry areas would be welcome. But the economy, based largely on agriculture, is geared to the dry season and depends on it for ripening and processing, of the widely grown grapes, for instance, and the production of raisins from them. Industries tend to rely heavily on the normal seasonal weather cycle; the interruption by the unusual disturbs.

REFERENCES

1. B. I. Miller, "A Study of the Filling of Hurricane Donna (1960) Over Land," *Monthly Weather Review*, XCIV (1964), 389–406.
2. A. L. Sugg and J. M. Pelissier, "The Hurricane Season of 1967," *Monthly Weather Review*, XCVI (1968), 242–50.
3. H. Riehl, *Jet Streams of the Atmosphere*, Technical Report 32 (Fort Collins: Colorado State University , Department of Atmospheric Science, 1962), 56–57.
4. F. K. Schwarz, "Unprecedented Rains in Virginia Associated with Remnants of Hurricane Camille," *Monthly Weather Review*, XCVIII (1970), 851–59.

PART V

Planning Coexistence with the Hurricane Hazard

CHAPTER 11

Threat Assessment and Risk Reduction

As devastating as a major hurricane can be, the long-term threat from hurricanes for any single habitation site, even at the most vulnerable coastline, is small. This is evident in Table 16. Yet a large responsibility devolves upon all levels of government to see that the development and use of coastal lands follows a pattern that reduces hurricane risks to a minimum. A risk that may appear small to the individual resident could, when integrated over a 100-km strike area, comprise a major problem for the state and federal governments to which thousands would turn for relief and rehabilitation after a hurricane strike. And it is the government that must restore beaches and structures in the public domain when they have been destroyed by a hurricane.

Table 16. Comparison of the Number of Hurricane Strikes with the Number of Times Fastest-Mile Winds in Excess of 50 Miles Per Hour Were Recorded at Official Anemometers of the National Weather Service (NWS)

	Number of cases, 1949 to 1978				
	Hurricane	*Fastest-mile winds at NWS stations*			
Location	*strikes*	*>50 mph*	*>60 mph*	*>70 mph*	*>100 mph*
Brownsville	3	2	1	0	0
Galveston	4	3	1	1	0
Tampa	3	5	3	0	0
Miami	5	8	5	2	1
Wilmington	6	6	3	2	0

NOTE: A *strike* is defined as an incident with hurricane-force winds affecting some portions of the area within 100 km of the anemometer. Tropical storm strikes are not counted.

The first step in developing viable plans for keeping hurricane risks at a minimum is to define the potential strike area and to determine the probable hurricane recurrence and severity for each

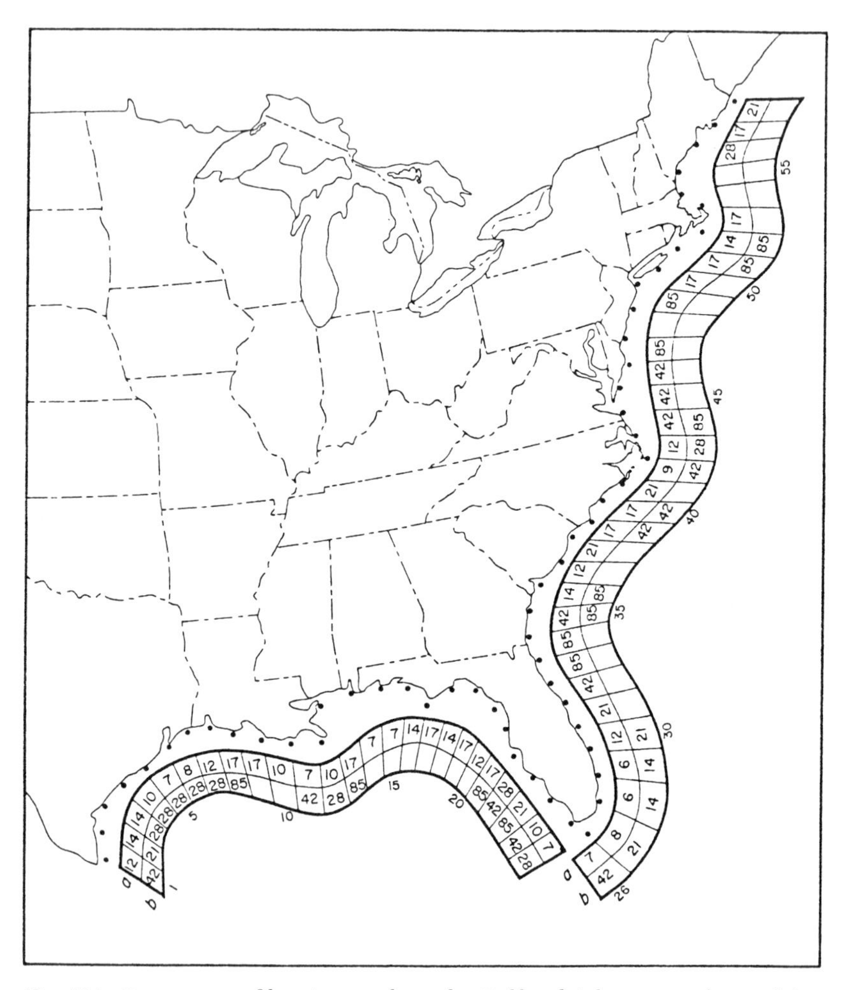

Fig. 124. Recurrence of hurricanes along the Gulf and Atlantic coastlines of the United States. Each of fifty-eight strike areas, 80 km in length, shows: *box a*, the number of years between occurrences of hurricanes—maximum winds greater than 120 kph—and, *box b*, severe hurricanes—maximum winds greater than 200 kph. Adapted from R. H. Simpson and M. B. Lawrence, *Atlantic Hurricane Frequencies*, Technical Memo NWS SR-58 (N.p.: Department of Commerce-NOAA, 1971), 5.

area. Figure 124 divides the United States's coastlines—Atlantic and Gulf of Mexico—into 80-km strike areas and for each area lists the recurrence, or return period, in years for *hurricanes* and for *great hurricanes*. Although few individual strike areas would expect a serious hurricane disaster more often than about once in 20 years, almost every year—in the long-term average—several of these fifty-eight strike areas should experience a hurricane crisis, if only from a near-miss.

Computing Return Periods

The usefulness of a computation of extreme values or of the recurrence of an extreme event depends upon the "goodness" of the data set used, that is, the representativeness of the samples and the size of the set. In Figure 124, the size of the strike area was defined as 80 km because this was the smallest coastline length for which there were enough hurricane strikes for a valid analysis, irrespective of the representativeness of the cases.

Frequency data were obtained by counting the numbers of tracks that carried hurricane-force winds across the coast in each segment and by assuming that a landfall in one segment would cause the hurricane-force winds in the right semicircle to intrude upon the adjacent segment. Then the figures in each box were fitted to a Poisson distribution.

With the recurrence data for strike areas in hand, the next step is to compute the probable extreme wind and inundation potential that may recur within 50, 100, 200, and 500 years. These are the basis for developing and adopting hurricane-resistant building standards and for deciding the allocation of public funds for structures to protect the beaches and shorelines against extreme erosion from hurricanes. Decisions of this kind, however, are rarely simple exercises in decision analysis, even with the recurrence data and probable extremes in hand. There is no guarantee that the extreme for the 100-year interval will not occur the very next hurricane season. Nevertheless, a systematic assessment of the threat is the starting point of all programs to reduce risks, and the discussion here will focus on the computation of expected extreme values for various return periods.

Some knowledge of hurricane occurrences in United States's coastal areas extends back to the seventeenth century. However, meteorologically useful descriptions of hurricane characteristics and extremes have been available for less than 100 years in most coastal areas, with a few notable exceptions.[1] Moreover, the number of occurrences for any one strike area is quite small, half the areas in Figure 124 having five or fewer cases. This places a heavy burden on any statistical analysis with so few cases available to determine valid and useful intervals for flood potentials or for extreme winds. Nevertheless, objective procedures must be generated to provide the best possible estimates, even though the results may need to be qualified in terms of probable errors of estimate.

Three questions must be answered with respect to a strike area: 1) What are the highest winds? 2) What are the peak tidal surges? 3) How far inland will inundation extend from hurricanes with appropriate return periods, say, of 25, 50, and 100 years? Finally, some estimate must be made of absolute extremes to be expected for a strike area.

The first question is a relatively straightforward task in which observed central pressures for the hurricane cases that affected the strike area are ranked, the cumulative and discrete probabilities computed, and, from this, the frequency and recurrence intervals determined. From the graph of recurrences, the minimum expected pressure is determined for the appropriate return periods and then converted to the maximum sustained-wind values.

Peak surges at the open coast and at bay shores are more complex to compute because they depend not only upon the pressure drop in the hurricane but also upon the local bathymetry, the velocity with which the center approaches the coast, the radius of maximum winds, and the phase of astronomical tides. Return periods, in this case, nearly always need to be computed using some form of the method of joint probabilities.[2] The inland spread of inundation is a separate problem that must be derived from the computation of open-coast surges and tailored to the orographic character of the strike area, the rainfall runoff, and the outfall from rivers and streams.[3]

To illustrate the application of such procedures, we shall con-

Table 17. Characteristics of Landfall Hurricanes That Brought a Sustained Hurricane-Force Wind to a 160-km Strike Area on the Central Texas Coast During the Period 1900–1978

Date	*Minimum pressure* P_o *(mb)*	*Radius of maximum wind R (mi)*	*Approach speed C (kt)*
9 Sep 1900	936	14	10
21 Jul 1909	959	19	12
17 Aug 1915	949	29	11
22 Jun 1921	954	17	11
28 Jun 1929	969	13	15
14 Aug 1932	942	12	15
23 Sep 1941	959	21	13
30 Aug 1942	951	18	14
27 Aug 1945	968	18	4
4 Oct 1949	963	20	11
11 Sep 1961	931	20	6

SOURCE: Ho, Francis P., R. W. Schwerdt, and H. B. Goodyear, *Some Climatological Characteristics of Hurricanes and Tropical Storms, Gulf and East Coast of the United States*, Technical Memo NWS 15 (N.p.: Department of Commerce—NOAA, 1975), 5.

sider a 160-km strike area along the central Texas coast from Matagorda (28.6° N, 96.5° W) to Galveston (29.2° N, 94.9° W). Table 17 chronologically lists the landfall strikes that brought hurricane-force winds to this zone during the period 1900–1978.* Included in the listing are approach speeds, C; radii of maximum winds, R; and central pressures, P_o. For these eleven cases, P_o values are listed in the order of their rank from 931 mb to 969 mb, and then an accumulative probability distribution is graphically obtained as in Figure 125.

The plotting positions for Figure 125 are computed from the formula

Eq. 11.1
$$P_a = \frac{r-0.5}{n} \times 0.54$$

where P_a is the cumulative probability, r is the rank number of the hurricane case, and n is the number of cases.[4] The coefficient,

* The threshold for sustained hurricane-force winds is defined in terms of the central pressure when it falls to a value of 982 mb.

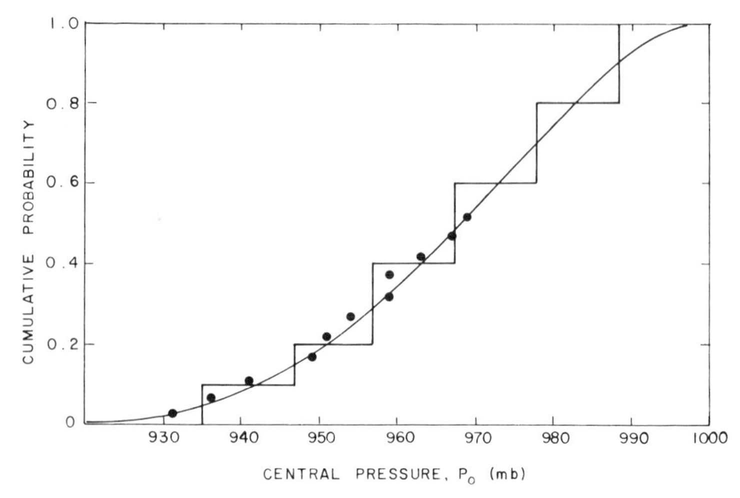

Fig. 125. Cumulative probabilities for occurrence of minimum pressure for hurricanes listed in Table 16 (see text).

0.54, accounts for the tropical cyclones of lesser strength that reached the strike area but were rejected in this listing as unsuited for the analysis of storm surge. It is the ratio of the number of tropical cyclones with hurricane-force winds to the total number of cyclone tracks in a strike area for the 79-year period. The curve, in this case, is fitted by eye, recalling that the highest central pressure in a tropical cyclone that will sustain gale-force winds at this latitude is about 1,000 mb. For this analysis, six convenient class intervals of central pressure are identified, and the expected frequency, F_e, for equal intervals (in this case, 10-mb intervals) is:

Eq. 11.2

$$F_e = \frac{nP_a}{N}$$

where N is the length of record. Finally the recurrence, r_e, a return period in years, is listed as the reciprocal of F_e. The expected frequency and recurrence intervals appear in Table 18.

From Figure 126, the central pressures for hurricanes that will recur in 25, 50, 100, 200, and 500 years are obtained; then using the conversion nomogram in Figure 127, the corresponding maximum sustained winds are obtained for these return periods. These appear in Table 19.

An estimation of peak-surge recurrences can be done in the same manner as for maximum winds, providing 1) dependable peak-surge heights were observed for a representative sample of

Table 18. Expected Frequency and Recurrence Intervals for Hurricanes as a Function of Central Pressure, P_0 (Based on a 79-Year Record)

P_o (mb)	p_a	F_e	r_e (years)
930	.02	.003	360
940	.08	.011	90
950	.19	.027	38
960	.35	.049	21
970	.54	.079	13
980	.76	.106	9
990	.93	.130	8

occurrences, 2) the bathymetry is homogeneous over a strike area, 3) a sample is restricted to hurricanes that make their landfalls almost normal to a coastline, and 4) the astronomical tidal range is small.* However, few strike areas qualify for a direct analysis of peak surges under these restrictions, and a return period obtained in this manner must be applied with the realization that probable errors may be large.

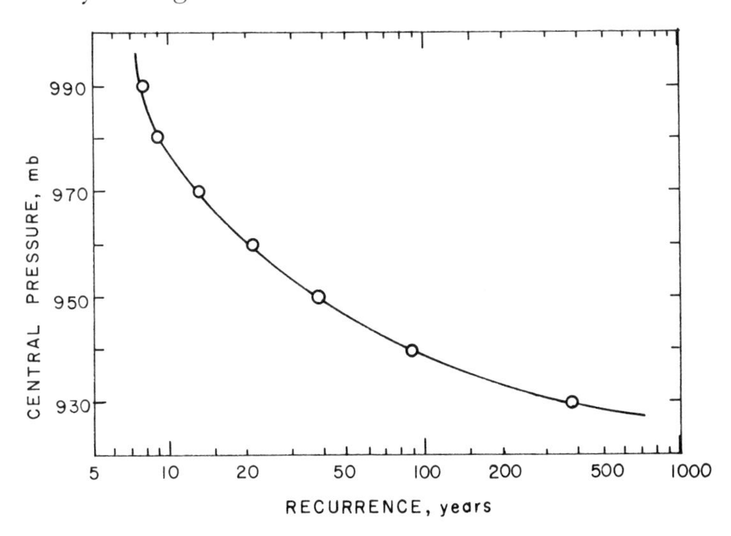

Fig. 126. Recurrence interval (years) for hurricanes when strength is measured by central pressure.

* The importance of these restrictions is clarified in the section on storm surges in Chapter 9.

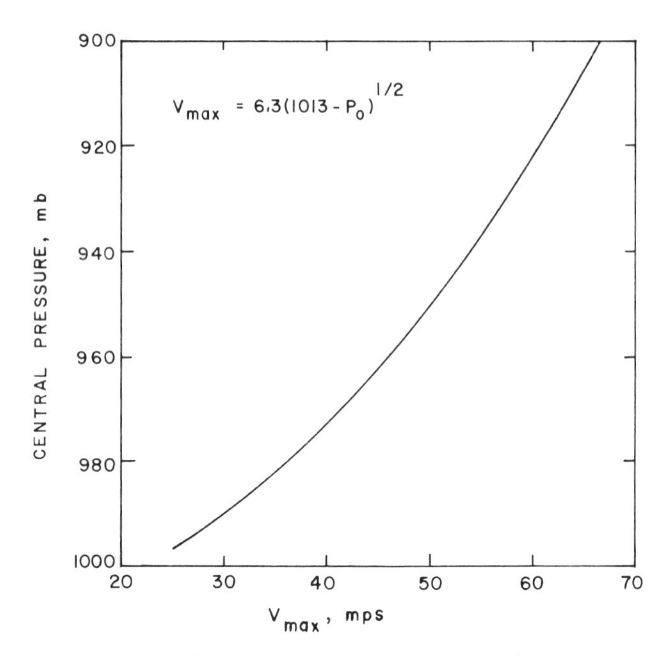

Fig. 127. Approximate values of maximum sustained wind speed in Atlantic hurricanes as a function of central pressure.

Table 19. Maximum Sustained Wind Speeds for Various Hurricane-Recurrence Intervals

Recurrence (years)	V_{max} *(sustained)* *(mps)*	*(mph)*
25	48.0	107
50	51.5	115
100	54.0	120
200	56.2	125
500	57.5	128

In the early 1970s, a hierarchy of computer models was developed for simulating hurricane storm surges. This opened the way for a hydrodynamic synthesis of the independent hurricane variables controlling peak-surge heights.[5] In this synthesis, the individual probabilities for each combination of variables must be obtained by computer processing and combined into a joint-probability statement of the peak surges for various return periods.

By this method, cumulative probabilities are computed for central pressure, P_0; for R, radius of maximum wind; and for C, speed of approach toward land. For the data sample in Table 17, the graph of cumulative-probability values for each variable is shown in Figures 125, 128, and 129, with six class intervals defined for each. The various combinations of these class intervals (6 × 6 × 6) provide a total of 216 peak-surge cases. After the discrete probabilities for each variable in each class interval are determined, the joint probability for each surge case is computed as the product of the 3 discrete probabilities for each of the 216 combinations. If a computer program is not available for machine computations, the nomograms in Figures 151 and 152 (in Appendix F) will provide good approximations. Proceeding as before to obtain distributions of P_a, P_0, R, and C, and computing the frequency and recurrence intervals, the expected peak surges for return periods of 25, 50, 100, and 500 years can be obtained.

Most sophisticated programs for computing the joint probabilities of peak surges incorporate a fourth variable, the probable phase of the astronomical tide, and a fifth, the angle of coastline crossing. The latter is more important for Florida and the United States east coast than for the Gulf of Mexico. The astronomical tidal range in most coastal areas is small (notable exceptions being

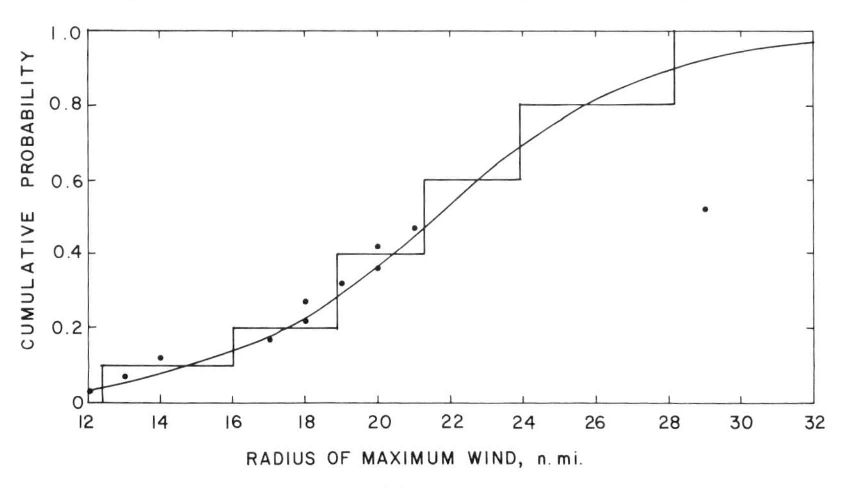

Fig. 128. Cumulative probabilities of hurricane radii of maximum wind (see Table 16).

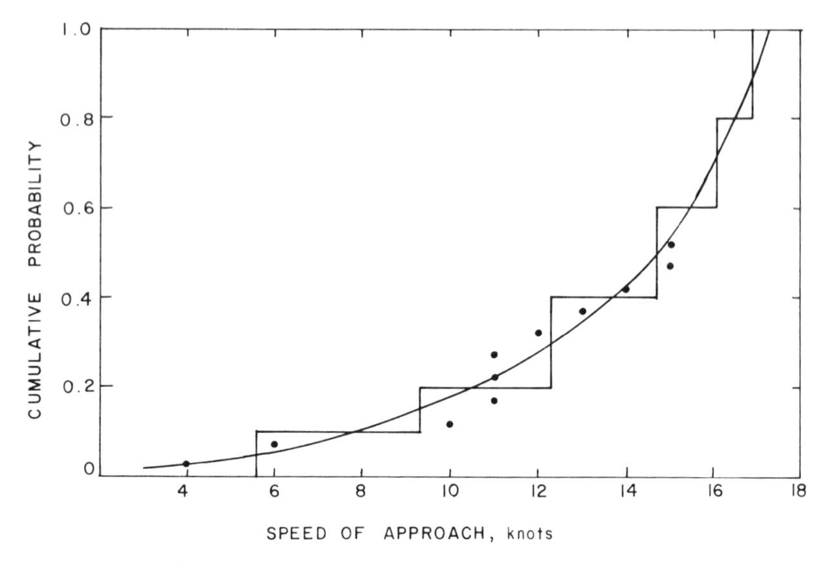

Fig. 129. Cumulative probabilities for rate of hurricane movement approaching landfall.

those for the Georgia, Carolina, and New England coasts) and does not exceed the secular changes in sea level. Obtaining the probability of secular changes is not a tractable problem in this kind of analysis.

The validity of the method of joint probabilities depends critically upon the mutual independence of the variables used in obtaining the joint probability. Although P_0, R, and C can, in some instances, be shown to be independent, it cannot be assumed a priori from physical reasoning that independence necessarily exists. Each new data set needs to be tested before this method is applied. From the scatter diagrams in Figure 130, it would appear that the P_0, R, and C values used in this example have low correlations and are sufficiently independent for use in the method of joint probabilities.

The greatest uncertainty concerns the question of whether the few cases available represent a random distribution of events. This needs more careful analyses if the computational results are to be credible. Such analyses require the use of procedures and tests that are beyond the scope of this discussion.

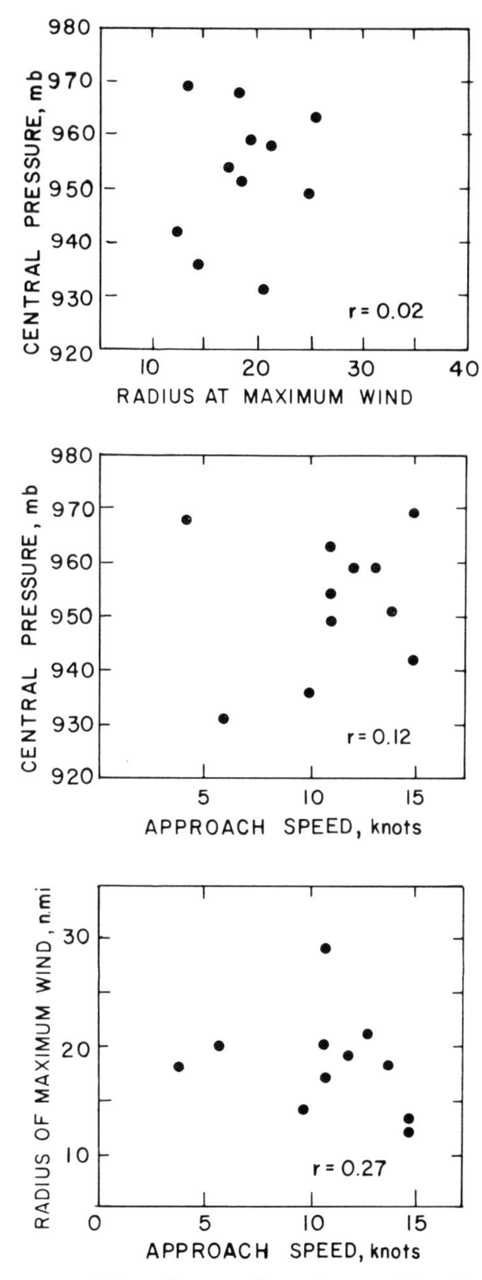

Fig. 130. Examination of the relative independence of the three variables: central pressure, radius of maximum wind, and approach speed, used in the computation of joint probabilities of storm surges; *r*, correlation coefficient.

Land-Use Planning and Regulation

Once a hurricane threat has been analyzed and assigned a strike area, the next question is how best to live with the threat while enjoying the benefits of coastal living and keeping risks to a minimum. This poses a hierarchy of problems—legal, political, and sociological—whereby individual and corporate rights and government responsibilities must interact if solutions are to be found that are compatible with the public welfare at large. Many noble efforts to initiate programs of land-use management or regulation have foundered either because of legal and political controversies over the regulatory rights of state and federal governments or because of a failure to comprehend or to account for human responses to the regulatory or management measures.

The need for a broader base of research to examine disaster potentials from (locally) rare events—hurricanes, earthquakes, and tornadoes—and the problems they pose, particularly those in the realm of the behavioral sciences, have been recognized. The need for research on human responses in the face of a disaster or risk of disaster was formally recognized by the United States in the 1960s when substantial government support was given to a series of research assessments on natural disasters. From this came a number of very useful analyses conducted by Gilbert White and his colleagues at the University of Colorado's Institute of Behavioral Science.[6] A series of monographs was prepared reporting the results of comprehensive studies of natural hazards, of research done and of research needs, of problems in land management and regulation for hazardous areas, and of the human responses to be expected in the face of such disasters as hurricanes, floods, tornadoes, and earthquakes, among others.[7] Similar studies are underway at other universities including Ohio State, University of Minnesota, and Texas A&M.[8]

These studies define the need to regulate and manage the use of coastal lands to preserve natural resources, vital public facilities, and the economic stability in potential strike areas and to protect against excessive loss of life and personal property. Figures 131 and 132 and Tables 20 and 21 provide impelling evidence that the needs for intelligent planning and management of coastal lands are not static; they are growing exponentially with time. Yet efforts to officially delimit and manage these areas tend to be confounded by the very forces that attract people to them and increase the population-at-risk. Aesthetic reasons often combine with economic reasons to create ever more intensive use of coastal lands that are vulnerable to hurricanes.

Those who own property outside urban areas where building codes are not enforced often insist on the freedom to build or improve their property without regard to risks from hurricane winds and flooding. But in exercising this freedom, they increase the risks faced by their close neighbors when, under hurricane stresses, poorly constructed residences collapse and their debris, borne by wind and surging waters, inflicts damage on better-built structures that otherwise would survive.

Some states have addressed this problem by indirection, plac-

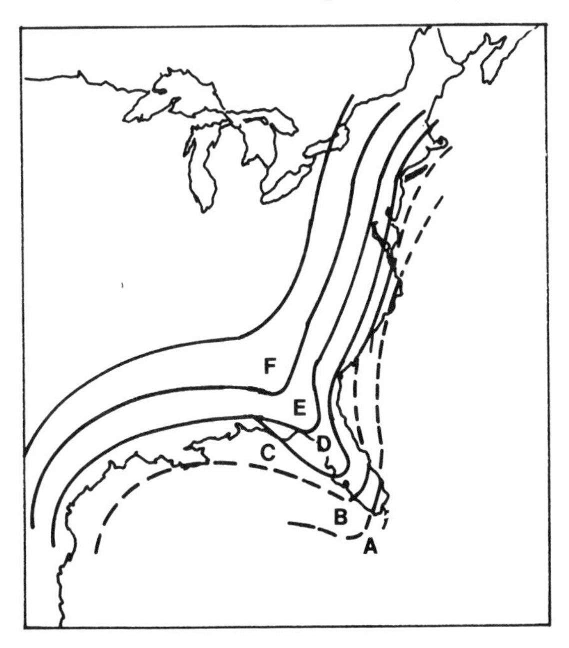

Fig. 131. Hurricane-hazard zones relative to the population- and property-at-risk data in Tables 20 and 21. W. Brinkman, *Hurricane Hazard in the U.S.: A Research Assessment*, National Science Foundation Monograph RA-E-75-007 (Boulder: Institute of Behavioral Science, University of Colorado, 1975), 12.

ing the burden upon the property owner to demonstrate that the land he proposes to develop is not subject to a natural hazard that requires more stringent building standards for safety. In other regions, incentives for the use of building standards and land-development measures in keeping with the natural-disaster potential are made by restricting the availability of insurance or by

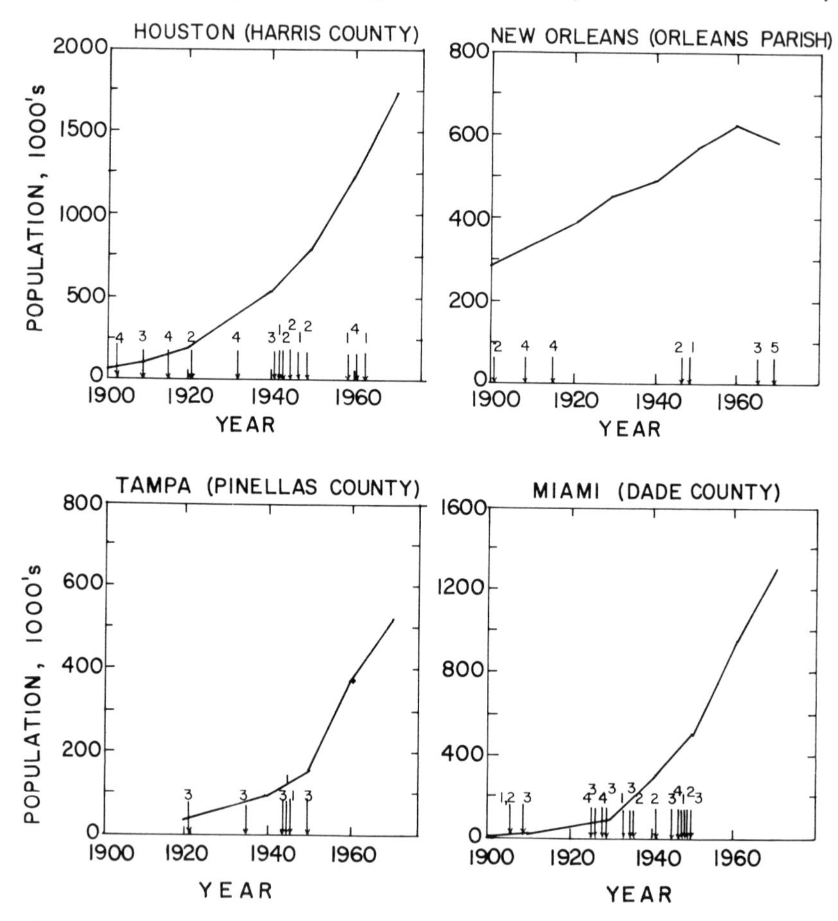

Fig. 132. Coastal population trends in Houston, New Orleans, Tampa Bay, and Miami areas. *Arrows* indicate individual occurrences and damage potential numbers (see Appendix C). P. J. Hebert and G. Taylor, *Hurricane Experience Levels of Coastal County Populations—Texas to Maine*, National Weather Service Southern Regional Technical Report (Silver Spring, Md.: Department of Commerce-NOAA, 1975), 32, 51, 55.

providing premium advantages for adhering to hurricane-resistant building standards.

In the mid-1970s, the state of Texas, through the initiative of its Coastal and Marine Council, inaugurated a comprehensive public program of hurricane awareness and hurricane-hazard assessment. This culminated in the development of a set of hurricane-resistant building standards upgrading the construction standards of the American National Standards Institute.[9] This document defines the classes and extremes of the hurricane hazard on the Texas coast, a Texas coast design hurricane, a means of delimiting the

Table 20. Estimated Population-at-Risk/Hurricane Wind

Hurricane-wind hazard zone	*At least one occurrence of peak gust in past 80 years in county-sized area*	*Population-at-risk*
A	125 mph or more	14,550,000
B	100 to 124 mph	37,230,000
C	75 to 99 mph	23,770,000
D	50 to 74 mph	20,490,000
Total: Zones A through D		96,040,000

SOURCE: W. Brinkman, *Hurricane Hazard in the U.S.: A Research Assessment*, National Science Foundation Monograph RA-E-75-007 (Boulder: Institute of Behavioral Science, University of Colorado, 1975), 13.

Table 21. Estimated Population-at-Risk/Storm Surge

Storm-surge hazard zone	*Return period*	*Population-at-risk*
A	less than 5 years	30,000
B	5–10 years	130,000
C	10–25 years	2,550,000
D	25–50 years	1,780,000
E	50–100 years	1,500,000
F	more than 100 years	610,000
Total: Zones A through F		6,600,000

SOURCE: W. Brinkman, *Hurricane Hazard in the U.S.: A Research Assessment*, National Science Foundation Monograph RA-E-75-007 (Boulder: Institute of Behavioral Science, University of Colorado, 1975), 11.

wind and flooding hazard in terms of the design hurricane, and the specific building standards needed to protect against each class of hazard. Although first attempts at legislation to implement the use of the recommended standards ran aground because of lobbying by special-interest groups, the voluntary application of these standards has been encouraged and their use rewarded by lowered insurance premiums applicable to this type of construction, together with other incentives to encourage proper use of coastal lands-in-hazard. The document also paves the way for constructive truth-in-selling and truth-in-lending legislation, all of which is calculated to impress upon the developer and user of coastal property the reality of natural hazards. As such, it stands as a pioneering effort and an example of a useful approach to land-use planning.

Beach Protection

Several regulatory measures are fundamental in protecting coastal lands and their development. First is the preservation of the beaches. Construction on, or the artificial reconfiguration of, beaches upsets the natural morphological processes that heal the erosive wounds from storms and maintain the stability of the beach, the berm, and the foredune structures that act as buffers against the advance of destructive waves inland when storm surges raise the sea level. The most important first step in beach protection is to establish a legal setback line for construction, development, and reconfiguration of beach structures. Such a setback line, in most cases, should extend from the position of mean high water at the beach to a sufficient distance inland to assure that natural dune grasses—the means of entrapping and retaining wind-borne sands that build and stabilize the dune structure—are not replaced with less-suitable ground cover. A comprehensive discussion of the problems and remedies in protecting coastal lands may be found in the three volumes of the *Shore Protection Manual* prepared by the United States Corps of Engineers Research Center.[10]

REFERENCES

1. D. Ludlum, *Early American Hurricanes, 1492–1870* (Boston: American Meteorological Society, 1963); J. C. Millas, *Hurricanes of the Caribbean Sea and Adjacent Regions During the Late Fifteenth, Sixteenth, and Seventeenth Centuries*, Preliminary and Final Reports to the U.S. Weather Bureau (Miami: Institute of Marine Science, University of Miami, June, 1962; June, 1963; June, 1964).
2. V. A. Myers, *Storm Tide Frequency on the South Carolina Coast*, Technical Report NWS 16 (Silver Spring, Md.: Department of Commerce-NOAA, 1975).
3. R. H. Simpson and J. C. Freeman, "Coastal Hazard Potentials," *Proceedings of the Symposium on Coastal Meteorology* (Boston: American Meteorological Society, 1976), 16–19.
4. E. J. Gumbel, *Statistics of Extremes* (New York: Columbia University Press, 1958).
5. V. A. Myers, *Joint Probability Method of Tide Frequency Analysis Applied to Atlantic City and Long Beach Island, N.J.*, Technical Memo WBTM Hydro-11 (Silver Spring, Md.: Department of Commerce, Environmental Sciences Services Administration, 1970), and *Storm Tide Frequency on the South Carolina Coast.*
6. G. F. White, *Natural Hazards, Local, National, Global* (London: Oxford University Press, 1974); G. F. White and J. E. Hass, *Assessment of Research on Natural Hazards* (Cambridge, Mass.: M.I.T. Press, 1975).
7. W. Brinkman, *Hurricane Hazard in the U.S.: A Research Assessment*, National Science Foundation Monograph RA-E-75-007 (Boulder: Institute of Behavioral Science, University of Colorado, 1975); E. J. Baker and J. G. McPhee, *Land Use Management and Regulation in Hazardous Areas: A Research Assessment*, National Science Foundation Monograph RA-E-75-008 (Boulder: Institute of Behavioral Science, University of Colorado, 1975).
8. E. L. Quarantelli and Russell R. Dynes, "Response to Social Crisis and Disaster," *Annual Review of Sociology*, III (1977), 23–49.
9. American National Standards Institute, *Building Requirements for Minimum Design Loads in Buildings and Other Structures*, Publication A58.1 (New York: American National Standards Institute, 1972), 1–60; Texas Coastal and Marine Council, *Model Minimum Resistant Building Standards for the Texas Gulf Coast* (Austin, Tex.: General Land Office, 1976).
10. Department of the Army, Corps of Engineers, Coastal Engineering Research Center, *Shore Protection Manual* (Washington, D.C.: Government Printing Office, 1973), Vols. I, II, III.

CHAPTER 12

Hurricane Awareness and Preparedness

On 27 June 1957, hurricane Audrey, in Category 4 on the hurricane scale (Appendix C), made a landfall near Cameron in western Louisiana shortly after daybreak—the first hurricane of consequence to strike that coastal area since 1918. Cameron had been under alerts or warnings eight times during the preceding 20 years, when hurricanes had brought heavy rains, and heavy seas had eroded the beaches. Disaster in these cases, however, had been reserved for landfalls elsewhere, often too distant for Cameron residents to be impressed with their narrow escapes.

More than five hundred people lost their lives in Audrey—*few* compared to the two thousand lost in the Okeechobee hurricane of 1928 and the six thousand lost in the Galveston hurricane of 1900. But with the improvements in science and technology, with aircraft reconnaissance and radar storm tracking, and with steady improvements in the skill of forecasts, such losses were shocking, incomprehensible, and for a culture that regards life as priceless, unacceptable. However, the facts and the conclusions from several careful studies of this disaster provide some insight to the need and the challenge to develop more effective programs for public awareness and preparedness.[1]

Audrey was well tracked by reconnaissance aircraft, its position and the rapid increases in strength were well advertised, and its landfall was predicted in space and time as accurately as scientific methodology of the day would permit. Coastal flooding was anticipated well in advance. The warnings and recommendations for evacuation made by the New Orleans forecast office provided time enough for coastal residents to escape before flooding and high winds cut escape routes. Yet for various reasons a majority of those threatened chose to remain in their homes, which had provided safe enough refuge for decades—in some instances, for nearly two generations. Many who made this decision died in Audrey. Those who survived experienced a cruel and harrowing experience that resulted in many grave injuries.

Physical science and technology had done their job fairly well, yet the warning system as a whole had failed. Why? From various studies of this disaster, not all with similar objectives and concerns, it seems evident that the factors contributing to failure fall into one of three categories, all of which involve some aspect of behavioral science: 1) human communication, 2) credibility of public advices, and 3) human responses in the face of an imminent emergency.

Most Cameron residents depended mainly upon a single radio station for warnings and advices of severe weather. This station was located in a metropolitan area some distance inland. Advisories on Audrey, issued every few hours from New Orleans, grew in length as the need for coastal warnings expanded and the hurricane threat increased. At one station, however, in order to make efficient use of air time during an impending crisis, the lengthy advisories were edited before being broadcast to include primarily those items of concern to the metropolitan area. Lost at times in the editing process were the explicit warnings to coastal residents to evacuate inland during the evening, before escape routes were cut by rising water. Although the complete advisories correctly predicted a landfall near daybreak, emphasis was given in the editing process to the (correct) prediction that the metropolitan area would receive its worst conditions "in the early forenoon." Some Cameron residents interpreted this to mean that there would be time enough by daybreak the next morning to decide whether to evacuate. Law-enforcement officials in Cameron Parish received the complete advisory by teletypewriter and a special telephone recommendation to initiate evacuations. All resources available in Cameron Parish were used to communicate the urgency of the situation and to motivate the evacuation. But the response was inadequate.

Credibility of warning advices and of their sources contributed to these uncertainties. No one doubted that a hurricane was nearby; yet having been warned many times of an approaching hurricane, and in many instances advised to move inland to safe shelter, the consequence of "crying wolf" too often exacted its toll on the credibility of advices—advices that emanated from anonymous sources through indirect channels. Experience has shown that the credibility of public advices may depend, in part, upon whether the advices are identified with anonymous sources or with an individual or a few highly visible individuals to whom the public can relate. Credibility tends to diminish as the anonymity of the source increases.

The human response in a hurricane crisis depends upon individual perception and evaluation of 1) the meaning of advices and warnings received, 2) the credibility of the advices, and 3) the individual's assessment of his own experience and judgment relative to the advices from official sources. Studies of human responses to natural hazards suggest that the response to crisis is unlikely to result in panic, regardless of uncertain or confusing information.[2] When doubt exists as to the credibility or meaning of advices, the

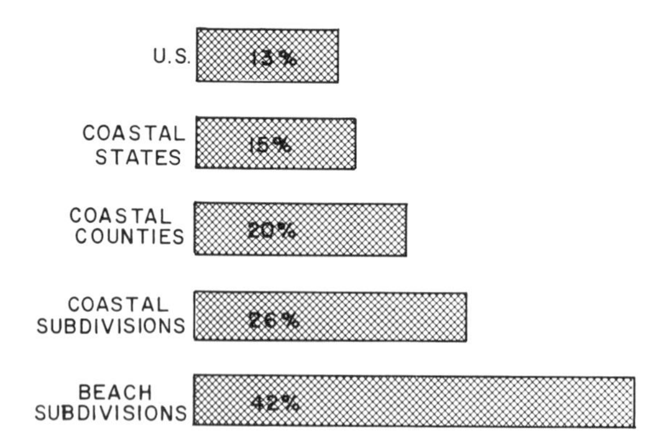

Fig. 133. Population increases in the United States during the period 1960–1970 compared with the percentage increases in coastal and beach areas. Adapted from W. Brinkman, *Hurricane Hazard in the U.S.: A Research Assessment*, National Science Foundation Monograph RA-E-75-007 (Boulder: Institute of Behavioral Science, University of Colorado, 1975), 14.

individual's response will depend upon the reliance he is willing to place on his own experience and judgment concerning the natural hazard.

Programs of public awareness and preparedness face a dual challenge: first, to supply technical information and generate public awareness of the hurricane's destructive potential and the emergency actions needed to protect life and property; second, to develop community-preparedness plans to assure an effective flow of information from sources whose credibility will be readily accepted as a basis for action.

The critical importance of effective hurricane-preparedness programs is illustrated in Figure 133, which reflects the result of migration to the seashores, where an exponential growth of population is occurring.[3] In many places, the population-at-risk has increased by more than an order of magnitude while the overall hurricane-experience level has been significantly lowered.

The Effect of Declining Hurricane Incidence in the United States

Still another factor that contributes to low levels of experience is the protracted decline in hurricane landfalls in the United States since 1948. Figure 134 shows that the annual expectancy for hurricane landfalls decreased from about 2.4 in 1948 to less than one a year by the mid-1970s, by far the lowest incidence level of the century. The incidence of severe hurricanes reaching the United States has also diminished. Although these statistics are interesting, they cannot be taken to mean that in all latitudes and regions hurricanes are diminishing in number (compare, *e.g.*, Figures 46 and 47).

This decline in the number of United States landfalls appears to be associated with the intrusion of mid-latitude westerlies into the tropics of the western Atlantic Ocean and Caribbean Sea, lowering the depth of the easterly trade winds and the moisture available to support the hurricane heat engine. Such an environment may account, also, for what appears to be an increasing tendency of tropical cyclones to lose strength or die over warm ocean waters before reaching a landfall.

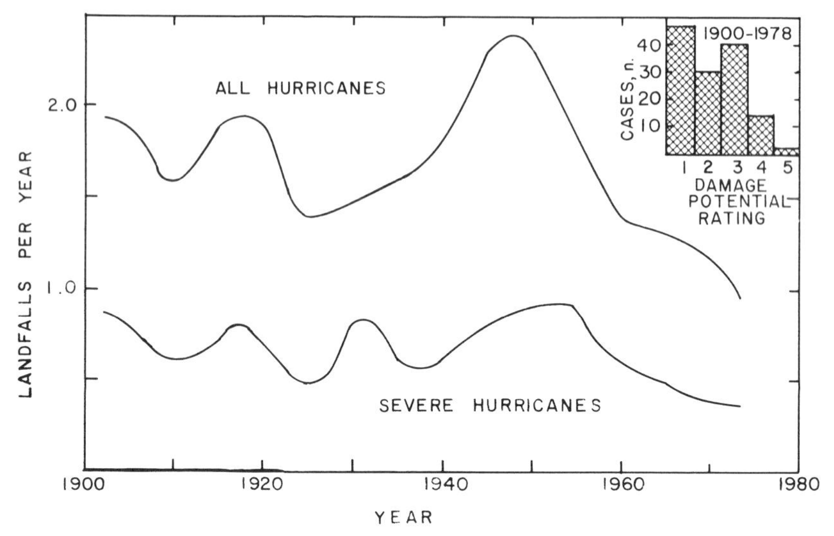

Fig. 134. Trend in number of hurricanes reaching the United States per year, given in 10-year averages in overlapping steps of 5 years each. Inset shows the frequency distribution of hurricanes by damage-potential categories (see Appendix C) for the period 1900–1978.

It is an open question whether this decline, sustained over nearly 3 decades and much sharper than that shown in Figure 47, merely represents a climatic aberration caused by the persistence of a planetary circulation that uniquely affects the western Atlantic, or results from some more obscure influence. Unfortunately, there are no dependable means of estimating the duration of this pronounced trend. However, if it is simply a climatic aberration, a recovery and sudden increase in hurricane landfalls in the early 1980s would not only place at risk an unprecedented amount of property but also a burgeoning population with dangerously low hurricane-experience levels.[4] Such conditions would set the stage for exceptional losses in a severe hurricane. Unless effective programs are mounted to defuse these deficiencies, the progressive reduction in lives lost reflected in Figure 135 could be rudely reversed.

In the United States, the Federal Reorganization Plan No. 3 of September 1978 combined all national programs for disaster planning and management under a new agency, the Federal Emergency Management Administration. This was a constructive step, conceptually. However, such reorganizations are per se rarely the answer to such difficult problems as those of emergency management, the most difficult of all being that of protecting the credibility of emergency advices and warnings so that prompt and effective action responses can be expected during emergencies.

Hurricane-Warning Advices

In the United States, the National Weather Service holds responsibility for the issuance of hurricane advisories, for explicit warnings to coastal areas and ships at sea, and for recommending emergency actions, including the evacuation of residents in areas subject to flooding. The skill in predicting the position of hurricane landfall is not high and, as we shall see, cannot be expected to improve significantly in the foreseeable future. Yet, if warnings are to remain credible, the amount of overwarning must be restricted to an extent consistent with safety. In an effort to deal with this dilemma constructively, three kinds of advices are used to communi-

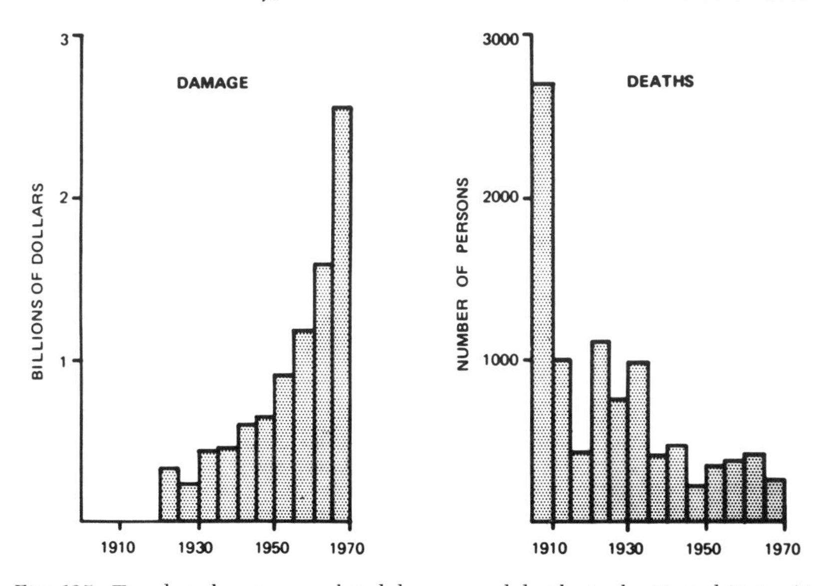

Fig. 135. Trends in hurricane-related damage and deaths in the United States (5-year averages). R. H. Simpson, "Hurricane Prediction," in *Geophysical Predictions* (Washington, D.C.: National Academy of Science, 1978), 142.

cate the degree and imminence of the threat. Each 6 hours—more frequently if needed—*advisories* are issued stating the current location of the hurricane center, the strength and movement of the system, and the changes expected during the next 24 hours. When the center approaches a coastline and the need for emergency actions is imminent, two additional action advices are headlined in the advisories:

1. Hurricane WATCH. An advice to a specific coastal segment that has a 50% probability of receiving a direct strike from a tropical cyclone of hurricane force within 36 hours. This advice has the purposes of alerting residents in the watch area to the potential need for evacuation or other emergency actions and to allow time for individual planning should such actions be required.
2. Hurricane WARNING. A warning of imminent danger for a specific segment of the coast and of the urgent need for immediate action to protect life and property.

The issuance of warnings and recommendations for specific emergency actions are usually timed to provide, when possible, a *maximum* of 12 to 18 hours for these actions. In practice this goal cannot always be met.

Emergency Evacuation of Coastal Residents

The lead time for evacuation of coastal residents when a hurricane threatens may vary from less than 10 hours to about 18 hours. In preparedness planning, however, no more than 10 hours for emergency actions can reasonably be expected to remain available once a specific *WARNING* has been posted. In Chapter 13, we will find that the outlook for the improvement of skills for predicting landfall is not promising, and it is unlikely that the research underway will provide a dependable means of extending this lead time. This poses a significant constraint that should be the initial point in hurricane-preparedness planning and longer-term coastal-zone management for hurricane-prone areas.

There are indeed some remarkable examples in which large numbers of people have been safely evacuated in short periods of time. In hurricane Carla (1961), more than 300,000 people were moved inland from exposed coastal areas; and in Camille (1969), evacuations that began before daybreak moved more than 75,000 inland before flooding and high winds cut escape routes in late afternoon.

In many areas, however, such massive relocations are not physically possible in so short a time frame. Notable examples are 1) the Florida Keys, where many stretches of highway accommodate only two lanes of traffic; 2) various segments of the Florida west coast, particularly where most high-capacity highways parallel the coastline, some segments of which are subject to early flooding; and 3) barrier islands, which have increasing residential development and populations that double or triple on holiday weekends; bridges and causeways could not accommodate the evacuation of more than half of those who would be threatened by a rapidly approaching hurricane.

Some officials, alarmed at these trends and with no suitable corrective measures in sight, have described hurricane scenarios that could claim the lives of more than ten thousand people in a single landfall area. Although such estimates may be conjecture, they do identify the urgent need for careful engineering of evacuation routes and procedures to assure that early rises in sea level will not cut those segments of escape routes subject to flooding.

In the early 1970s, an alternative concept of *in situ* evacuation was proposed in a study by the Federal Executive Board of Miami, Florida.[5] This study recommended that all suitable high-rise structures, irrespective of their exposures near a shoreline, be certified for use as shelters during hurricane emergencies. The use of such refuges has sometimes been referred to as "vertical evacuation." It was suggested that evacuees be encouraged to occupy interior hallways and common-use rooms located on those floors that were safe from flooding. Other proposals recommended that in new communities vulnerable to flooding, public buildings, schools, and libraries be constructed on elevated earthen platforms, using building standards and designs that would assure a safe shelter for nearby residents. Among other recommendations, it was proposed that the developer of a large, new coastal community be required to intersperse his single-level residences or garden apartments with high-rise units that could serve the purposes of vertical evacuation during emergencies. These measures would alleviate some of the burden, and eliminate some of the pitfalls, of horizontal evacuation.

However, the kind of land-use regulations required for effective *in situ* evacuation procedures are always confronted with political opposition having a wide range of motivations. The irony of the democratic process in the United States is that such programs as this, which address the long-term public welfare, rarely succeed except during the brief period that follows a hurricane disaster. The longer the interval between these disasters, the more aggravated are the risks and the potential for still greater disaster.

REFERENCES

1. F. L. Bates *et al*., *The Social and Psychological Consequences of a Natural Disaster: A Longitudinal Study of Hurricane Audrey*, National Research Council Disaster Study 18 (Washington, D.C.: National Academy of Science, 1959); National Academy of Science, National Research Council, *Disaster Research Group Field Studies of Disaster Behavior: An Inventory*, Disaster Study No. 14 (Washington, D.C.: National Academy of Science, 1961).
2. National Academy of Science, *Disaster Research Group Field Studies of Disaster Behavior*.
3. W. Brinkman, *Hurricane Hazard in the U.S.: A Research Assessment*, National Science Foundation Monograph RA-E-75-007 (Boulder: Institute of Behavioral Science, University of Colorado, 1975).
4. P. J. Hebert and G. Taylor, *Hurricane Experience Levels of Coastal County Populations—Texas to Maine*, National Weather Service Southern Regional Technical Report (Silver Spring, Md.: Department of Commerce-NOAA, 1975).
5. R. H. Simpson, "Evacuation of Coastal Residents During Hurricanes: A Pilot Study," Report of the Miami Federal Executive Board to the U.S. Office of Management and Budget, Washington, D.C., (1973).

CHAPTER 13

Prediction and Warning

From Chapter 12, it is apparent that the impact of a hurricane—at landfall or on a vessel at sea—will depend, in part, upon the awareness, preparedness, and human responses to warning advices. However, with public-preparedness machinery in good working order, and with full awareness of the hurricane and its threat, proper actions to protect life and property depend precariously upon the timeliness and accuracy of each component of the forecast. However, because hurricane prediction is not an exact science despite great strides that have been made to improve its accuracy, preparedness actions at all levels of responsibility must take into account the limitations of prediction procedures and how they may impact upon the emergency actions taken. It is the purpose of this chapter to provide a background of information on prediction procedures, how they have evolved, and the kinds of errors that may occur and require consideration in deciding emergency actions.

Figure 136 shows the consequences of an error in predicting landfall. In this case an error in predicting direction of movement—one that can reasonably be expected in a 24-hour forecast—causes a critical change in locating peak storm surges and highest winds, and in identifying beach areas from which residents must be evacuated. A similar error in advices to vessels at sea could lead the skipper of a slow-moving tug or pleasure craft to choose an avoidance procedure that would expose him to the dangerous, instead of the so-called navigable, semicircle of the hurricane. This

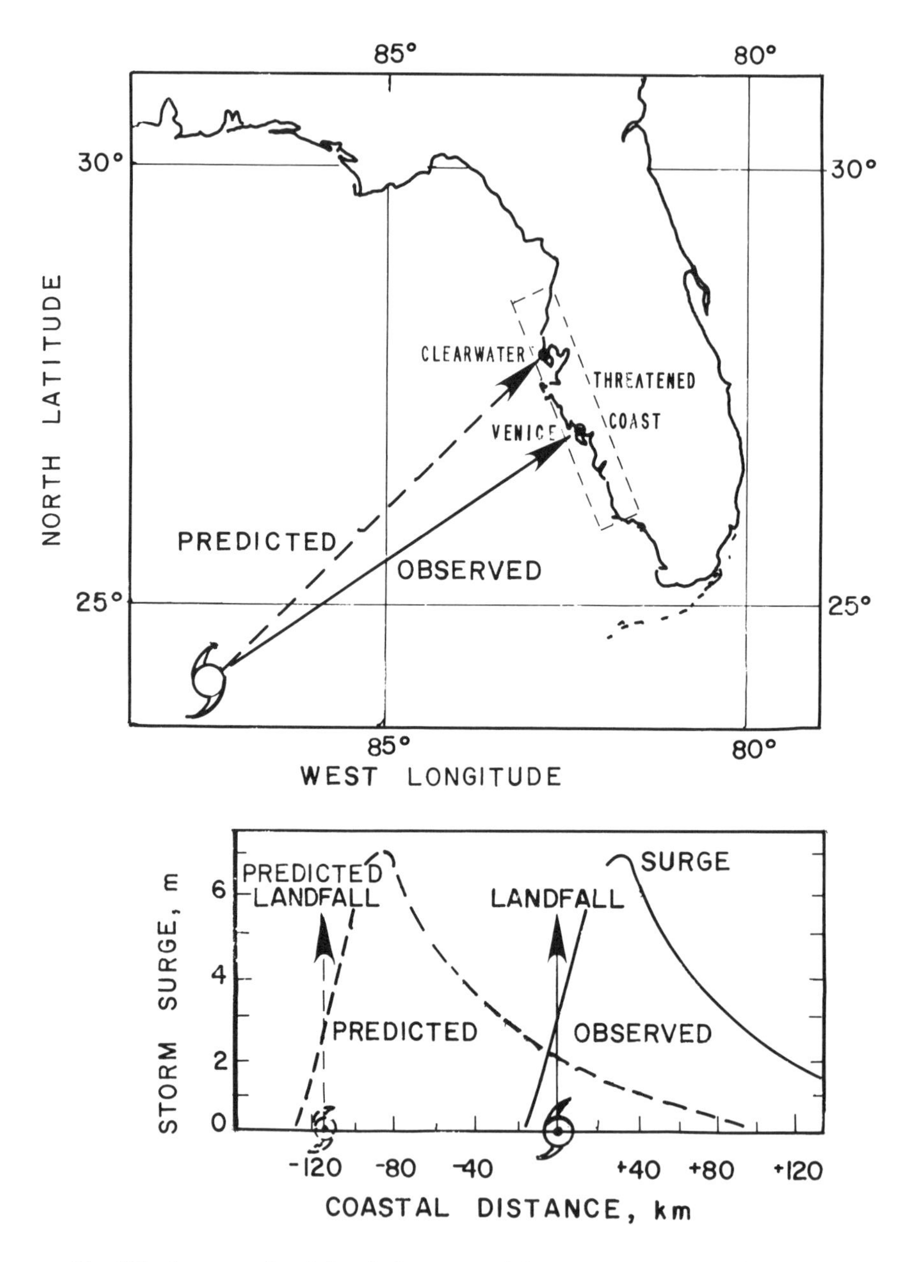

Fig. 136. An example of the shift in coastal danger zone: *upper*, the predicted direction of hurricane movement, 24 hours before landfall, differs from the observed track by 10°; *lower*, the resulting change in coastal hazard requiring evacuation of residents.

illustrates the need to understand what hurricane forecasting involves and the limits that present scientific knowledge places upon predictability.

Two Critical Forecast Problems

The advisory for hurricanes approaching the coast must contain at least five elements: 1) the current position, strength, and movement; 2) the predicted time and position at landfall; 3) the strength at landfall; 4) peak hurricane tides and areas of inundation; and 5) specific warnings for the coastal areas at risk. Of these, the most critical are the predictions of position and strength at landfall. If these two are known, the remainder can usually be observed or derived from data supplied by the customary hurricane-monitoring facilities—aircraft, weather satellites, and radar. A small shift in expected landfall position, as shown in Figure 136, can quickly change the requirements for evacuation and for other emergency-preparedness measures. A rapid increase in hurricane strength during its last 12 hours at sea can sometimes double the peak-surge heights and escalate the urgency for evacuation.

Despite significant progress during the 1960s and 1970s in understanding the hurricane, its structure, and its sources of energy, the skill in predicting movement and changes in strength seemed to reach a plateau in the 1970s.[1] In fact, since the early 1960s, progress has been slow in predicting marked changes in strength more than a few hours in advance, and most successful predictions of marked changes are based upon diagnostic reasoning. The physical reasons for such remarkable last-minute changes as occurred, for example, in Celia (1970), when central pressure fell from 988 mb to 943 mb during the last 14 hours before landfall, and in Carmen (1974), when pressure rose from 935 mb to 980 mb during the last 18 hours, remain obscure even after extensive postanalyses.

In recognition of these limitations, the forecaster has no rational alternative but to overwarn, and the preparedness official no alternative but to overevacuate, despite the losses of credibility that come from "crying wolf." However, some insight to the probable accuracy of predictions and warnings should constructively influence preparedness and evasive actions in the selection of a "course of least regret."

Predicting Hurricane Movement

Success in predicting hurricane movement depends upon the initial direction of movement, the stage of development, and the strength of the steering current. Errors are generally smallest for mature westward-moving hurricanes and largest for hurricanes accelerating north or northeastward. Figure 137 shows that the average error in predicting movement is about 80 km after 12 hours, increasing almost linearly to 380 km after 48 hours. The 24-hour forecast—normally the basis for explicit warnings and initial preparedness decisions in 1980—had an average error of 160 km with a standard deviation of 40 km. Once a landfall target has been specified and warnings are up, however, the movement of the center is monitored continuously by aircraft, weather satellites, and radar; and the forecaster can adjust the landfall position and the extent of tidal inundations in accordance with the hour-by-hour changes that are observed during the final hours as the center approaches.

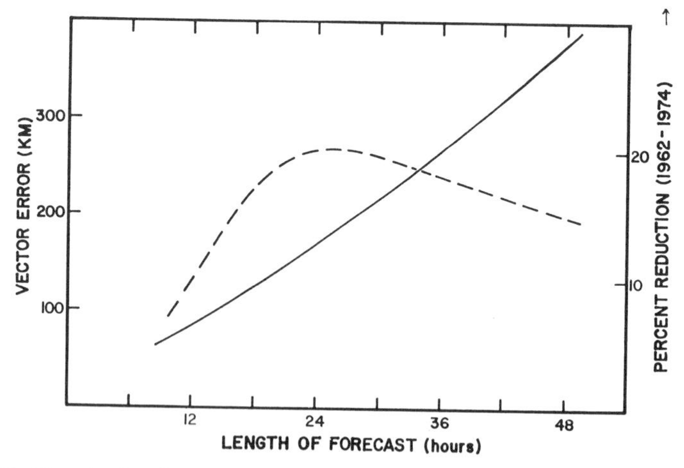

Fig. 137. Magnitude of vector errors in predicting hurricane movement for forecast periods (*solid line*) up to 48 hours, and the percent reduction in error magnitude from 1962–1974 (*broken line*). R. H. Simpson, "Hurricane Prediction," in *Geophysical Predictions* (Washington, D.C.: National Academy of Science, 1978), 142.

Changes in Strength

The prediction of changes in strength continues to depend primarily upon diagnostic reasoning guided by successive measurements of central pressure by aircraft and by the changes in hurricane cloud patterns observed by weather satellites that can be related both directly and empirically to changes in strength of the system.[2] Nevertheless, very rapid changes cannot be reliably anticipated far enough in advance to significantly influence preparedness measures, once the warnings have been posted.

Unique Nature of the Forecast Problem

Forecasting for hurricanes requires about the same basic observations and tools of analysis as for baroclinic storms of higher latitude. However, the hurricane presents the forecaster with a unique array of problems.

1. The hurricane is a macroscale event with a mesoscale impact (hundreds of kilometers versus tens of kilometers).
2. It is basically an oceanic storm conceived in a barotropic environment, from the outset dependent upon fluxes of heat from a warm ocean surface to maintain its destructive potential.
3. Having achieved maturity, the hurricane may outrun its barotropic environment and draw upon baroclinic sources to circulate moisture-poor air through its heat engine rapidly enough to supply the latent heat required to sustain its structure and fury, even while moving over cold water and sometimes great distances inland.[3]
4. Its major destructive potential at the coast is due to a combination of inundation from hurricane tides and of high winds, most of the damage and loss of life being confined to a coastal area within 80–100 km of landfall.
5. Important destructive by-products may occur far inland, including flash and river flooding and tornadoes that usually occur before the arrival of sustained gale- or hurricane-force winds.

Since a hurricane clusters most of its destructive potential quite close to its low-pressure center, the forecaster must seek greater precision in projecting the track of this center than is required in forecasting most other severe weather events.

Background of Prediction Methods

Movement. Before the advent of dynamic prediction models, hurricane movement was regarded conceptually as the response of a vortex to a steering current.[4] Most forecast decision making was centered around the identification of the steering level and the reasoning about changes that could modify the steering and future track of the system. This process is still pursued diagnostically by forecasters to test the credibility of model results.

In the 1940s, Grady Norton used the wind direction and speed at the top of the hurricane as an index to the movement of the vortex.[5] However, the three-dimensional hurricane circulation must move dynamically as a complex response to the resultant forces imposed throughout the entire storm volume by the environment interacting with the vortex. An unresolved problem is the degree to which the circulation within the vortex may influence the displacement of the center.

One of the earliest models to provide objective predictions of movement was proposed by H. Riehl in 1956.[6] Riehl considered that the best available index to steering the hurricane is the geostrophic flow of the environment at the level of nondivergence (4–6 km). He computed zonal and meridional components of geostrophic wind from 500-mb analyses using a rectangular grid superimposed on the vortex. These data were used as inputs to a regression based upon historic storm cases to obtain the westward and northward components of displacement for the ensuing 24-hour period. The method worked quite well in a research environment, but operationally it suffered from subjectivity of hurried hand-analyses for the 500-mb surface.

During the late 1950s and early 1960s, the search for methods less sensitive to subjective analysis led to the use of statistical screening procedures to select predictors from surface charts.[7] This produced a model that, although it took no cognizance of upper-level circulation, displayed good skill with westward-moving hurricanes and blazed the trail for the development of a hierarchy of similar but more powerful models for use at the National Hurricane Center.[8] These incorporated the best conceptual features of the Riehl model and the early screening models. The application of these models, none of which drew explicitly upon the results of research on hurricane dynamic and energy processes, was primarily responsible for a significant increase in forecast skills at the National Hurricane Center in the 1960s.[9] The impact of research on hurricane structure and energy processes was felt mainly in the heuristic reasoning applied by forecasters to test the credibility and acceptability of machine-guidance products, including track predictions from various models. However, with ever more inscrutable models being used, the forecaster now finds it more difficult to apply dynamical reasoning to decide which of the model results is more credible and should become the basis for his forecast.

Development. Tropical rain disturbances release abundant supplies of latent heat but develop into dangerous windstorms only when cumulus convective processes succeed in concentrating the air warmed by latent heat together with that warmed anomalously by subsidence (a by-product of convection) into a narrow deep layer of the troposphere. Strong wind shear near the core of the rain disturbance interferes with the deep-layer storage of heat required to generate and maintain the warm core and the surface-pressure gradients that drive the heat engine. The forecaster uses this knowledge diagnostically as a means of distinguishing disturbances that are likely to develop from those that will not, in terms of the amount of shear between lower and upper levels in the disturbance core. The simulation of development is complex because of the need to incorporate, explicitly or by parameterization, the smaller-scale motions that distribute heat generated by the cumuli throughout the warm core.[10]

The explosive development of disturbances and rapid growth of hurricanes into extreme events are currently beyond the reach of operational models and remain unresolved problems. These unresolved problems are the more important because significant changes in strength or size of the hurricane strongly influence the height of peak storm surges, the extent of coastal inundation, and the requirement for evacuation.

Prediction Models

Three classes of models for predicting hurricane movement were in use in 1980: 1) kinematic analog models, 2) dynamic analog models, and 3) pure dynamical models. The first draws upon the climatology of hurricane tracks and of persistence of movement to produce a most-probable displacement of the center. The output is a function of initial position, past movement, and calendar time of occurrence. The computation does not take into account the environment or its influences on the hurricane. The second extracts from historical cases the dynamical properties of the near or the larger-scale environment that correlate with some aspect of hurricane movement. These are combined in a multiple regression statement as analogs to the migration of the vortex. The third class, not concerned with history, combines basic principles of fluid motion, the thermodynamics of an ideal gas, and the application of conservation relationships to predict the behavior and displacement of the hurricane vortex.

The first two classes suffer from incomplete hurricane climatology, especially with regard to the cases with critical changes in movement and strength. Dynamical models encounter at least three kinds of problems. The first, and probably the most important, is that of initialization—the description of the initial state of the atmosphere when computations begin—especially the description of processes in the vigorous inner core of the vortex. The second is the fact that higher (space) resolution is needed to describe what is going on in the inner core than is needed for the large-scale environment. While the environment can usually be described adequately with grid data at 300-km spacings, the active vortex, especially approaching the region of maximum winds, may require a grid spacing of 5–10 km.[11] Since a 5- to 10-km resolution over an entire domain is too costly both in time and resources, some investigators have turned their attention to a system of nested grids of varying mesh length. Work on such a system began in the mid-1970s at the NOAA laboratories at Miami and is being pursued by a number of other research groups. Third, an adequate simulation of the heating function generated by cumulus convection has not been adequately resolved.[12]

Dynamic Analogs. The first completely objective procedure for predicting hurricane movement, using machine analyses of current weather data, was developed for use at the National Hurricane Center in 1964 and is known as NHC-64. This model and its updated successor, NHC-67, employ predictors obtained from analyses of circulations at 1,000, 700, and 500 mb over a large synoptic-scale domain. The predictors are based upon statistical screening of data from historical hurricane cases. The method computes latitudinal and meridional components of motion, then generates storm positions for successive 12-hour intervals up to 72 hours in advance. This method provides more conservative estimates of the poleward component of motion than do other models. For this reason, it was still in use at the National Hurricane Center in the late 1970s and has been adapted for use in the western Pacific Ocean by Japan and the People's Republic of China.

In the late 1960s, the United States Navy developed a statistically constrained dynamical method that returned to the concept of a steering level.[13] This method makes machine analyses at standard pressure surfaces (850, 700, 500, and 300 mb) and then filters out the perturbations with scale sizes of hundreds of kilometers, including the hurricane vortex, which is reduced to a point entity. This point is moved in accordance with the initial large-scale geostrophic flow for 72 to 96 hours. Changes in the large-scale flow and the bias imposed by the geostrophic assumptions are adjusted by subtracting the observed vector error after 12 hours of movement from the computed position for subsequent 12-hour intervals.

Statistical Analogs. In 1967, in response to a requirement of the National Aeronautics and Space Administration for probabilistic information on hurricane movements, an analog model was developed known as HURRAN (Hurricane Analog).[14] The only current information required by this model is the position, direction, and speed of movement of the system for the preceding 12 hours. Drawing upon a 100-year record of hurricane tracks stored on tape, the model computes the most probable track for a 72-hour period based on the movement of historic hurricanes that had occurred at the same time of year and whose positions and movement vectors were similar to the present case. The output is a probabilistic statement of track positions at 12-hour intervals up to 72 hours ahead. For each position, a probability ellipse is defined within which the storm center has a 50% probability of residing at that time period (Figure 138).

The remarkable skill shown by this purely statistical model led to its adoption as an objective tool for identifying coastal sectors placed under a hurricane watch. The hurricane watch is posted 36 hours before landfall for a sector of coastline enclosed between the tangents connecting the extremities of the ellipses for the 24- and 48-hour track positions as the 36-hour position (Figure 138) reaches the coast. This identifies the coastal sector that has at least a 50% probability of a direct hurricane strike within 36 hours.

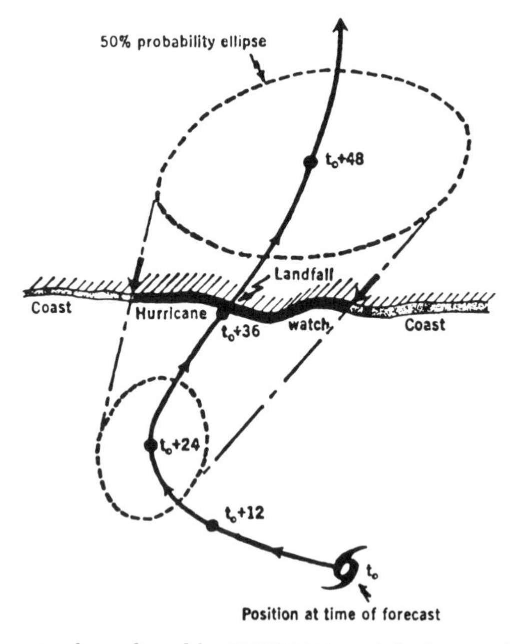

Fig. 138. Hurricane track predicted by HURRAN model, showing how the 50% probability ellipses for 24 hours and 48 hours are used to define the coastal segment in which a hurricane watch is to be posted. R. H. Simpson, "Hurricane Prediction," in *Geophysical Predictions* (Washington, D.C.: National Academy of Science, 1978), 147.

The principal shortcoming of HURRAN and other analog methods is limited usefulness during highly anomalous movements, since too few analogs are available for computing track positions on such occasions. This handicap was alleviated when an auxiliary method known as CLIPER was developed.[15] This method draws its predictors solely from climatology and persistence (of past motion). The output, similar to that of HURRAN, is a probabilistic statement, including a family of probability ellipses. In combination, HURRAN and CLIPER provided more reliable predictions of zonal movement than did the alternative prediction methods, while NHC-67 provided more reliable predictions of meridional movement. The NHC-72 model combined HURRAN, CLIPER, and NHC-67 into a single procedure that accounts for an astonishing amount of the variance in both zonal and meridional components of movement as shown in Figure 139.

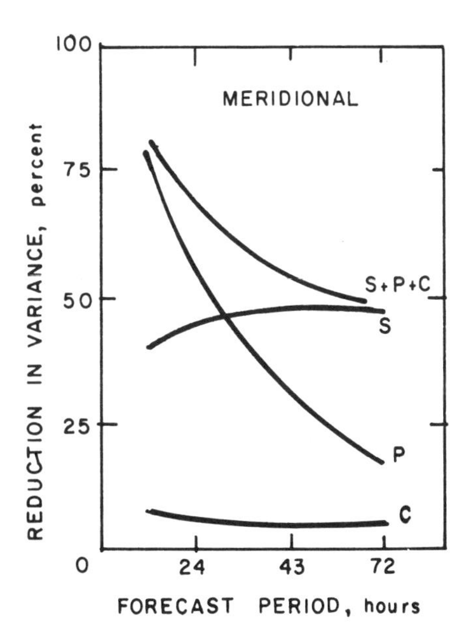

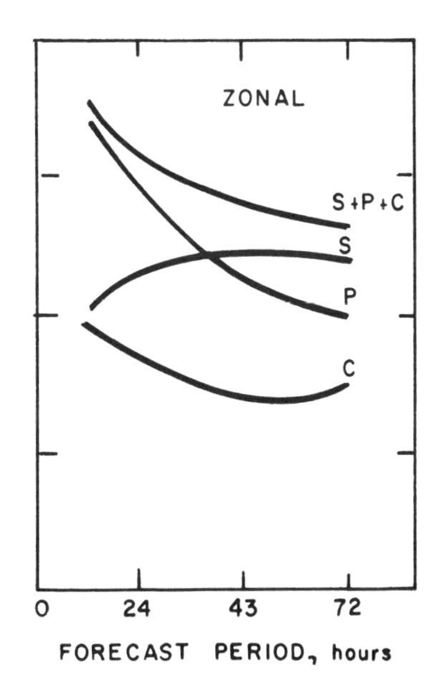

Fig. 139. Reduction in variance of meridional and zonal components of hurricane-movement forecasts due to predictors from *P*, persistence; *C*, climatology; and *S*, circulation dynamics. Adapted from R. H. Simpson, "Hurricane Prediction," in *Geophysical Predictions* (Washington, D.C.: National Academy of Science, 1978), 148.

The most recent improvement in statistical dynamic models has been the incorporation of predictors by Charles Neumann and Miles Lawrence from prognostic charts of the 500-mb circulation.[16] This method is known as NHC-73 (the number referring to the year of implementation).

The United States Navy and some private meteorologists have used similar statistical procedures to answer a different kind of question.[17] Instead of asking What coastal sector is most likely to be struck? this method asks What is the probability that a specific ship haven or shore installation will suffer a direct strike? Both methods provide valuable probability assessments as a basis for operational decisions.

Considering the skills achieved with statistically founded models in probabilistic format, the principal challenge remaining for the more sophisticated dynamical models is to predict anomalous movements, explosive development of the hurricane at sea, and heavy precipitation after a hurricane passes inland.

The first dynamical model to be successfully applied operationally was developed at the Massachusetts Institute of Technology by Fred Sanders and Robert Burpee.[18] This barotropic model (known as SANBAR) computes pressure-weighted mean winds for the layer 1,000 to 100 mb, from which stream functions are generated and used as inputs to the prediction model. A grid resolution of 165 km over a geographically fixed domain is used. In the initialization process, the vortex is replaced with an ideal vortex modified to provide initial steering that is consistent with the observed motion of the system. Curiously, the performance of this model, although competitive with the CLIPER, HURRAN, and NHC-67 models, gave more erratic results for the first 24 hours of prediction but often provided superior results for 48 and 72 hours, except in cases in which the vortex moved into an area where boundary influences adversely affected the results.

The first primitive equation (three-dimensional) model was developed by Stanley Rosenthal and Richard Anthes in 1969.[19] They demonstrated that, while axisymmetric hurricane models are adequate for studying many aspects of hurricane dynamics, a number of significant structural features, including the spiral rainbands, cannot be simulated without accounting for asymmetries in circulation. This work led Anthes to conclude that the growth of asymmetries in hurricanes is correlated with intensification.

The first asymmetric model that used real data for inputs was developed by Banner Miller. Using a 75-km grid resolution, accurate predictions of movement were made for hurricanes Alma (1962) and Celia (1970) and of development in Celia, although the explosive deepening of Celia was not predicted.[20]

Subsequently, investigators succeeded in obtaining useful predictions of movement and at least the right sign of development trends.[21] The principal obstacle to the use of each of these methods operationally has been a combination of the initial-value problem and the vast amounts of computer time required, even for a 24-hour prediction.

In 1973, after a number of other investigators had conducted successful experiments using primitive equation models, the National Weather Service inaugurated a research program to develop an operational three-dimensional hurricane model.[22] The horizontal grid resolution adopted was 60 km over a domain that moved relative to the earth's surface. A unique feature was the initialization procedure that began with the removal of the observed vortex and the generation of a dynamically consistent vortex that was meshed with the environment at a radius of 11° latitude. This presented some problems for westward-moving hurricanes in low latitudes where the artificially spun-up vortex was oversized and the predicted motion was to the right of the observed track. The operational results from this model (NMC—Hovermale) compared favorably with the simple statistical and dynamic analogs after several years of experimental usage, but at the end of the 1970s, its reliability in predicting irregular movements remained uncertain. However, the model did show superiority in predicting the distribution of heavy rains as a hurricane moved inland.

Comparison of Prediction Skills

Nearly all prediction models perform well when persistence is a good predictor and poorly when nonextrapolatable movement or explosive development occurs. Fortunately, a high percentage of forecasts involve regular and persistent movement. However, this makes it more difficult to make a fair assessment of a method's skill in handling irregular movements.

The best comparative test is one that makes separate comparisons for persistent-movement and for irregular-movement cases. Again, since the latter cases constitute only a small percentage of the total cases, the sample size is often too small for the results to be convincing. Table 22 gives some idea of the variation in magnitude of vector errors between regular landfall cases and irregular-movement cases. These are official forecasts for time periods 12, 24, and 48 hours beyond the time of observations upon which predictions were based. During periods when movements may be described as irregular, the error escalates substantially, reflecting the inadequacy of guidance products on such occasions. Table 23 shows the average error for 12-, 24-, and 48-hour movement computed by three statistically founded models and one dynamical model.

Unfortunately such models, like other prediction procedures, accumulate most of their skill and credibility during periods when extrapolation is a good predictor; few perform well when the hurricane displays surprising behavior. None are designed to predict changes in strength except the more powerful dynamical models, and these have never succeeded in anticipating rapid or explosive increases or losses in strength. The anticipation of such changes will probably have to await more effective means for simulating the dynamical processes in the vortex, including such features as the changes in spiral-rainband configuration and asymmetries in the

Table 22. Average Error for the Official Hurricane Movement Forecasts (in kilometers), 1972–1975

Length of forecast (hours)	*All forecasts*			*Landfall approaches*			*Irregular movements*		
	ϵ (*km*)	*Number of cases*	σ_s (*km*)	ϵ (*km*)	*Number of cases*	σ_s (*km*)	ϵ (*km*)	*Number of cases*	σ_s (*km*)
12	89	398	37						
24	179	363	112	171	21	92	267	36	150
48	410	240	143						

NOTE: ϵ = vector error; σ_s = standard deviation

SOURCE: Adapted from R. H. Simpson, "Hurricane Prediction," in *Geophysical Predictions* (Washington, D.C.: National Academy of Science, 1978), 149.

outflow cloudiness, which have been shown to correlate with the onset of intensification.[23]

The Role of Monitoring Systems

To protect against the surprises that most hurricanes hold in store, the forecaster must reach out for every shred of evidence that he can use diagnostically to anticipate critical changes. In times of crisis the effectiveness of hurricane-warning systems through the years has always remained more a function of the quality of monitoring systems than of the prediction methods that guide the judgment of the forecaster. The primary monitoring systems, listed in the order of their probable value to the forecaster in generating coastal warnings, are the reconnaissance aircraft, the weather satellite, radar, observations from ships at sea, and ocean-buoy systems.

Reconnaissance aircraft provide the forecaster with about the only means—diagnostic ones—of directly applying what he has learned from the years of research on hurricane structure and the energy processes that drive it. Because of their limited endurance and range, aircraft cannot effectively monitor changes of the environment or the net interaction between vortex and environment. As of 1980 it was the *only* system capable of directly monitoring the consequences of that interaction as well as the changes due to transports from the environment while tracking the changes in movement of the system.

Table 23. Average Error in Model Computations of Movement (in kilometers), 1972–1975

Length of prediction (hours)	*NHC-67*		*NHC-72*		*HURRAN*		*SANBAR*	
	ϵ *(km)*	*Number of cases*	ϵ *(km)*	*Number of cases*	ϵ *(km)*	*Number of cases*	ϵ *(km)*	*Number of cases*
12	92	250	85	208	84	153	97	136
24	199	216	187	180	221	135	187	119
48	539	150	539	128	489	124	478	85

NOTE: ϵ = vector error

SOURCE: Adapted from R. H. Simpson, "Hurricane Prediction," in *Geophysical Predictions* (Washington, D.C.: National Academy of Science, 1978), 150.

The Synchronous Meteorological Satellite, the most effective facility for monitoring changes in the large-scale environment, is also a dependable means of keeping tabs on the direction and speed of vortex movement. Because it continuously looks down on the same geographical features, its pictures, available as often as five times an hour, can be composited into a time-lapse movie loop from which the movement of systems can be accurately measured. Changes in size and configuration of a hurricane central dense overcast (CDO) provide an empirical means of inferring trends in strength. Someday it may supply a dynamical means for predicting trends in strength as a function of changes occurring in the outflow and in the rainbands. At present, however, the CDO—essentially an exhaust product—blocks the satellite view of the dynamical processes in the energy-releasing layers below. This deficiency has a fair chance of being alleviated in the upcoming new generation of satellites, which will use microwave sounders to sense temperature distributions throughout the clouded hurricane volume. Until then, however, it remains for the reconnaissance aircraft to obtain the dynamical data from the energy-releasing layers, which are important for diagnostic reasoning and analysis and for initializing dynamical models.

Radar can "see" the hurricane only in terms of the precipitation patterns. Nominally, the system can be viewed in sufficient detail to monitor its movement and to draw some inferences as to development when viewed from a maximum distance of about 280 km—the distance at which the beam from a radar at the surface passes through the 6-km level of precipitation. At this distance, a hurricane moving as slowly as 12 kt is within 12 to 13 hours of the coastline. By then, if all is going well, the warning decisions will have been made, the preparedness die cast, and evacuation procedures under way. Thus, radar is a monitoring tool whose primary function is to "fine tune" the approach track, the probable position and strength of the hurricane at landfall, and to provide the basis for deciding adjustments in emergency measures dictated by "eleventh-hour" changes in movement and strength.

The number of observations from ships at sea near the hurricane diminish rapidly following the first hurricane advisory and play only a small role in the predictions upon which coastal warnings are based. Ocean buoys provide only a handful of observations at present; but even if a dense network were in operation, their primary operational impact would very likely be the monitoring of changes in growth potential or actual changes in strength of hurricanes approaching landfall. In view of the limited capabilities of modern prediction methods and the restricted applicability of the observing systems we have discussed, the ability of the forecaster to detect and warn of irregular hurricane tracks and changes of strength will, in the foreseeable future, probably continue to depend mainly upon a steady flow of observations both from reconnaissance aircraft and from the meteorological satellite.

The Outlook for Increasing Forecast Skills

Good progress has been made with dynamical models in simulating hurricane behavior; yet, by comparison with the results

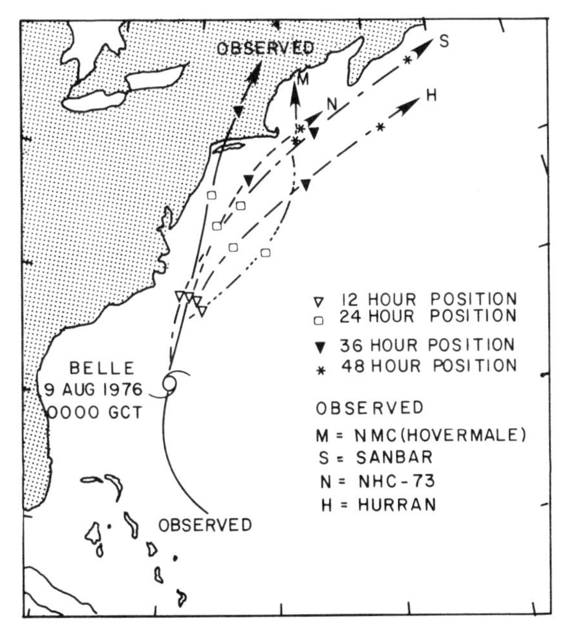

Fig. 140. Observed track of hurricane Belle, 1976, compared with predicted tracks from four operational models supplied to the forecaster to guide his decisions on the need and location of coastal warnings. R. H. Simpson, "Hurricane Prediction," in *Geophysical Predictions* (Washington, D.C.: National Academy of Science, 1978), 145.

achieved with simpler and less costly probabilistic models, the progress is less impressive. Neither, however, lends confidence to the forecaster that he can avoid serious errors in predicting landfall. When he must decide where to post warnings on the basis of the guidance from four independent prediction sources—Figure 140 being a typical example—he faces a scientific dilemma, because he has no direct means of determining which is closest to the truth. His only resort is to fall back on heuristic kinematic reasoning to assure that warning advices do not fail disastrously. One alternative that remains unexploited, however, is to develop and apply more sophisticated procedures for using the abundant data from aircraft, satellites, and radar for diagnostic tests of the credibility of results from deterministic models, and for objective decision making in selecting the model output that will best guide his forecast.[24]

REFERENCES

1. W. M. Frank, "The Structure and Energetics of the Tropical Cyclone: I. Storm Structure, II. Dynamics and Energetics," *Monthly Weather Review*, CV (1977), 1119–50; Harry F. Hawkins and S. M. Imbembo, "Structure of a Small, Intense Hurricane," *Monthly Weather Review*, CIV (1976), 418–42; D. J. Shea and W. M. Gray, "The Hurricane's Inner Core Region: I. Symmetric and Asymmetric Structure; II. Thermal Stability and Dynamic Characteristics," *Journal of Atmospheric Science*, XXX (1973), 1544–76; R. A. Anthes, "Hurricane Model Experiments with a New Cumulus Parameterization Scheme," *Monthly Weather Review*, CV (1977), 287–300; R. H. Simpson, "Hurricane Prediction," in *Geophysical Predictions* (Washington, D.C.: National Academy of Science, 1978), 142–52; Charles Neumann, *A Guide to Atlantic and Eastern Pacific Models for the Prediction of Tropical Cyclones*, Technical Report NWS-NHC-11 (N.p.: Department of Commerce-NOAA, 1978); C. J. Neumann, J. Pelissier, and J. Hovermale, "Can We Expect Significant Improvement in Hurricane Prediction?" Paper Presented at the Twelfth National Conference on Tropical Meteorology and Hurricanes (New Orleans), April, 1979, in press.
2. W. E. Shenk and E. B. Rodgers, "Nimbus 3/ATS 3 Observations of the Evolution of Hurricane Camille," *Journal of Applied Meteorology*, XVII (1978), 458–76; Edward Rodgers *et al.* "The Benefits of Using Short-Interval Satellite Images to Derive Winds for Tropical Cyclones," *Monthly Weather Review*, CVII (1979), 575–84; V. F. Dvorak and S. Wright, "Tropical Cyclone Intensity Analysis Using Enhanced Infrared Satellite Data," *Proceedings of the Eleventh Technical Conference on Hurricanes and Tropical Meteorology* (Boston: American Meteorological Society, 1977), 268–83; Vernon F. Dvorak, "Tropical Cyclone Intensity Analyses and Forecasting from Satellite Imagery," *Monthly Weather Review*, CIII (1975), 420–30.
3. E. Palmén, "Vertical Circulation and Release of Kinetic Energy During the Development of Hurricane Hazel into an Extratropical Storm," *Tellus*, X (1958), 1–23; R. H. Simpson and R. A. Pielke, "Hurricane Development and Movement," *Applied Mechanics Review*, (May, 1976), 601–609.

4. Simpson and Pielke, "Hurricane Development and Movement."
5. G. Norton, "A Soliloquy on Hurricane Forecasting," (Manuscript at National Hurricane Center, Miami, 1949).
6. H. Riehl, W. H. Haggard, and R. W. Sanborn, "On the Prediction of 24-Hour Hurricane Motion," *Journal of Meteorology*, XIII (1956), 415–30.
7. K. W. Veigas, R. G. Miller, and G. H. Howe, "Probabilistic Prediction of Hurricane Movements by Synoptic Climatology," Travelers Weather Research Center's Occasional Papers in Meteorology, II (1959); R. G. Miller, "The Screening Procedure," Travelers Weather Research Center Final Report, Contract No. AF19 (G) 4–1590, (1958), 89–96.
8. B. I. Miller and P. Moore, "A Comparison of Hurricane Steering Levels," *Bulletin of the American Meteorological Society*, XLI (1960), 59–63; B. I. Miller and P. Chase, "Prediction of Hurricane Motion by Statistical Methods," *Monthly Weather Review*, (1966), 399–405.
9. E. Dunn, R. C. Gentry, and B. Lewis, "An Eight-Year Experiment in Improving Forecasts of Hurricane Motion," *Monthly Weather Review*, (1968), 708–13.
10. S. L. Rosenthal, "The Sensitivity of Simulated Hurricane Development to Cumulus Parameterization Details," *Monthly Weather Review*, CVII (1979), 193–97; Shea and Gray, "The Hurricane's Inner-Core Region, II."
11. Robert W. Jones, "Three-dimensional Hurricane Model Experiments with Release of Latent Heat on the Resolvable Scale," Paper Presented at the Twelfth Conference on Hurricanes and Tropical Meteorology (New Orleans) April, 1979, in press.
12. S. L. Rosenthal, "The Sensitivity of Simulated Hurricane Development to Cumulus Parameterization Details."
13. R. Renard *et al.*, "Forecasting the Motion of North Atlantic Tropical Cyclones by the Objective MOHATT Scheme," *Monthly Weather Review*, C (1973), 206–14.
14. J. R. Hope and C. J. Neumann, "An Operational Technique for Relating the Movement of Existing Tropical Cyclones to Past Tracks," *Monthly Weather Review*, XCIII (1970), 925–33.
15. C. J. Neumann, *An Alternate to the HURRAN Tropical Cyclone Forecast System*, Technical Memorandum NWS SR-52 (N.p.: Department of Commerce, 1972).
16. C. J. Neumann and M. Lawrence, "An Experiment in the Statistical-Dynamical Prediction of Tropical Cyclone Motion," *Monthly Weather Review*, CIV (1975), 925–33.
17. Samson Brand, "The Navy Typhoon Haven Program," Paper Presented at the Twelfth Conference on Hurricanes and Tropical Meteorology (New Orleans) April, 1979, in press.
18. R. Sanders and R. Burpee, "Experiments in Barotropic Hurricane Track Forecasting," *Journal of Applied Meteorology*, VI (1968), 313–23.
19. R. A. Anthes, "The Development of Asymmetries in a Three-Dimensional Numerical Model of the Tropical Cyclone," *Monthly Weather Review*, C (1972), 461–76.
20. B. I. Miller, P. O. Chase, and B. R. Jarvinen, "Numerical Prediction of Tropical Weather Systems," *Monthly Weather Review*, C (1972), 825–35.
21. B. F. Ceselski, "Cumulus Convection in Weak and Strong Tropical Disturbances," *Journal of Atmospheric Science*, XXXI (1974), 1241–55; G. W. Leg and R. L. Elsberry, "Forecasts of Typhoon Irma Using a Nested Grid Model," *Monthly Weather Review*, CIV (1976), 1154–63.
22. Anthes, "The Development of Asymmetries in a Three-Dimensional Numerical Model of the Tropical Cyclone"; T. N. Krishnamurti and M. Kanamitsu, "A Study of a Coastal Easterly Wave," *Tellus*, XXV (1973), 568–78; J. B. Hovermale, "First-Season Storm-Movement Characteristics of the NHC Objective Hurricane Forecast Model," *Proceedings of the Twelfth Annual Hurricane Warning and Evaluation Conference (Coral Gables)* (N.p.: Department of Commerce-NOAA, December, 1975).
23. Anthes, "The Development of Asymmetries in a Three-Dimensional Numerical Model of the Tropical Cyclone."

24. J. E. George and W. M. Gray, "Tropical Cyclone Recurvature and Nonrecurvature as Related to Surrounding Wind-Height Fields," *Journal of Applied Meteorology*, XVI (1977), 34–42, and "Tropical Cyclone Motion and Surrounding Parameter Relationships," *Journal of Applied Meteorology*, XV (1976), 1252–64.

CHAPTER 14

Scenarios of Actions and Impacts

Few natural events generate as much sustained and intense drama as the hurricane, but the drama assumes quite different forms depending upon the perspective from which the hurricane impact is viewed—professional concern, concern for the safety of one's family and for the protection of one's property, or, from a distance,

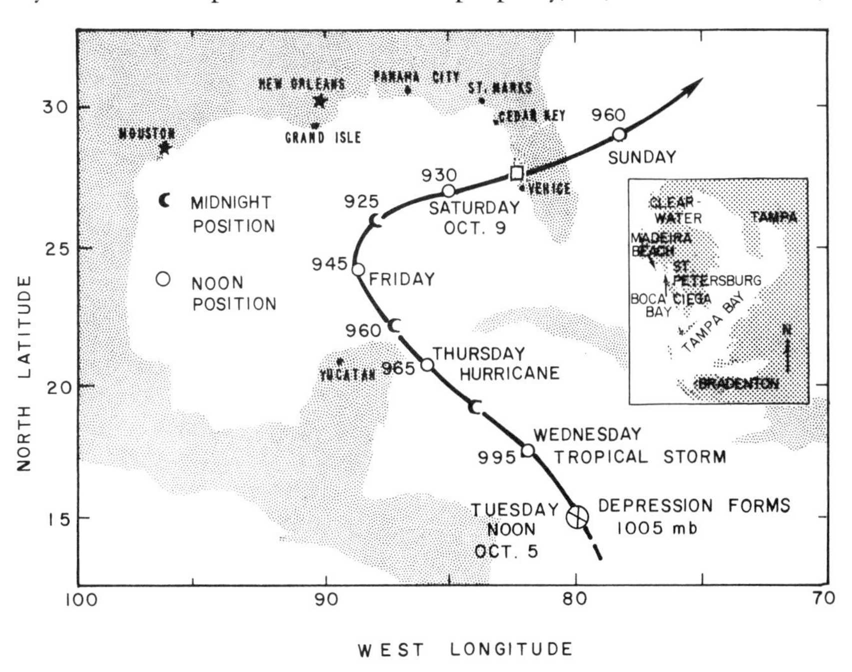

Fig. 141. Track and development history of hurricane Helen, a fictitious October hurricane, which is the basis of the scenarios in Chapter 14.

concern for close friends or relatives who seem to be threatened. This chapter views the drama from several perspectives, not for the sake of drama but as a means of illustrating the potential impact of the hurricane as it may be influenced by one's initial understanding, perception, and interpretation of the event as it unfolds and by one's response to warning advices.

The first scenario reflects the concern of the hurricane forecaster who must correctly identify a landfall and warn the public and, concomitantly, the concern of the emergency-preparedness official and law-enforcement agency where responsibilities reside for planning and for assisting with the evacuation of residents from exposed coastal areas. The second scenario reflects the perceptions and responses of the coastal resident who must make decisions about preparedness actions and the adjustment of his personal plans to be ready for swift action if the hurricane does strike. The third scenario is an example of reaction to a near-miss as a hurricane threatens an island area.

National Hurricane Center
Miami, Florida
Tuesday, 5 October, 11 A.M.

The hurricane specialist has just received a summary message from the reconnaissance aircraft investigating a tropical depression in the western Caribbean Sea. The center was found near 15° N, 80° W, with a low pressure of 1,005 mb. Strongest winds, 33 kt, were located about 60 nmi northeast of the center. A giant canopy of cirrostratus clouds extended as far north as Grand Cayman and Jamaica, and beneath it was a succession of small cumulus cloud lines—a few with heavy rain—spiraling gently toward the depression center. During the 4 hours of reconnaissance in this weather system, the center migrated slowly toward north-northwest (Figure 88).

This organized cloud system, a veritable factory of nearly every cloud type to be found in the international cloud atlas, had formed when the equatorial trough had surged northward across Panama into the Caribbean Sea on the preceding Friday. There a weak low-pressure center had languished without development until the parent trough reformed in its normal position south of the isthmus. By noon on Monday, 4 October, the geostationary satellite had detected the formation of large rainbands extending northward from Panama in a spiral pattern east and north of the pressure center. That afternoon a ship bound from New Orleans to the Canal Zone had encountered torrential rain near the fourteenth parallel, and its barometer had fallen 4 mb in less than 3 hours. Although the strongest sustained winds recorded had been only 25 kt, they had been from the northwest—sufficient evidence for the forecaster to confirm his suspicions that the wind circulation at the surface had systematically closed around the area of falling pressure. The time had come to implement procedures for continuous monitoring and surveillance of a tropical cyclone. By midafternoon, plans were completed to dispatch a weather reconnaissance plane early Tuesday morning (5 October) to investigate the structure and changes occurring in this young cyclone.

As it approached the sprawling rain disturbance on Tuesday morning, the reconnaissance plane radioed a succession of messages to the hurricane center, and finally a summary statement from the airborne meteorologist. This information, together with an hour-by-hour evaluation of changes in the large-scale cloudiness patterns viewed by satellite, was enough to prompt the hurricane specialist on duty to announce to the public and the ships at sea that the depression which had formed over the weekend appeared to be developing into a tropical storm. At 12 noon a hurricane center bulletin with this advice was relayed by cable and radioteletypewriter to all points in the Caribbean and southeastern United States. It stated that the depression, with increasing strength, was moving towards the north-northwest at 8 to 10 kt, and was expected to become a tropical storm and turn northwestward toward the Yucatan Channel within 12 to 24 hours. This indication was based upon a quick appraisal of the probable movement from climatology and of the steering currents in the atmospheric layer surrounding the pressure center. By 4 P.M., the first results from numerical-model computations were transmitted from the National Meteorological Center in Maryland to the hurricane center. These indicated that the incipient storm system should move northwestward at a speed of about 10 kt for the following 48 hours. Two of the four models used in these computations projected a track across the extreme western tip of Cuba; the third, a track through the center of the Yucatan Channel; and the fourth, a track across the northeast tip of Yucatan.

Through the afternoon and evening, evidence from satellite pictures showed that the depression continued to gain strength, and the decision was reached to name the cyclone tropical storm Helen. At 5 A.M., Wednesday, the first official advisory on the eighth tropical storm of the season was released to the public. The time had come for the hurricane center to calculate the risks: first, the rapidity with which Helen might gain hurricane strength; and second, how soon it might affect landmasses where specific early warnings would be needed.

The hurricane specialist noted sharp rises of temperature in the upper troposphere near the storm center and the development of a circulation pattern favorable for whisking the warm outflow from the top of the cloud system north and eastward into colder regions. Looking further afield for evidence of separate sources of energy that could cause explosive increases in Helen's wind strength—the kind which may convert a young storm into a superhurricane in less than a day—the hurricane specialist found none. He decided that indications supported the conclusion that Helen would grow at a steady pace and reach hurricane force within 24 hours or less, a little faster than normal for a storm in this location at this time of year. The closest vulnerable point in the United States was Key West, more than 450 nmi to the north and nearly 2 days' travel time. Moreover, Helen was moving toward the Yucatan Channel on a course that, projected, would extend to the Louisiana coastline in 4 to 5 days. At this point, Helen had the earmarks of a fairly well-behaved hurricane that would not be difficult to predict. Ample time should be available to warn coastal areas.

Later Wednesday morning another aircraft reconnaissance confirmed that Helen was on track but with minimum pressure, 995 mb, falling steadily, and with maximum winds of 50 kt. Satellite pictures had shown some contraction of the large shield of dense overcast that covered the storm center; yet there was no visual evidence of the formation of an eye. However, on the last penetration by the reconnaissance aircraft at 3 P.M., sustained winds had increased to a maximum of 60 kt. Faced with these trends, the hurricane specialist agreed that Helen would undoubtedly reach full hurricane strength—64 kt—by midnight. Anticipating this, the 6 P.M. advisory stated that Helen was of hurricane strength and would threaten extreme western Cuba and northeastern Yucatan with rising tides and dangerous winds by midday Thursday.

By early Wednesday evening infrared pictures from the weather satellite revealed important changes. There was further contraction of the large cirrus cloud shield. Near the center, which earlier had been obscured by high clouds, a distinct, small, nearly cloudless eye appeared. The satellite analyst advised the forecaster that the changes in cloud patterns indicated that the central pressure was falling rapidly. This conclusion was confirmed by the reconnaissance aircraft on its first penetration Thursday morning, when a low pressure of 965 mb was recorded together with winds of nearly 90 kt. Nevertheless, Helen continued on track near 20° N and 85° W moving to the northwest at a speed of 12 kt.

On Thursday, the results of computations from the four numerical prediction models became available to the forecaster shortly before noon. They made his day less pleasant! The first of the models projected a track at about the same speed northwestward to the central Gulf of Mexico but showed the center turning toward the central Texas coast after 30 hours. The second two projected a track towards landfall on the Louisiana coast 2 days later. The fourth called for recurvature to the northeast after 24 hours, with landfall Saturday near Panama City, Florida. With this range of potential landfalls proposed by his guidance materials, the forecaster faced the task of deciding, on the basis of diagnostic reasoning, which of the prediction models was closest to the truth. However, there were still at least 12 hours before he would have to decide the area he must alert by posting a hurricane watch. Additional information would be available by that time.

Just after noon on Thursday, the hurricane specialist noted an unusual development appearing in the upper troposphere and still another on the 500-mb chart (6-km level), both in the broad belt of westerlies over the United States. At 500 mb, pressures were rising from Oklahoma to the Great Lakes, and a trough over the eastern seaboard was moving out to sea. At 200 mb, the corresponding trough had left behind a circulation center in a pool of cool air centered over Mississippi and Alabama. Again, at 500 mb, in the Pacific Northwest, a very pronounced trough of low pressure was moving inland from the Pacific, encroaching upon a ridge of high pressure extending from Arizona to Minnesota. The hemispheric prediction model indicated that the west coast trough would move eastward but that most of its energy would remain near the Canadian border, first passing north of the ridge and then cascading cool air southward and bringing strong west winds as far as Georgia and Alabama within 48 hours.

The steering flow in the immediate environment of Helen porvided no hint of a decided change in the projected track, nor did the short-period numerical prediction for the southeastern United States. It was clear, however, that steering currents would weaken as Helen approached the central Gulf of Mexico and that her future movement would be uncertain.

Faced with this dilemma, the hurricane specialist concluded that, unless a different trend showed up within his 12 hours of grace, he would resist posting a hurricane watch for the Texas and west Louisiana coasts and place his bets on a landfall somewhere east of New Orleans. This was based on his conviction that the trough in the Pacific northwest would extend its influence eastward rapidly enough to cascade energy down the eastern side of the high-pressure ridge and bring west or southwest winds to the northern Gulf of Mexico in time to divert Helen to a north or northeastward track before it could make a landfall. Having reached this decision, he still worried about what would happen in the meanwhile to that pool of cool air at 200 mb cut off from its source and languishing over the central Gulf states.

As Helen approached the Gulf of Mexico, long-distance telephone calls from many official sources claimed much of the time of the hurricane center's director. After initiating calls of his own to national organizations with responsibilities for hurricane preparedness, and conferring with weather-service directors in Cuba and Mexico about the more immediate threat to their respective coasts, he responded to official inquiries from governors' offices in three Gulf coast states and from the mayor of Merida. He even made a brief live comment for use by a television station in faraway Sidney, Australia, where international distribution of advices on Helen had generated great curiosity as to the portion of the United States most likely to experience disaster.

At dusk on Thursday, the eye of Helen passed over the extreme tip of Yucatan and moved into the Gulf of Mexico; the reconnaissance aircraft reported a central pressure of 960 mb with maximum winds of 95 kt. The 6 P.M. advisory stated that Helen continued to move northwestward at 12 kt with little change in direction or speed expected within 18 to 24 hours, even though further strengthening could be expected. By midafternoon the computations of movement by the four prediction models had arrived; one had predicted a landfall near Grand Isle, Louisiana; the others clustered landfalls between Pensacola and Apalachicola, Florida. Strangely, however, the one model that had earlier called for a landfall near Panama City now predicted the landfall at Grand Isle.

By this time, Helen was the subject of constant comment and speculation by coastal residents from Florida to Texas. Few believed, however, that Helen would ever become a direct menace to *their* localities. Shortly after midnight on Thursday, (7 October), two developments impressed the hurricane specialist enough that he called the director of the center to pass along the information: First, the latest 200-mb chart showed that the pool of cool air that had settled down over Mississippi and Louisiana had begun to sink and meld with its environment, a clear signal that fresh kinetic energy was being released. This could effect the development of Helen. Second, on the 500-mb chart the trough in the Pacific Northwest had advanced across Canada and was approaching the Great Lakes. Clearly the stage was set for it to cascade its energy southward and with it the west winds that would surely recurve Helen to the northeast probably earlier than appeared likely the day before.

At 3 A.M. Friday, the four prediction models showed little change in the movement predicted 12 hours earlier; however, one model called for a landfall as far east as Cedar Key, Florida. In a special advisory at 5 A.M., the hurricane center issued its first hurricane watch for the coastal area extending from Grand Isle, Louisiana, to Saint Marks, Florida. Helen was designated a dangerous hurricane threatening further increases in strength. The bulletin also stated that she would probably reach landfall Saturday afternoon or evening.

At 7 A.M., a terse message from the reconnaissance aircraft reported that pressures in Helen had dropped to 945 mb with maximum winds of 120 kt. The movement, though still towards the northwest, had slowed to 8 kt. Preparedness organizations in the four states implicated in the watch area moved into action, readying hurricane shelters and checking supplies, procedures, evacuation plans, and other emergency measures that would be needed when the precise area of warning was posted. The pace of activities quickened at the hurricane center: schedules were revised to provide additional staff around the clock and to supply live telecasts from the center at Miami as Helen approached landfall. The weather satellite revealed a continuation of the strengthening trends, confirming the worst fears of the preceding night when the disappearance of the pool of cool air at 200 mb was first noted. Helen might well become a superhurricane!

At 9 A.M. Friday, the center director initiated several calls—to his Washington headquarters, to the Red Cross, to several national agencies concerned with preparedness, and to the Florida governor's office. Each was told that Helen would probably make landfall as a superhurricane and would pose extraordinary requirements both for preparations and subsequently for relief and rehabilitation. The Red Cross was advised that Atlanta would be the closest safe depot for stockpiling disaster relief supplies already en route to the southeastern United States.

By late afternoon on Friday, time-lapse movies from successive satellite pictures gave the first indication that Helen was turning as predicted to a northward track. By 9 P.M. it was obvious that the hurricane had continued turning to a north-northeast heading and was moving faster. At 500 mb, the trough at higher latitudes had extended its influence to the north-central Gulf of Mexico, and winds were southwesterly as far south as New Orleans and Pensacola.

The forecaster could now be virtually certain that Louisiana, Mississippi, and Alabama would not feel the fury of Helen. But how much of the west-central coast of Florida would be affected? Specific hurricane warnings had to be posted for some of the watch area, and soon! Within less than 24 hours, Helen would find her mark and be ashore!

In a special advisory at 10 P.M., the changing course of Helen was announced; the hurricane watch was extended southward on the west Florida coast to Fort Myers. It was indicated that hurricane warnings would be posted for some sections of the Florida coast before daybreak and that Helen would reach the coast by midnight Saturday. Shortly before 11 P.M., the movement computations from the four models were transmitted from the central computer in Maryland. The results once again were discomfiting. The model that had last predicted a landfall at Grand Isle was now identifying a landfall near Sarasota, Florida. Two others selected Panama City, Florida; and the fourth, Saint Marks. By this time, however, it was clear that the southward penetration of the 500-mb trough had occurred in greater force than expected; Helen was now certain to turn farther to the east than expected and accelerate toward the coast. The instantaneous steering of the deep-layer mean-flow chart implied a landfall between Saint Marks and Cedar Key, Florida.

The chips were down; the hurricane specialist must decide on the location and extent of hurricane warnings. The consensus of the hurricane staff was that the greatest threat lay between the Tampa Bay area and Saint Marks, Florida. But should the initial warning include Tampa Bay, which would need to evacuate more than 100,000 people if a hurricane emergency were declared? The course of least regret demanded that this populous area be included in the hurricane warning area; if it were not included, and it then developed that warnings were indeed needed with less than 6 hours left for emergency action, evacuation could not be completed and many lives would be lost.

In the midst of the decision crisis, a brief message from the reconnaissance plane, which had reached the storm just before midnight, reported that pressures in Helen had dropped to 925 mb and sustained winds were up to 140 kt. Helen's center at the time of penetration had been moving on a heading of 40° but appeared to be continuing its turn toward the east.

At 3 A.M., hurricane warnings were posted from Saint Marks to Sarasota, Florida, with gale warnings westward to Panama City and southward to Naples. By daybreak, the Red Cross and civil preparedness agencies had opened hurricane shelters in the areas where hurricane warnings were displayed. Repeated announcements by television and radio stations urged coastal residents in low-lying areas to select a safe shelter inland and move there as quickly as possible.

Throughout the night infrared pictures from the weather satellite confirmed Helen's further turn eastward, and shortly after daybreak its projected course indicated a landfall near Clearwater, Florida. The 6 A.M. advisory extended hurricane warnings southward to Fort Myers and gale warnings to Key West. Hurricane tides in Tampa Bay were predicted to reach a peak height of 4.5 to 5.5 m with 3 to 3.5 m in the Sarasota area.

The noon advisory from the hurricane center billed Helen as a superhurricane with a damage potential approaching five on a scale of one to five. There was no precedent in history for such a hurricane affecting this coastal area. All residents living in low-lying areas within 2 km of the coast were advised to seek safe shelter before rising tides and high winds cut off escape routes.

The die was cast now. Radar would track the center inland, the satellite would note significant changes in the cloudiness character that might indicate further alterations in strength, and reconnaissance aircraft would fly continually back and forth through the center as Helen approached the coast. However, little time was left for emergency actions not already underway.

By midafternoon, waves were splashing across, and partially inundating, some sections of all the causeways in the Tampa Bay area. Evacuation by these routes could no longer proceed safely. By 10 P.M., 2 hours before Helen reached the coast at Clearwater, flooding was affecting many areas of Clearwater and Saint Petersburg beach; portions of Sarasota were isolated by rising tides. However, as Helen swept ever faster toward the shore, central pressures slowly rose and the lowest pressure measured when the center reached the coast just before midnight was only 935 mb. Maximum winds at the open beach were 120 kt, diminishing inland to no more than 100 kt within several kilometers of the shoreline.

Because no doubt had been left in the minds of coastal residents that Helen was a superhurricane posing a greater threat than any previous hurricane in that area, because more than 12 daylight hours of hurricane warning had been given, and because residents had responded to the pleas of evacuation officials to leave exposed areas and seek safe shelter, the death toll was less than one hundred. In postanalyses of the events, however, it became clear that had the hurricane forecaster not selected the course of least regret and extended hurricane warnings from the very first to include all the Tampa Bay area, insufficient time would have remained to evacuate all who needed to leave. The death toll, it was concluded, could have been an order of magnitude higher.

Saint Petersburg, Florida
Tuesday, 5 October, 12 noon

Jerry Jensen, a successful businessman, a manufacturer of marine electronics, and a yachtsman, kept his 40-foot sloop at a private dock behind his sumptuous residence located just off Boca Ciega Bay a short distance inland from the Gulf of Mexico. Jerry was en route to a luncheon appointment with his business colleague and fellow yachtsman, Ben Thomas, to coordinate plans for a long weekend aboard their yachts. They had planned a rendezvous with other yachtsmen at Venice, Florida. As he pulled into the parking lot at Saint Petersburg beach at noon, he listened on his car radio to an announcement from the National Hurricane Center that a depression which had formed in the southwest Caribbean Sea the past weekend was gaining strength and that conditions favored further development to tropical storm strength within the next 24 hours. The depression was on a northerly course at 10 kt.

In the Tampa Bay area it was a beautiful day, warm and sunny with no hint that fall was at hand except that the air felt a bit drier than it had in August. At lunch, Jerry and Ben discussed the effect that the depression might have on their weekend. However, since an October hurricane had not affected the Gulf of Mexico for more than 10 years and the southwestern coast of Florida had been free of a hurricane strike since Donna in 1960, they both agreed that their plans should remain intact while they kept a careful watch on developments. They planned to sail nearshore in the Gulf of Mexico to Venice on Thursday and return home on Sunday. They agreed that, if necessary, they would delay their departure until Friday morning. Back at work, both men called home and advised their families that the depression might cause some delay in departure but that the trip was still on.

Wednesday morning, 6 October, Jerry awoke to still another magnificent, clear morning with gentle winds out of the east-northeast. However, when he turned on the television for the morning news, the local meteorologist was in the process of reading the first advisory on tropical storm Helen, which had been christened at 5 A.M. Helen was moving northwestward at 10 kt towards the Yucatan Channel with conditions favoring still further increases in strength.

After breakfast Jerry plotted the successive positions of the depression given in earlier bulletins and, projecting the track, it appeared to him that the storm—if it remained on the same course—should reach the Louisiana coast. However, he recalled from a meteorology course, which he had taken from the Coast Guard Auxiliary, that October hurricanes moving out of the Caribbean into the Gulf of Mexico usually turn from northwest to a north or northeasterly track before they reach landfall on the Gulf coast. He would have to watch this storm very carefully before deciding to go ahead with the weekend plans.

Several times during the day on Wednesday, Jerry tuned in the continuous weather broadcast of the National Oceanic and Atmospheric Adminstration (NOAA) to check on the latest bulletin concerning the movement of Helen. The noon television news summarized the developments and reported that reconnaissance aircraft that morning had measured a central pressure of 995 mb with maximum sustained winds of 50 kt. Helen was moving on a steady northwesterly course still at 10 kt and with no change in speed or direction anticipated during the next 24 hours, but with the likelihood she might reach hurricane strength within that period.

At Saint Petersburg, the sun set over a placid, undisturbed Gulf of Mexico with northeasterly winds of about 15 kt and a nearly cloudless sky except for high cirrus racing northward at a good clip. What would the decision be? Should Jerry leave on schedule with his family Thursday morning or play the waiting game a little longer? Helen continued to gain strength but still seemed to be heading for the west coast of Louisiana, and the hurricane center foresaw no change in speed or direction. Even if she changed course and headed directly for Florida, Helen was at least 48 hours away; and, if worse came to worst, he could put in at Sarasota and secure his sloop in the narrow inland waterway. After Jerry had consulted at length with Ben Thomas, both families agreed that the course of least regret would be to delay further for more assurance that Helen would not recurve and threaten their trip—if not their residences on Boca Ciega Bay.

Thursday morning, a thin veil of cirrostratus covered the sky in the Tampa Bay area, but the wind remained gentle and northeasterly; the Gulf of Mexico near the shore was virtually unruffled. Each hour, local television stations as well as NOAA's continuous weather broadcast updated the progress of Helen, which had reached hurricane force Wednesday evening.

By midmorning, the thin cirrostratus observed at dawn had disappeared, the sky over the Tampa Bay area was brilliantly blue, and the wind was unchanged, although a few puffs of small cumulus clouds in narrow lines had appeared over the water.

The news broadcaster at noon on Thursday read the latest advisory, which announced that the pressures in Helen had dropped to 965 mb and maximum winds had increased to nearly 90 kt. However, she was still on course, moving further to the west than indicated earlier, and was predicted to reach the tip of Yucatan by dusk that day. No change in direction was indicated, but some increase in speed of movement was expected during the next 24 hours. This would bring Helen to the center of the Gulf of Mexico on a northwesterly track by noon Friday.

After plotting and extrapolating the straight track that had brought Helen toward the tip of Yucatan, Jerry and Ben decided that, unless there were changes in Helen's track, or an unusual development during the night, they would cast off for Venice at daybreak Friday. Checks were made of the NOAA continuous weather broadcasts periodically. At 11 P.M. a bulletin confirmed that Helen had cleared Yucatan and was in the Gulf of Mexico with the same course and speed and very little change in central pressure and maximum winds.

At 5 A.M. Friday there had been no rain or threat of rain in the Clearwater area, and while the stars were obscured by a high layer of cirrus, the lower layers contained only small stratocumulus clouds moving more rapidly now from the east. However, surface winds were holding at 12–15 kt, occasionally rising to 18 kt. The NOAA radio indicated no new developments in Helen. So at daybreak, the Jensen family boarded their sloop, which had been provisioned for the trip the day before, and cast off from their dock. By VHF radio, Jerry raised Ben Thomas and confirmed that he was also en route for their rendezvous at John's Pass. Even as the sloop moved into the open bay approaching the pass, the wind veered to the southeast as it gusted up to 25 kt. By the time the rendezvous had been made with Ben Thomas, and their mainsails and working jibs were raised, the wind was blowing a steady 18 to 20 kt, and there was a heavy chop to the water. Clearly this was going to be a nasty beat to windward all the way to Venice, especially over an open sea, where the winds could be expected to be higher than over the bay. They would be taking green water over their bows. In fact, if the wind veered further, it might be necessary to turn seaward and then tack once or twice before reaching Venice. As they moved out into the open Gulf of Mexico under power and sail and headed south, a long row of small cumulus clouds, some precipitating, could be seen in the distance trailing off to the northwest. No other rain was in sight, although an overcast of cirrostratus clouds almost obscured the sun's disk. After an hour at sea, the wind, which remained east-southeasterly, had increased to a steady 25 kt with gusts to 35 kt, necessitating a reef in the mainsail. After 2 hours, a double check of their navigation revealed that they had averaged only 4 kt against the heavy seas since entering the open Gulf of Mexico. At this rate, Venice seemed very far away. By ten o'clock, the two sloops picked up a special bulletin broadcast from a Tampa commercial station which quoted from a National Hurricane Center advisory that pressures appeared to be falling rapidly in hurricane Helen and winds were in excess of 100 kt. While apparently still moving on a northwestward course, there had been a slight turn to the right during the last 6 hours. The NOAA continuous broadcast was becoming more difficult to receive now and information on the progress of Helen was obtained mainly from commercial radio stations.

By this time, both sloops had been forced to divert to a more seaward course in order to make efficient use of their sails. This had increased the boat speed from 4 to 6 ½ kt, but after 45 minutes the seas had increased to more than 2 m and winds were sustained at 30 kt with occasional gusts to 40 kt.

At noon a Sarasota radio broadcasted an advisory from the hurricane center which contained the alarming news that Helen was becoming a superhurricane; central pressure had dropped to 945 mb and she was slowly turning to a more northerly heading. After both Jerry and Ben received this message, it took only a brief ship-to-ship radio conversation to decide that they had had enough; they must now return home immediately.

Neither boat was in immediate danger and both skippers were competent. The increase in wind speed with distance seaward was alarming, and two long tacks landward would be required to reach their destination. This was not an inviting outlook, for it meant they would arrive at Venice long after dark. Moreover, with the changes in the course and strength of Helen, all bets were off as to the location of her ultimate landfall. The hurricane watch posted for the coastal area from Grand Isle, Louisiana to Saint Marks, Florida, was not unduly alarming, but with no more mobility than that afforded by sails on a 40-foot sloop, the better part of wisdom was to take no chances. Besides at a time like this it *had* to be better ashore than afloat!

En route back to their Boca Ciega Bay homes—a much more comfortable journey than the one outbound—they were overtaken by a northward moving band of innocuous-appearing cumulus clouds that brought with it a surprisingly heavy burst of rain and wind that briefly gusted above 40 kt. Arriving home shortly after 5 P.M., they found that winds in the protected area of their dock—only about 2 km inland from the open coast—seemed almost calm in comparison to the beating they had taken by winds and seas in the open ocean.

The 6 P.M. advisory on Friday stressed the need for alertness and readiness for fast action in view of the uncertainty of Helen's landfall position. Jerry checked the small barograph he always kept as a showpiece in his home. It was rarely useful in helping him understand or predict the weather, since often the more severe and rapid changes occurred without significant changes in the pressure trace on the barograph sheet. Only in higher latitudes are barometers, in the absence of other weather data, useful in anticipating the nature of weather changes. This time, however, he noted that the diurnal changes, which nearly always superimpose a wavy trace with two rises and falls each day, had changed little for nearly 24 hours but had now begun a steady slow fall.

Jerry had resolved to watch the weather at least until midnight, feeling ill at ease about the changes and uncertainties in Helen's track and the fact that she had developed into a superhurricane. He was aware that in October and November hurricanes could change course very rapidly and, upon turning eastward, could double or triple their speed in only a few hours, leaving less time for preparation than would have been available at speeds on the previous heading.

The 11 P.M. news carried a special announcement from the hurricane center stating that Helen was turning further eastward, pressures in the hurricane were still falling, and the greatest danger to coastal areas had shifted eastward. At 12:30 A.M., Jerry retired realizing that he must catch some sleep before the difficult task of preparing for whatever Helen might bring on Saturday.

At 5 A.M., when Jerry's alarm went off, all seemed peaceful and there was no evidence of rain showers during the night. However, when he went outdoors to survey the situation, he could see that his yacht was riding at his dock more than half a meter higher than normal. There was a heavy overcast of high clouds but no sign of rain in the immediate vicinity. The wind was south-southeast at 12 to 15 kt at his residence. He decided to drive over to have a look at the ocean. As he approached, he could see the surf was very heavy at the beach, and winds were much stronger there than they had been at his home. In the distance offshore, a long line of tall cumulus clouds extended from southeast to northwest and was approaching the shore. The surf was running up on the beach, and erosion had already begun in a few spots. Tides at the open beach were running nearly a meter above normal. On the way back home, an earlier announcement was repeated on his car radio that the Tampa Bay area was under a hurricane warning and that Helen was expected to make landfall a short distance north of the Tampa Bay area Saturday evening. Evacuation of beach and exposed areas had been advised, and all residents in the area of hurricane warnings were urged to complete their preparations for the worst hurricane ever to attack west-central Florida.

Jerry examined his barograph once more. The trace, which had continued to show slowly falling pressure through the night, had leveled off and was steady now. He realized later that he was experiencing what is known as the "bar of the storm," a leveling-off of pressure before the arrival of the rapid pressure falls that sustain damaging winds.

Just before 7 A.M., the first of a succession of spiral rainbands—the one Jerry had seen just offshore—passed over Saint Petersburg beach, bringing torrential rain and wind gusts to 40 kt. This was but a very brief sample of what was to come at more frequent intervals and more severely as the day went on. Jerry described to the family his plans to prepare for Helen, and individual responsibilities were assigned to each one. Because peak hurricane tides were expected to be more than 5 m, it was obvious they could not remain at home. With this kind of flooding the living room of their house, at a grade level of 3 m, would have more than 1½ m of water in it. If tides were higher, the chances of personal survival would rapidly diminish, even if the house itself survived the extreme winds and the battering of waves and debris. The most important decision they had to make was where they would go to find a safe shelter and what escape route they would use once they had secured the house and taken precautions to protect irreplaceable items.

By this time, the television and radio stations were warning that tides could be expected to rise rapidly enough to jeopardize many escape routes by midafternoon and urged that everyone planning to cross the causeways of Tampa Bay do so before noon. At 10 A.M., Jerry decided to call Ben Thomas to compare notes on evacuation plans and the selection of an appropriate shelter. Radios had been announcing the locations of Red Cross shelters; television stations were displaying maps giving the location of recommended shelters and escape routes; some high-rise structures not far from the beach were listed as safe shelters. Following a suggestion generated by the hurricane-preparedness committee of which he was a member, Ben proposed that both families plan to seek safe shelter in a high-rise office building nearer home. He pointed out that the new, well-constructed 8-story building on Madeira Beach where he had an office on the sixth floor had been certified for use in vertical evacuations during hurricane emergencies. Both families decided to take advantage of this alternative. So at 2 P.M. the Jensens packed food and emergency supplies into their auto and drove to the new Marston Building to join the Thomas family. Before closing his residence door, Jerry took a final look at his barograph, which had experienced a number of rapid rises and falls as the heavy rainbands passed overhead. However, it now indicated a decided fall, and a freshening of sustained winds, now more than 25 kt, was the obvious response.

From its location two blocks inland Ben's business suite offered a view northwestward out over the Gulf of Mexico. The die had now been cast; both families had done everything possible to prepare for the worst that Helen might bring. It turned out to be far worse, however, than they had believed possible. During the next 10 hours as Helen approached shore, a succession of rainbands moved inland over Madeira Beach, each one with stronger gusting winds and heavier gushing rain than the last, each followed by sustained winds of higher speed. The wind remained from a southeasterly direction almost parallel to the coastline until after dark. By 5 P.M., several of the low-lying streets near the oceanfront were awash with a combination of sea and rainwater. At dusk, the rain was heavy and continuous. The wind, sustained at 45 kt, had superimposed upon it a succession of long-period gusts in which speeds first increased to 65 or 70 kt, then suddenly dropped to what seemed to be a near-calm as the influence of the gust disappeared and winds gradually returned to their sustained values. By ten o'clock, the roar of the wind had become a high-pitched screech, and it was no longer safe to be near any exterior window. The two families moved out of the business suite into interior hallways for further protection against a possible loss of glass openings on the exposed side.

Just before midnight, Helen made her landfall just north of Clearwater, and Madeira Beach found itself in the southern part of the hurricane eye. The sudden cessation of the wind screech and violent shaking of the building provided such a contrast that there seemed to be a dead calm—an utter silence—though this was a dangerous illusion. Cautiously moving to one of the windows, Jerry could see a few stars overhead, but six floors below, every street was inundated to a depth that could not be estimated. A few structures had come off their foundations and were floating in the swirling waters below. There was unbelievable devastation everywhere. The roofs of several weaker structures had been blown off and the walls had partially collapsed.

At 12:30 A.M. the uneasy quiet was broken; and with the sensation one might expect with the approach and passage of a squadron of jet transport aircraft a few hundred feet overhead, the reality of devastating winds from a superhurricane returned, this time advancing from a northwesterly direction. During the remainder of Saturday night, Helen swept across the Florida peninsula, devastating the most valuable citrus crop in the nation. By dawn on Sunday morning, however, Helen had moved out to sea, tidal waters were rapidly receding, and the full impact of the disaster was apparent. Jerry found that his car had been washed more than a block away, where it nestled against the wall of a masonry building. Of course it could no longer provide him transportation. As he walked back to his residence several kilometers away, carefully avoiding downed electrical wires and hazards of many kinds, he was appalled at the extent of destruction and inundation. His home had remained intact, but rising tides had left their watermarks head-high on the living room walls, and his barometer, left on a high closet shelf, had gone completely off scale as the lowest pressures had arrived. Had he remembered to reset the scale 50 mb higher before he left home, he would have had a valuable recording of the hurricane pressure trace of the worst hurricane ever to hit the Tampa Bay area.

The yacht, which he had carefully bridled to the tall pilings at his dock, protecting the gunwales by rows of old automobile tires lashed to the lifelines, had finally been torn free of its moorings when the waters began surging back out to sea. It had been swept over the seawall and deposited in the patio, the stern floating placidly in the deep end of his swimming pool, the keel resting against the opposite end of the pool, and the stem jutting skyward, an incongruous gesture of survival.

The Jensens and the Thomases had called their shots closely and avoided personal injury, but they suffered severe, unavoidable losses of personal property. The losses were largely offset by the combination of flood and windstorm insurance they had maintained, though with no cause to file a single claim in the previous 10 years of residency on Boca Ciega Bay.

San Juan, Puerto Rico
September 1945

On 13 September 1928, a hurricane of extreme intensity passed squarely over the whole island of Puerto Rico. This date, the feast day of San Felipe, has never been forgotten by the population of the island. The palm tree in Figure 142, surviving but pierced by a wooden board, gives a vivid picture of the wind force on that day. Small wonder, therefore, that nervousness rose to a fever pitch when, on 12 September 1945, a hurricane was located northeast of the Lesser Antilles and moving straight west. This event happened long before the days of radar and satellite surveillance.

In those days, the Institute of Tropical Meteorology, founded and maintained by the University of Chicago on the campus of the

Fig. 142. Royal palm pierced by pine board (10 feet × 3 inches × 1 inch) during San Felipe hurricane at Puerto Rico, 13 September 1928. From Ivan Ray Tannehill, *Hurricanes: Their Nature and History*, 9th rev. ed. (copyright © 1956 by Princeton University Press), Fig. 66, p. 126; reprinted by permission of Princeton University Press.

University of Puerto Rico in the San Juan suburb of Rio Piedras, was operating in full swing. There was much lecturing on hurricanes, but no one present in 1945 had ever experienced one. Hence, curiosity was at a pinnacle at the institute.

It was known, of course, that hurricanes are strongest on the right side looking down its track in the direction of motion (not wind) and that, from all previous experience, this side would not come to Puerto Rico from a hurricane already situated at 19° N when crossing longitude 60° W. But the hurricane, whose intensity was not well known, appeared to be moving west and, for some hours, even south of west, making one of those curious short-period oscillations shown in Figure 51. Thus it was likely that gale winds and potential flooding would reach the north coast. Except for the storm's coming closest to Puerto Rico on the most fateful of all days in the year, the day of San Felipe, everything might have progressed calmly. Businesses closed during the afternoon of 12 September and everyone went home for hurricane preparation; the hammering of general boarding-up was heard all over the San Juan area.

Since extrapolation placed the hurricane center north of the island during the night and morning of 13 September, a group of us went down in late afternoon to spend the night at the weather bureau, high on the shore of San Juan and just east of the main part of the city. We settled in with great expectations. The wind started to blow from the west and northwest with increasing force, the sky was overcast, and the barometer fell slowly. In the late evening, 25- to 30-kt gusts were recorded, but then a curious thing happened. The wind died down completely, only to resume blowing from the south to southwest during the time the center was still estimated to be well to the northeast.

We felt ourselves now to be entering the potentially most active part of the hurricane's southern side. However, the barometer started rising, and during the next several hours, winds remained a gentle 15 to 20 kt. It gradually dawned on us that we had seen it all. Various members of the group started to drift on home, with the last of the hard-core optimists leaving about three in the morning.

Since the cyclonic circulation was completely closed, with west winds of 20 kt on the coast, the center could not have been more than about 80 km north of the island. The next morning Air Force hurricane reconnaissance found 100-kt winds on the northern side. It was clear by then that we had experienced the immature stage, with a very large asymmetry between the northern and southern sides.

The hurricane went on to score a bulls'-eye on southeastern Florida on 15–16 September, causing great damage. Although 13 September 1945 has been forgotten in Puerto Rico, the memory called San Felipe lives on.

CHAPTER 15

Direct Reduction of the Hurricane Threat

The increasing population-at-risk and annual property losses from hurricanes—more than $800 million as of 1978—should engender a sense of urgency in the search for a means of mitigating the destructive potential of hurricanes. Indeed, a report in June 1978 to the Congress of the United States by the National Weather Modification Advisory Board documented the need to mount a large research effort for this purpose.[1]

A brute-force approach to such a task is clearly untenable. The huge energy transactions in a typical, moderate-sized hurricane instigated by condensation heating (see Chapters 1, 4, and 6) far exceed any energy force that man can bring to bear. Even when one considers the more powerful hydrogen bombs, the energy that man can thereby release—even if it could directly oppose the operation of the hurricane heat engine, which it *cannot*—is small. The condensation heat energy required to fuel the heat engine of the mature hurricane will one day equate to that of about 8,000 megatons of fusion energy. Any viable hypothesis for hurricane moderation must be based upon the "Achilles' heel" concept, whereby a relatively small amount of energy is applied strategically to alter the energy-transformation processes, the local rates of kinetic-energy production, and the maximum winds that can be generated by the hurricane.

Although a number of different means have been proposed for reducing hurricane wind speeds, the only concept to have undergone full-scale experimentation by the end of the 1970s is one that uses cloud seeding to release heat of fusion in the outer eye wall and inner rainbands. The first seeding conducted in a tropical cyclone was done by Project Cirrus in 1947.[2] It was apparently intended to be more of an exploratory trial of a seeding method than a scientific experiment in weather modification. In this case, dry ice was dropped into stratiform clouds near the eye wall of a hurricane east of Jacksonville, Florida, on 13 October 1947. Another brief and unsuccessful test of a seeding method was conducted in hurricane Daisy on 27 August 1958 using an airborne device for generating silver-iodide smoke plumes.

In 1961, the first of a series of field experiments to test a physical hypothesis for the reduction or moderation of damaging winds in a tropical cyclone by a cloud-seeding procedure was carried out in hurricane Esther north of Puerto Rico. This series of experiments became known as Project STORMFURY. Table 24 lists the hurricanes in which moderation experiments were conducted by Project STORMFURY and the results that were observed.

The term *moderation* is appropriate to describe the objectives of cloud seeding in tropical cyclones. To reduce the destructive winds would reduce wind damage exponentially. However, to attempt to steer hurricanes away from coastal targets or to break up hurricane-seedling disturbances before they can acquire destructive winds is a less reasonable, if not implausible, objective. To alter the tracks or the recurrences of tropical disturbances and

Table 24. Project STORMFURY Cloud-Seeding Results

Hurricane	*Seeding dates*	*Type of seeding*	*Results*		
			Eye wall	*Winds*	*Pressure*
Esther	9/16/61	single	broke open	diminished	rose
	9/17/61	single	null*		no change
Beulah	9/13/63	single	null*		fell
	9/24/63	single	expanded	diminished (14%)	rose
Debbie	8/18/69	multiple	expanded	diminished (30%)	rose
	8/20/69	multiple	expanded	diminished (15%)	little change
Ginger†	9/26/71	multiple	none	diminished	no change
	9/28/71	multiple	none	diminished	no change

*Seeding flares were incorrectly placed.
†Exploratory seeding; conducted in rainbands alone.
SOURCE: R. H. Simpson *et al.*, *TYMOD: Typhoon Moderation*, Final Report Prepared for the Government of the Philippines (Arlington: Virginia Technology, 1978), B.8.

cyclones would pose the risk of reducing annual rainfall amounts or altering rainfall distributions in some coastal regions.[3] The potential impact upon agriculture and water resources, admittedly difficult to evaluate, would not make this approach attractive, if viable.

The STORMFURY Experiments

The first full-scale experiment aimed at reducing peak winds of the hurricane was conducted in hurricane Esther (1961), based upon a hypothesis by Robert and Joanne Simpson and inspired by unique observations of H. Riehl in a reconnaissance flight into hurricane Donna (1960).[4] Riehl noted that nearly all the outflow cloudiness stemmed from an aggregation of convection in the right-front quadrant of the eye wall. The series of related experiments that followed became known as Project STORMFURY. The initial hypothesis sought to change the distribution of mass and the position of maximum pressure gradients. This was to have been accomplished by using silver-iodide-crystal generators to seed clouds in the eye wall from the radius of maximum winds outward. The hypothesis anticipated that the release of latent heat of fusion would upset the balance of forces in the vortex, causing the ring of maximum winds to expand and the eye wall to reform at a greater distance from the center. This expansion, the hypothesis argued, would cause a reduction in maximum wind speeds in accordance with the conservation principle for absolute angular momentum, as discussed in Chapters 4 and 8.

An innovation, first used in Project STORMFURY, was the pyrotechnic generator for release of silver-iodide smoke. These generators, initially packaged as rocket canisters (later small flares), were dispersed from aircraft flying through the tops of hurricane clouds and were timed to burn through the depth of the layer of supercooled clouds. The generators were first developed and tested at the China Lake Naval Weapons Center in California.[5]

The second series of experiments, conducted in hurricane Beulah (1963), provided results that seemed to support those from the Esther experiment in 1961. While many important questions about the experiment and the hypothesis remained unanswered, there was enough encouragement to continue the experiment under the joint sponsorship of the United States Weather Bureau and the United States Navy.[6] Figure 143 shows the expansion of the eye wall, the reduction of wind speeds, and the changes in radial pressure gradients that were observed after the seeding in Beulah.

Because no hurricanes appeared in areas where the experiment could be effectively carried out, Project STORMFURY did not conduct another experiment until 1969, when hurricane Debbie was seeded on 2 successive days.[7] This time the seeding was done a total of five times at 90-minute intervals. The results were also more impressive than in earlier or subsequent experiments: peak winds fell 30% on the first seeding day and 16% on the second. Figure 144 shows the successive reduction in wind speeds that occurred the first seeding day. The only other seeding experiment during the 1970s was in hurricane Ginger in September 1971. However, the hypothesis in this case was inapplicable because, at the time of seeding, no identifiable eye wall or apparent source of supercooled liquid water was present. Seeding was confined to the rainbands, and no systematic change in structure was observed following these seedings. Based upon the results of the first three experiments and extensive numerical modeling, a number of changes were made in the hypothesis.

Basic Cloud-Seeding Concepts for Hurricane Moderation

Figure 145 illustrates the revised conceptual basis for moderating the hurricane's destructive potential. Instead of an outward migration of the eye wall, as suggested by the initial hypothesis, the revised concept anticipates the replacement of the initial eye wall with a new one of larger diameter, created by the dynamic growth of clouds in the inner rainbands or outer eye wall. As these clouds reach the outflow level, the new eye wall becomes the primary conduit for vertical transport of mass to the outflow, preempting the vertical transports in the original eye. Since the vertical transport in the new eye wall occurs at greater distance from the center,

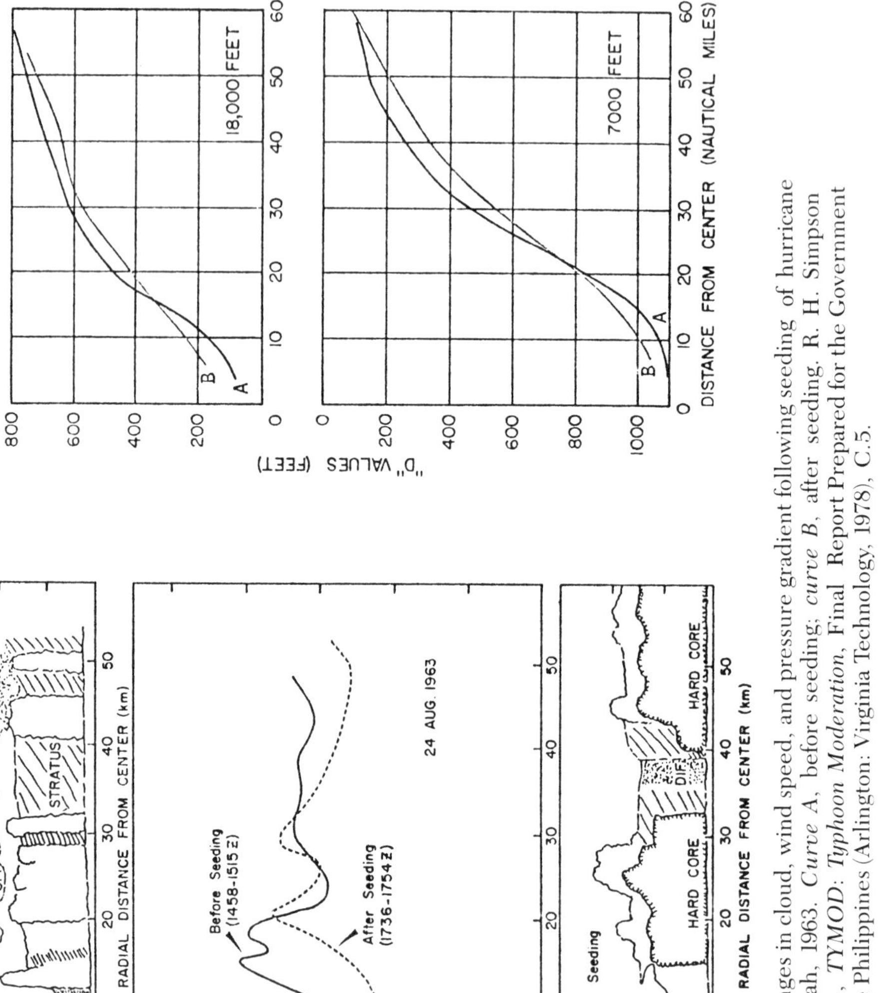

Fig. 143. Changes in cloud, wind speed, and pressure gradient following seeding of hurricane Beulah, 1963. *Curve A*, before seeding; *curve B*, after seeding. R. H. Simpson *et al.*, *TYMOD: Typhoon Moderation*, Final Report Prepared for the Government of the Philippines (Arlington: Virginia Technology, 1978), C.5.

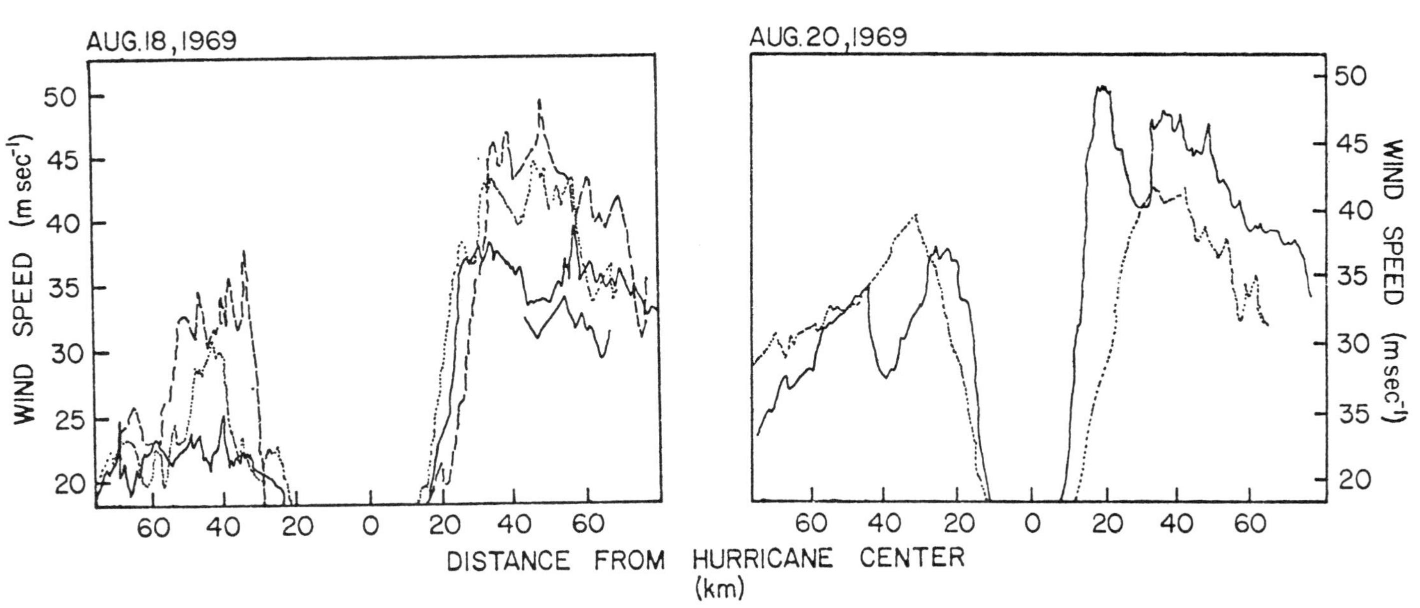

Fig. 144. Profiles of wind speed (3,600-m altitude) in hurricane Debbie after five successive soundings on each of 2 days. For 18 August, *broken line*, wind speed before seeding; *dotted line*, after third seeding; *solid line*, 4 hours after fifth seeding. For 20 August, *solid line*, before seeding; *dotted line*, 6 hours after fifth seeding. Adapted from R. C. Gentry, "Hurricane Modification," in W. Hess (ed.), *Weather and Climate Modification* (New York: John Wiley & Sons, 1974), 497–521.

the peak tangential wind speeds will be reduced. This follows from the argument in Chapters 4 and 8 that the conservation of angular momentum transported from the environment toward the center causes an increase in tangential wind speeds as a function of decreasing distance from the center. If seeding causes the maximum winds to occur a greater distance from the center, the maximum wind speed will be less.

Although this reasoning is at the heart of the STORMFURY hypothesis, it is obviously not a sufficient statement of the adjustment processes upon which the ultimate benefits of seeding depend.[8] Changes in the radial profiles of momentum in the hurricane must be supported by corresponding adjustments in the radial distribu-

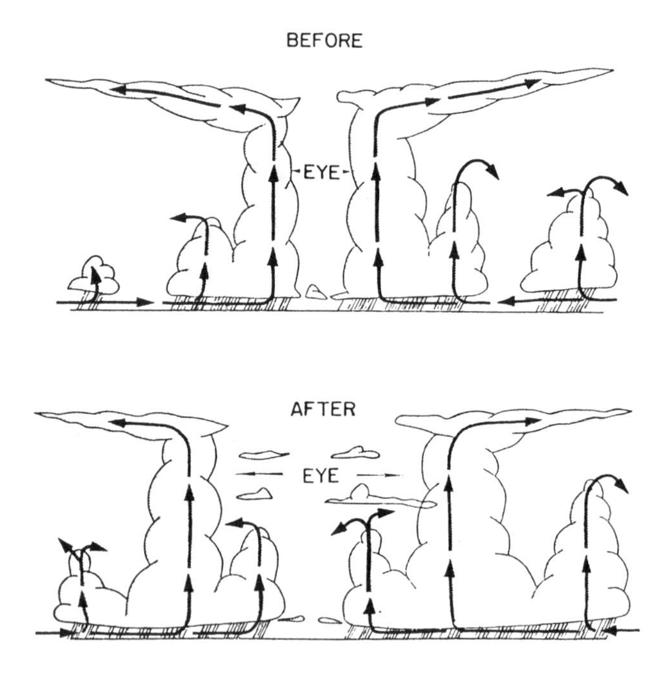

Fig. 145. Vertical cross sections through a hurricane eye wall and rain bands before and after seeding. Dynamic growth of seeded clouds in the inner rain bands provides new conduits for conducting mass to the outflow layer and causes decay of the old eye wall. R. H. Simpson *et al.*, *TYMOD*: *Typhoon Moderation*, Final Report Prepared for the Government of the Philippines (Arlington: Virginia Technology, 1978), A.6.

tion of mass, or pressure gradients, if the change is to be other than impulsive and short lived.

Since the dynamic growth of a new eye wall of larger diameter necessarily changes the position in which most of the latent heat is released, a modification of the warm core or the horizontal mean-temperature gradients, and consequently of the surface pressure gradients, must ensue. The reestablishment of dynamic stability following seeding, therefore, depends upon the maintenance of the new eye wall over a period sufficient for thermal adjustments to be completed. If this is not achieved, there is some evidence from numerical simulations that the net result of seeding could be an oscillatory expansion and contraction of the ring of maximum winds, resulting in stronger peak winds than before seeding began.[9] To avoid this possibility, a similar program, designed for use by the Philippine government in a typhoon-moderation experiment (Project TYMOD), proposed that seeding continue for at least 24 hours as a typhoon approaches landfall.[10]

Potential Benefits and Unanswered Questions

The damage potential of hurricanes at landfall is related not only to the peak winds but also to flooding and battering due to the hurricane tide. The reduction in wind postulated to result from cloud seeding might appear to be small. However, as shown in Chapter 8, the reduction in pressure forces on structures is proportional to the square of the wind speed, and of the wind energy to the cube of the wind speed. Hence a reduction of no more than 15% in maximum wind speeds will reduce the destructive potential from wind pressures by at least 30% and, for such reductions as observed in the Debbie experiment, by more than 50%. This should have a substantial impact on the losses due to wind.

However, the impact of cloud seeding on storm-surge heights and flooding potential involves some uncertainty. As shown in Chapter 9, the storm-surge peak height is uniquely related to the central pressure and radius of maximum wind. It reduces with rising central pressure and, in general, increases with the expansion of the ring of maximum wind. However, unless the central pressure rises, peak storm-surge heights would increase slightly as the RMW increases. Maximum peak surges occur at an RMW of about 50 km; for larger radii, however, the peak surge diminishes. The STORMFURY hypothesis does not address the question of changes in central pressure. However, from analytical reasoning and some numerical simulations, the central pressure must eventually rise if a sustained reduction in maximum winds is achieved. Moreover, if, by repetitive seeding, the eye wall continues to reform at progressively greater distances from the center, it can be argued that the radius of maximum wind will ultimately exceed the critical value of 50 km, and storm surges will diminish.

A fundamental problem in validating the responses to cloud seeding is that of separating natural changes from those induced by seeding. Statisticians have demonstrated the need for careful randomization and control procedures if the statistical analyses of the seeding results are to determine the strength of responses to seeding and the potential cost-effectiveness of an operational program.

There remain a number of uncertain links in the chain of reasoning used in the hypothesis. It has been assumed a priori that there are always large supplies of supercooled cloud liquid water in the hurricane ready to be frozen upon seeding with silver iodide. This assumption needs further investigation.

The question of how seeding may affect hurricane rainfall rates and distributions also needs further research. The answer to this question depends upon whether the number of convective conduits (cumulonimbi) for transporting mass in the eye wall increases in proportion to the volume of the eye wall. If so, as the wall expands in response to seeding, the total rainfall generated in the eye wall would be spread over a larger area and local rainfall rates and, thereby, the potential for flash floods, would be reduced. Because the total amount of rain remains essentially unchanged, however, the contribution to river flooding would not be altered.

Other Methods for Hurricane Moderation

The Weather Modification Advisory Board's report to Congress in 1978 enumerated a number of alternate methods, not involving cloud seeding, that might be used in future experiments.[11] First, hurricanes often generate cold wakes in the ocean, mainly by bringing cooler water to the surface from the thermocline and below. Theory and observation show that the amount of this cooling depends on the storm's intensity, its speed of motion, and the ocean stratification. Evidence suggests that hurricanes tend to lose strength when they encounter a water surface cooler than the critical value of about 26°C. Several occasions have been documented in which a hurricane has moved over the wake of cold water left by another hurricane that passed several days earlier. In these instances, when the cool water was encountered by the second hurricane, a marked loss in strength followed. Could an artificial means of cooling the sea surface help in reducing the impact of hurricanes? Is it plausible to attempt to cool several thousand square kilometers of ocean surface in the Gulf of Mexico or along the eastern seaboard?

Cooling an ocean surface layer might be accomplished as a by-product of other advertent modifications of the ocean environment. Two possibilities have been suggested: 1) the generation of artificial upwelling for the production of power and for open-sea mariculture, and 2) the importation of icebergs nominally as a freshwater resource but secondarily to cool the water surface in strategic areas and thereby reduce the strength of a tropical cyclone transiting the area. In 1978, at least three artificial upwelling techniques were being considered: ocean thermal-energy conversion, pumping by using wave energy, and pumping with conventional forms of energy.

Ocean Thermal-Energy Conversion (OTEC). The extraction of energy from the temperature difference between the warm surface waters and the deep cold water—planned for pilot tests in the early 1980s—brings to the surface large quantities of cold water, 15°C to 20°C colder than the surface. Initial estimates had speculated that the colder water, once brought to the surface, would eventually sink back. However, two mechanisms act to lower the density of the below-thermocline water enough to keep it at the surface: 1) The water from below the thermocline would have lower salinity than that at the surface, and 2) the dissolved gases kept in solution under pressure in deep water would be liberated, creating a bubble mixture that would change the effective density of cold water brought to the surface. These mechanisms together would cause the cold water to remain on the surface.

Icebergs. In 1977, a proposal to tow antarctic icebergs into low latitudes as a source of freshwater gained much attention, and in this connection, a few scientists gave serious consideration to the possibilities of using such a source to reduce the strength of tropical cyclones transiting the area.[12] Aside from the scientific feasibility, however, the technical problems of towing, the economic considerations, and finally questions of environmental impact made this suggestion for hurricane moderation less interesting and of limited applicability.

Reduction of Evaporation from the Ocean. Another method, first proposed in the early 1960s—one unlikely to have an adverse environmental impact—would seek the reduction of evaporation from the ocean in the area where air is actively flowing into the storm, *i.e.*, along a central sector of the streamlines relative to the moving core.[13]

From Figures 74 and 75, a steady flow of heat from the ocean to the surface layer of air flowing into the vortex is of critical importance if the air near the storm core is to ascend at higher temperatures than in the outer vortex and to establish the warm central core that supports the low central pressures and destructive winds. If evaporation could be reduced, say, 25% in the surface layer where pressure drops rapidly toward the center, the mixing ratio at a pressure of 950 mb would be 20 g/kg and the cloud base would be at an altitude of 300 m. Temperatures in the air ascending from this base are compared in Figure 74 with those starting at the same pressure but with a mixing ratio of 21 g/kg. At low levels, these differences are small, but they become substantial above 400 mb in the critical outflow layer (Tab. 25). Such changes would reduce the rate of mass circulation—in, up, and out—in the hurricane and reduce maximum wind speeds.

Table 25. Change of Temperature in Hurricane Core If Ocean Evaporation into Inflow Is Reduced by 25 Percent

Pressure (*mb*)	*Initial values* A (°C)	B (°C)	ΔT (°C)
800	18	18	0
700	14	13	−1
600	8	7	−1
500	1	0	−1
400*	−8	−9	−1
300	−21	−23	−2
200	−41	−45	−4
150	−55	−62	−7

NOTE: For column A, pressure = 950 mb and mixing ratio = 21 g/kg in hurricane core; for column B, pressure = 950 mb, mixing ratio = 20 g/kg.
* Ice phase included below freezing.

To reduce evaporation, it has been proposed (and tested on some water reservoirs) that a monomolecular layer of hexadecanol be spread over the water surface. Such a layer would reduce evaporation by more than 80%; a multimolecular layer would be less effective. Although a form of the chemical compound has been developed that maintains its monomolecular cover on reservoirs with winds up to 35 kt, it remains to be shown that such a cover could be maintained in the open sea with high waves and gale-force winds. Further development and testing would very likely be needed before an adequate chemical cover for this purpose could be readied. Logistically, it would also pose quite a problem to deliver and spread the chemical over the appropriate area. One study estimated that it would take a fleet of thirty C-130 cargo aircraft to deploy the amount of chemical over the required area with an effective schedule of delivery.

Use of Carbon Black. Still another method, proposed by W. M. Gray, is to introduce particulates such as carbon black into the hurricane circulation to strategically alter the net radiation flux in a manner that would change the static stability in the vortex.[14] Done strategically, this might cause cumulus clouds to grow to the outflow level at greater distances from the center. It was argued that this would have the net effect of expanding the diameter of the warm core and reducing the surface pressure gradients and thereby the maximum sustained wind speed that could be achieved. This concept, tested with numerical simulation models, gave results that were not inconsistent with the hypothesis. However, the hypothesis has never been tested in a real tropical cyclone, in part because the present United States experiment, Project STORMFURY, remains incomplete, and funding for two parallel projects has not been available.

Outlook for Direct Intervention

There is little question that an operational procedure that could be counted upon to reduce the potential wind damage from hurricanes or typhoons by as little as 20% would be quite cost-effective. Whether such dependability can be achieved, even approximately, remains an open scientific question.

However, with the new technology that has become available—primarily such indirect sensing systems as airborne Doppler radar and lidar—the answers to these scientific questions are far more tractable than when Project STORMFURY was inaugurated. The greatest roadblock in evaluating such experiments remains the separation of responses to cloud seeding (or other experimental treatments) from the natural changes continually occurring in the hurricane. With proper application of indirect sensing probes deployed by reconnaissance aircraft, a means is within reach for continually monitoring the changes in circulation at all levels in the vortex. Using such probes, there is a reasonable likelihood that the sources of all changes can be traced, and cause and effect defined.[15]

The question that needs an answer is not *whether* we can determine if direct reductions of the hurricane threat are possible and, if so, cost-effective, but rather *should we* invest the resources to obtain the answer, considering the likelihood of success. Scientists will disagree on the answer to this question. Moreover, politicians and their constituencies are unlikely to agree on the priority for multiyear funding for such an experiment, until or unless a major hurricane brings virtually unparalleled disaster to a coastal area of the United States.

REFERENCES

1. U.S. Department of Commerce, Weather Modification Advisory Board, *Proposals for a National Policy and Program* (Silver Spring, Md.: Department of Commerce, June 30, 1978), 99–120, Vol. 1 of *The Management of Weather Resources*.
2. General Electric Research Laboratory, *Project Cirrus*, Final Report Under Contract W-36-039-SC-38141, RL566, (Albany, N.Y.: N.p., 1951).
3. H. Landsberg, "Manmade Climatic Changes" in F. Singer (ed.), *Changing Global Environment* (Boston: D. Ridel, 1975), 197–234; G. W. Cry, "Effects of Tropical Cyclone Rainfall on the Distribution over the Eastern and Southern United States," Professional Paper 1 (Department of Commerce, ESSA, 1967).
4. R. H. Simpson, M. R. Ahrens, and R. D. Decker, *A Cloud-Seeding Experiment in Hurricane Esther, 1961*, National Hurricane Research Project Report 60 (Washington, D.C.: Department of Commerce-NOAA, 1963); H. Riehl to Robert H. Simpson, personal communication.
5. P. St. Amand and G. W. Henderson, *Project Cyclops: An Experiment in Hurricane Modification*, Technical Report (Ordinance Test Station, China Lake, Calif.: U.S. Navy, 1962), 3.
6. R. H. Simpson and J. Simpson, "Experiments in Hurricane Modification," *Scientific American*, CCXI (1964), 27–38; R. C. Gentry, "Hurricane Modification," in W. N. Hess (ed.), *Weather and Climate Modification* (London: Wiley & Sons, 1974), 497–521.
7. Gentry, "Hurricane Modification."
8. S. L. Rosenthal, "Computer Simulation of Hurricane Development and Structure," in W. N. Hess (ed.), *Weather and Climate Modification* (London: Wiley & Sons, 1974), 522–51.
9. R. Jones, *Oscillation Movement of the Eye Wall Following Cloud Seeding*, National Hurricane and Environmental Meteorological Laboratory Technical Report (Miami: Department of Commerce-NOAA, 1977), n. p.
10. R. H. Simpson *et al.*, *TYMOD: Typhoon Moderation*, Final Report Prepared for the Government of the Philippines (Arlington: Virginia Technology, 1978).
11. U.S. Department of Commerce, *Proposals for a National Policy and Program*, I, 99–120.
12. Joanne Simpson, "Iceberg Utilization: Comparison with Cloud Seeding and Potential Weather Impacts," in A. A. Husseiny (ed.), *Iceberg Utilization* (New York: Paragon Press, 1978), 624–39.
13. R. H. Simpson and J. Simpson, "Why Experiment on Tropical Hurricanes?" *Transactions of the New York Academy of Science*, XXVIII (1966), 1045–1062.
14. W. M. Gray, "Feasibility of Beneficial Hurricane Modification by Carbon Dust Seeding," Occasional Paper 196 (Department of Atmospheric Sciences, Colorado State University, Ft. Collins, 1973).
15. J. Simpson, "What Weather Modification Needs: A Scientist's View," *Journal of Applied Meteorology*, LXXIV (1978), 365–70.

APPENDIX A

Glossary

The definitions here are limited in scope to the context in which they are used in this volume.

ADVECTION. The transfer of a property, such as heat or moisture, by horizontal motion.

AERODYNAMIC LIFT. A force acting perpendicularly to the flow of air around an object, *e.g.*, that which results from the relative movement of air over an airplane wing.

AIR MASS. A large body of air with fairly homogeneous properties of temperature and moisture.

ANALOG. A weather circulation with properties that resemble those of a weather map from which a prediction is to be made. The observed weather sequence from analog maps provides the basis for an analog prediction. The analog procedure may be *statistical*, summarizing the kinematic properties of many analog cases, or *dynamical*, summarizing the distribution of forces.

ANEMOMETER. An instrument for measuring wind speed.

ANTICYCLONE. A high-pressure system in which winds move *clockwise* around the pressure center in the Northern Hemisphere, *counterclockwise* in the Southern Hemisphere; generally characterized by fair weather.

BACKSHORE. Upper shore zone beyond the reach of ordinary waves and tides, landward from the beach crest.

BACKWASH. The seaward return of flood waters that have accumulated inland due to a hurricane tide.

BAR (OF A HURRICANE). The outer limit of the central cloud mass, where a transition to hurricane clouds, rain, and winds often takes place suddenly.

BAROCLINIC ENVIRONMENT. An atmosphere characterized by significant variations in temperature, horizontally, with density a function of both pressure and temperature.

BAROTROPIC ENVIRONMENT. An atmosphere in which temperature variations, horizontally, are small or absent, with density a function of pressure alone.

BARRIER ISLAND. An island lying offshore that separates and protects the mainland from the open sea.

BATHYMETRIC CHART. A mapping of ocean bottom contours.

BATHYMETRY. The study of water-depth patterns.

BERM (BEACH). A nearly horizontal part of the beach or backshore formed by the deposit of material by the action of higher waves.

CENTRAL PRESSURE (OF A HURRICANE). The minimum surface pressure at the cyclone center; a good measure of a hurricane's strength.

CIRCULATION (GENERAL OR PLANETARY). The seasonal mean-wind regimes prevailing over the globe.

CIRRUS CLOUDS. Ice-crystal clouds, usually thin and found only in the upper troposphere or where temperatures are far below freezing.

CLIMATOLOGICAL HURRICANE TRACK. An average of historical tracks, usually computed on a monthly or seasonal basis.

CLOUD NUCLEI. Minute particles that have the property of adsorbing the water vapor and becoming the nuclei around which cloud droplets (condensation nuclei) or ice crystals (freezing nuclei) may grow.

CLOUD PHYSICS. A study of the physical properties of clouds, including cloud nuclei and the processes that produce precipitation.

CLOUD SEEDING. The artificial insemination of cloud nuclei to accelerate the freezing or condensation processes in clouds, often to enhance precipitation but, in other instances, to alter the storm dynamics and reduce the weather extremes normally generated.

CONDENSATION LEVEL. The level at which moist rising air is cooled to the temperature of saturation.

CONTINUITY OF MASS. The principle stating that, in a circulating mass of air, its bulk remains constant, is neither created nor destroyed.

CONVECTION. The mechanically or thermally produced upward or downward movement of a limited part of the atmosphere, essential to the formation of clouds, in particular, cumulus clouds.

CONVECTIVE PROCESS. The vertical motions induced within a cloud, including the subsidence in its immediate environment; processes that transfer heat and momentum from one level to another.

CONVERGENCE. A horizontal accumulation of air over a given region due to nonuniform winds.

CORIOLIS FORCE. A force related to the earth's rotation (*see* Chapter 4); often called an *apparent force* because it does not exist in absolute coordinates, for instance, when the earth is viewed from fixed stars.

CYCLONE. A low-pressure system in which winds move *counterclockwise* around the pressure center in the Northern Hemisphere, *clockwise* in the Southern Hemisphere; usually attended by foul weather and strong wind speeds. (*See also* TROPICAL CYCLONE and EXTRATROPICAL CYCLONE.)

DEW-POINT TEMPERATURE. The temperature to which air must be cooled (without change in water vapor or pressure) to reach saturation.

DISTURBANCE. A disruption of the normal or prevailing flow or circulation; usually associated with abnormal cloudiness and precipitation.

DIVERGENCE. The opposite of convergence (above); the horizontal depletion of mass over a given area due to nonuniform winds.

DOLDRUMS. A nautical term for the region of light winds between the trade winds of the two hemispheres (the equatorial low-pressure trough), in which sailing vessels may be becalmed for days.

DOMAIN. A region over which physical processes are studied and computations carried out.

DRY ICE. Solid carbon dioxide, often used in granulated form for cloud seeding.

DYNAMIC INSTABILITY. A characteristic distribution of wind velocities in which a small imposed perturbation causes wave disturbances to increase in amplitude and individual particles to accelerate continuously away from an initial position; often occurs in a barotropic environment.

EKMAN LAYER (OF THE OCEAN). The portion of ocean nearest the surface in which water is set in motion by wind stresses at the surface, diminishing with depth and usually vanishing at a depth of about 50 fathoms (91 m).

ENERGY. *See* text, p. 8. *Kinetic energy (of motion)* is proportional to the square of the wind velocity; *latent heat*, proportional to water vapor content of air; *potential energy (available for conversion to kinetic energy)*, proportional to the elevation of the center of gravity in a cooler air mass above the center of gravity in the environment.

EXTRATROPICAL CYCLONE. Any synoptic-scale cyclone that is not a tropical cyclone. It usually originates in middle or higher latitudes in a baroclinic environment. Its foul weather and strong winds affect a much broader area than those of a tropical cyclone, with worst conditions occurring at a greater distance from the low-pressure center.

FEEDBACK. The return of a portion of an output quantity to the input of a system.

FETCH. The sea distance over which an air current moves in a nearly straight line to generate waves on the ocean surface.

FOREDUNE. The seafronting, primary dune immediately landward of the berm or backshore.

FRONT. A discontinuity or narrow transition zone that separates the characteristic properties of two air masses at the earth's surface.

GEOSTROPHIC (EARTH-TURNING) BALANCE. In the atmosphere, the balance between Coriolis and pressure forces that causes a wind to move parallel to the isobars without change in speed (*see* Chapter 4, p. 62).

GRADIENT. The change of a given property over a fixed distance, *e.g.*, a rise in temperature of 3° C over a distance of 100 km.

GRID RESOLUTION. The detail with which a property can be described using data spaced in accordance with a grid of given mesh length.

HABITATION LAYER. *See* text, p. 193.

HEAT, SENSIBLE, LATENT. *See* ENERGY.

HUMIDITY. *Absolute humidity* is the water-vapor density or mass of water vapor in a given volume; *relative*, the ratio of observed to saturation water-vapor content; *specific*, the mass of water vapor in a given mass of air.

HURRICANE. A regional name for tropical cyclones with maximum winds over 63 kt (32 mps) in the Atlantic and eastern North Pacific Oceans.

HURRICANE TIDE. *See* text, p. 244.

HYDROGRAPH. A graphical plot of changes in water levels, or of river stage or discharge, with time.

INVERSE BAROMETER (PERTAINING TO STORM SURGE). The component of rise in sea level resulting from pressure gradients in a storm system.

INVISCID. Frictionless.

ISOBAR. A line, or isopleth, of constant pressure on a weather map.

ISOPLETH. A line of equal or constant value of a given quantity.

ISOTACH. A line of constant wind speed on a weather map.

ISOTHERM. A line of constant temperature on a weather map.

JET STREAM. A narrow current of air with high velocity relative to its environment; mainly in the high troposphere near 300 mb to 200 mb but occasionally in the layer below the 1,600-m level. Wind speeds may reach 200 kt (100 mps) in the upper troposphere, less than half this value in the lower troposphere.

LANDFALL (OF A HURRICANE). The position at a seacoast where the center of a hurricane passes from sea to land.

LIDAR (LIGHT INTENSITY DETECTION AND RANGING). An atmospheric probe system, analogous to radar, using radiation in the near-visible spectrum.

LONGSHORE CURRENT. A shoal-water current moving essentially parallel to the coastline; generally caused by waves breaking along a line at an acute angle to the coast.

MACROSCALE EVENT. A circulation system or disturbance whose dimensions exceed 1,000 km.

MARIGRAM. A graphical record of variations in tides.

MERIDIONAL. Directed north-south, usually positive in the poleward direction.

MESOSCALE EVENT. A circulation system or disturbance whose dimensions are generally less than 100 km.

MIXING LENGTH. The mean travel distance over which an air eddy maintains its identity.

MIXING RATIO (ATMOSPHERIC). The amount of water vapor in a given mass of dry air.

MODEL. A numerical scheme for describing the properties of an atmospheric circulation system—usually a disturbance or cyclone—its development, and/or movement. A *dynamical model* is based upon the fundamental physics that govern the changes in properties or movement; a *statistical*, upon a statistical evaluation of the behavior in historic cases.

MOMENTUM. *Angular momentum* is the product of the moment of inertia and the angular velocity of the body; *linear*, the mass of a body times its velocity (*See* text, p. 155).

MONOMOLECULAR LAYER. A film whose thickness is that of a single molecule.

MONSOON. A wind whose direction reverses seasonally, typically over the South China Sea and the eastern Indian Ocean. A process in which the monsoon wind blows generally from an atmospheric cold source to a warm source.

OROGRAPHIC. Related to, or caused by, terrain variations.

OROGRAPHY. The nature of a region with regard to its elevated terrain.

PARAMETER. A derived constant or function, often arbitrarily assigned, symbolizing a physical process.

PARAMETERIZATION. The representation of a physical process in the harness of one or more parameters.

PROBABILITY ELLIPSE. A statistically derived elliptical area inside of which an uncertain event will have a probability of occurring equal to that assigned to the ellipse; for example, the 50% probability ellipse inside of which a hurricane center will be located after, say, 24 hours of movement (*See* Fig. 138).

RADAR (RADIO DETECTION AND RANGING). In meteorology, used to detect and display distributions and intensity of precipitation. *Pulse Doppler* is a radar system that measures the speed with which precipitation targets approach or recede from the radar antenna and, thereby, the wind component transporting the precipitation.

RADIATION. The process by which heat energy is propogated from a material body, the heat flux being proportional to the fourth power of the body's temperature; in earth-atmosphere processes, *long-wave radiation* is from terrestrial and atmospheric sources, mainly the infrared spectrum; *short-wave*, from solar sources, mainly in the visible or near-visible spectrum.

RADIOSONDE. A balloon-borne atmospheric sounding device for measuring vertical profiles of pressure, temperature, and humidity.

RAM PRESSURE (INSIDE A STRUCTURE). The increase of internal pressure over ambient (external) values that occurs when an opening in the windward face of the structure exposes the interior to the ram action of wind.

RAWINSONDE. A radiosonde also equipped to measure vertical profiles of the horizontal wind velocity.

RECURVATURE (OF A HURRICANE). Usually the change from movement toward the west, under the influence of trade winds, to a north or northeastward movement, under the influence of westerlies from higher latitudes.

RESULTANT FORCE. The net remaining force after all acting forces have been vectorally added.

RESURGENCE. A resonant recurrence of long gravity waves in a water basin that causes tidal perturbations; related to seiche.

RIDGE. An area of pronounced anticyclonic flow where pressure generally reaches a maximum.

SATELLITE (WEATHER). An earth-orbiting capsule instrumented to maintain a surveillance of cloud systems and, in some instances, to measure mean temperatures of the atmosphere, the clouds, the oceans, and some terrestrial features. A *synchronous satellite* orbits in an equatorial plane synchronous with the earth's rotation and thus continuously views the same geography and offers the opportunity to develop time-lapse movies of cloud movement and development over this geographical area.

SATURATION VAPOR PRESSURE. The atmospheric vapor pressure over a flat body of water that, at a given temperature, is in equilibrium (no fluxes) with the liquid water surface. The equilibrium pressure is a function of temperature alone.

SEICHE. Motion, sometimes oscillatory, of long (gravity) waves in a large body of water in response to an excitation force (surface-wind stresses or sudden pressure changes) that may cause sudden rises in tidal levels in a coastal or estuarine region.

SHEAR LINE. A synoptic-scale line across which wind velocity components parallel to the line change discontinously in the cyclonic sense.

SINGULAR POINT (IN A WIND FIELD). A point in a flow field at which wind speed vanishes.

STABILITY. A general property of atmospheric systems that forces the return to an equilibrium state after a disturbance is imposed. (*See also* DYNAMIC INSTABILITY.) *Thermal stability* requires that a parcel of air, displaced vertically, returns to its initial position.

STORM SURGE. *See* text, p. 232.
STRATOSPHERE. A deep atmospheric layer with nearly constant or increasing temperature with height whose base lies near the 16-km level in the tropics.
STREAMLINE. A line each element of which is tangent to the instantaneous wind direction.
SUBTROPICS. The intermediate region between tropical and middle latitudes, generally in the latitude belt of 25°–35°, that varies with season and region.
SUPERCOOLED WATER. Liquid cloud droplets or precipitation at temperatures colder than freezing.
SYNOPTIC SCALE. *See* MACROSCALE EVENT

TEMPERATURE INVERSION. An atmospheric layer, usually thin, where temperature increases upward.
THERMOCLINE. A layer of pronounced vertical temperature gradient in the ocean (temperatures decreasing downward), usually near the water surface.
TRADE WINDS. Steady easterly winds blowing across the subtropical and tropical oceans, generally with an equatorward component.
TRAJECTORY, OR TRACK, (OF A HURRICANE). (a) The path taken by the center of a tropical cyclone; (b) the path of individual air particles spiraling about the cyclone center.
TROPICAL CYCLONE. A synoptic-scale near-circular cyclone generally originating over tropical oceans, distinguished by torrential rains and damaging winds that reach a maximum very near the pressure center.
TROPICS. The latitude belt in which weather typical of the tropics occurs. Geographically, it is the global region between the tropics of Cancer and Capricorn; its weather is variable with geographic region and season.
TROPOSPHERE. The well-mixed weather-bearing layer of the atmosphere, extending from the earth's surface to the base of the stratosphere, through which temperatures decrease with height.
TROUGH. The counterpart of a ridge; an area of pronounced cyclonic flow where pressure generally reaches a minimum.
TURBULENCE. Small-scale, irregular, and random motions acting in and through a mean flow.
TYPHOON. The regional term for a tropical cyclone in the western North Pacific Ocean.

UNITS OF MEASURE. S.I., metric, English. *See* Appendix B.

VAPOR PRESSURE. The partial atmospheric pressure exerted by the water vapor in air.

VORTEX (ATMOSPHERIC). A rotational system (possessing vorticity) with a discrete center. Circulation may be either cyclonic or anticyclonic.

VORTICITY. The (local) rotation (of air) about a discrete center; may be reflected in the cyclonic or anticyclonic turning of streamlines or by wind shear perpendicular to an axis of rotation. *Absolute vorticity* is the sum of the local or relative vorticity and that due to the earth's rotation.

WIND. Air motion relative to the earth.

Cyclostrophic motion is supported by a balance between pressure and centrifugal forces.

Extreme (hurricane) winds have highest gust speed, generally about 1.3 times the maximum sustained wind.

Geostrophic motion is supported by a balance between pressure gradient and Coriolis forces.

Gradient motion is supported by a balance between pressure gradient, Coriolis forces, and centrifugal forces.

Maximum sustained (hurricanes, severe storms) wind is the fastest mile of wind passing the anemometer.

Radial component is the component of motion in a radial direction with respect to a center of circulation, positive outward.

Sustained wind is the average wind for a 1-minute interval (U.S.) or for a 10-minute interval (international).

Tangential component is the component of motion encircling a center of circulation (a rotational component), positive in a cyclonic sense.

WIND SHEAR. The change in wind velocity relative to a horizontal or vertical axis; *e.g.*, vertical shear of the horizontal wind, the change in wind velocity with height.

APPENDIX B

Units of Measure and Conversion Factors

Table 26. Unit Conversion Factors

Unit	*Conversions*	*Unit*	*Conversions*
	Length		*Mass*
1 in	2.540 cm	1 kg	2.2 lb
1 ft	30.480 m		*Force*
1 m	3.281 ft	1 N	10^5 dy
	Distance		*Pressure*
1 nmi	1.151 mi	1 mb	10^2 Pa
	1.852 km		0.0295 in (hg)
1 mi	1.609 km	1 lb (sq ft)$^{-1}$	4.88 kg m^{-2}
1° lat	59.996 nmi		*Speed*
	69.055 mi	1 mps	1.94 kt
	111.136 km		3.59 kph
	Depth	1° lat (6 hr)$^{-1}$	10 kt
1 fa	6 ft		
	1.829 m		

Table 27. Systems of Measure: Units, Symbols, and Definitions

Quantity	*S.I. unit*	*Early metric*	*Maritime*	*English*
Length	meter (m)	centimeter (cm)	foot (ft)	foot (ft)
Distance	meter (m)	kilometer (km)	nautical mile (nmi)	mile (mi)
Depth	meter (m)	meter (m)	fathom (fa)	foot (ft)
Mass	kilogram (kg)	gram (g)	pound (lb)	pound (lb)
Time	second (s)	second (s)	second (s)	second (s)
Speed	meter per second (mps)	centimeter per second (cm s^{-1}) kilometer per hour (km hr^{-1})	knot (kt) nmi hr^{-1}	miles per hour (mph)
Temperature				
sensible	degree Celsius (C)	degree Celsius (°C)	——	degree Fahrenheit (°F)
potential	degree Kelvin (°K)	degree Kelvin (°K)	——	degree Kelvin (°K)
Force	newton (N) kg m s^{-2}	dyne (dy) g cm s^{-2}	poundal (pl)	poundal (pl)
Pressure	pascal (Pa) Nm^{-2}	millibar (mb) 10^3 dy cm^{-2}	inches, mercury (in)	inches, mercury (in)
Energy				
motion	joule (J) kg m^2 s^{-2}	joule (J) 10^7 g cm^2 s^{-2}	footpoundal (fp)	footpoundal (fp)
heat	joule (J) kg m^2 s^{-2}	calorie (cal) (4,187 J)	——	British Thermal Unit (BTU) (252 cal)
Power	watt (W) J s^{-1}	watt (W) 10^7 g cm^2 s^{-3}	horsepower (hp) 550 fp s^{-1}	horsepower (hp) 550 fp s^{-1}

APPENDIX C

The Saffir / Simpson Damage-Potential Scale

The Saffir / Simpson Damage-Potential Scale is used by the National Weather Service to give public-safety officials a continuing assessment of the potential for wind and storm-surge damage from a hurricane in progress.* Scale numbers are made available to public-safety officials when a hurricane is within 72 hours of landfall.

Scale numbers range from 1 to 5. Scale No. 1 begins with hurricanes in which the maximum sustained winds are at least 74 mph (119 kph) or which will produce a storm surge 4 to 5 feet (1.4 m) above normal water level. Scale No. 5 applies to those in which the maximum sustained winds are 155 mph (249 kph) or more, or which have the potential of producing a storm surge more than 18 feet (5.5 m) above normal. Atmospheric-pressure ranges have been adapted to this scale, and pressure ranges associated with each are listed in Table 28.

The scale numbers are not forecasts but are based on observed conditions at a given time in a hurricane's lifespan. They represent an estimate of what the storm would do to a coastal area if it were to strike without change in size or strength. Scale assessments are revised regularly as new observations are made, and public-safety organizations are kept informed of the hurricane's disaster potential.

The damage-potential scale indicates probable property damage and evacuation recommendations as listed below:

1. Winds of 74 to 95 mph (119 to 153 kph). Damage primarily to shrubbery, trees, foliage, and unanchored mobile homes. No real damage

* Developed by Herbert Saffir, consulting engineer, Dade County, Florida, and Robert H. Simpson, former director of the National Hurricane Center.

to other structures. Some damage to poorly constructed signs. And/or storm surge 4 to 5 feet (~ 1.5 m) above normal. Low-lying coastal roads inundated, minor pier damage, some small craft torn from moorings in exposed anchorage.

2. Winds of 96 to 110 mph (154 to 178 kph). Considerable damage to shrubbery and tree foliage; some trees blown down. Major damage to exposed mobile homes. Extensive damage to poorly constructed signs. Some damage to roofing materials of buildings; some window and door damage. No major damage to buildings. And/or storm surge 6 to 8 feet (~ 2 to 2.5 m) above normal. Coastal roads and low-lying escape routes inland cut by rising water 2 to 4 hours before arrival of hurricane center. Considerable damage to piers; marinas flooded. Small craft torn from moorings in unprotected anchorages. Evacuation of some shoreline residences and low-lying island areas required.
3. Winds of 111 to 130 mph (179 to 209 kph). Foliage torn from trees; large trees blown down. Practically all poorly constructed signs blown down. Some damage to roofing materials of buildings; some window and door damage. Some structural damage to small buildings. Mobile homes destroyed. And/or storm surge 9 to 12 feet (~ 2.6 to 3.9 m) above normal. Serious flooding at coast; many smaller structures near coast destroyed; larger structures near coast damaged by battering waves and floating debris. Low-lying escape routes inland cut by rising water 3 to 5 hours before hurricane center arrives. Flat terrain 5 feet (1.5 m) or less above sea level flooded inland 8 miles (~ 13 km) or more. Evacuation of low-lying residences within several blocks of shoreline possibly required.
4. Winds of 131 to 155 mph (211 to 249 kph). Shrubs and trees blown down; all signs down. Extensive damage to roofing materials, windows, and doors. Complete failure of roofs on many small residences. Complete destruction of mobile homes. And/or storm surge 13 to 18 feet (~ 4 to 5.5 m) above normal. Flat terrain 10 feet (~ 3 m) or less above sea level flooded inland as far as 6 miles (~ 10 km). Major damage to lower floors of structures near shore due to flooding and battering by waves and floating debris. Low-lying escape routes inland cut by rising water 3 to 5 hours before hurricane center arrives. Major erosion of beaches. Massive evacuation of all residences within 500 yards (~ 455 m) of shore possibly required, and evacuation of single-story residences on low ground within 2 miles (~ 3 km) of shore required.

5. Winds greater than 155 mph (249 kph). Shrubs and trees blown down; considerable damage to roofs of buildings; all signs down. Very severe and extensive damage to windows and doors. Complete failure of roofs on many residences and industrial buildings; extensive shattering of glass in windows and doors. Some complete building failures. Small buildings overturned or blown away. Complete destruction of mobile homes. And/or storm surge greater than 18 feet (~ 5.5 m) above normal. Major damage to lower floors of all structures less than 15 feet (~ 4.5 m) above sea level within 500 yards (~ 455 m) of shore. Low-lying escape routes inland cut by rising water 3 to 5 hours before hurricane center arrives. Massive evacuation of residential areas on low ground within 5 to 10 miles (~ 8 to 16 km) of shore possibly required.

Table 28. Saffir/Simpson Damage-Potential Scale Ranges

Scale number (category)	*Central pressure*		*Winds (mph)*	*Surge (ft)*	*Damage*
	Millibars	*Inches*			
1	≥980	≥28.94	74–95	4–5	Minimal
2	965–979	28.50–28.91	96–110	6–8	Moderate
3	945–964	27.91–28.47	111–130	9–12	Extensive
4	920–944	27.17–27.88	131–155	13–18	Extreme
5	<920	<27.17	>155	>18	Catastrophic

SOURCE: P. J. Hebert and G. Taylor, *Hurricane Experience Levels of Coastal County Populations—Texas to Maine*, National Weather Service Southern Regional Technical Report (Silver Spring, Md.: Department of Commerce—NOAA, 1975), 2.

Table 29. Chronological List of Hurricanes That Affected the United States During the Period 1900–1974

	States affected	*Highest damage potential*	*Minimum pressure (mb)*
1900 Sep	Texas	4	931
1901 Jul	North Carolina	1	——
Aug	Louisiana, Mississippi	2	972
1903 Sep	Florida	2	976
Sep	New Jersey, New York, Connecticut	1	990

Table 29—Continued

	States affected	*Highest damage potential*	*Minimum pressure (mb)*
1906 Jun	Florida	1	——
Sep	South Carolina, North Carolina	3	947
Sep	Mississippi, Alabama	3	958
Oct	Florida	2	967
1908 Jul	North Carolina	1	——
1909 Jul	Texas	3	958
Aug	Texas	2	——
Sep	Louisiana	4	931
Oct	Florida	3	957
1910 Sep	Texas	2	965
Oct	Florida	3	955
1911 Aug	Florida, Alabama	1	——
Aug	Georgia, South Carolina	2	——
1912 Oct	Texas	1	——
1913 Jun	Texas	1	——
Sep	North Carolina	1	——
1915 Aug	Texas	4	945
Sep	Florida	1	988
Sep	Louisiana	4	931
1916 Jul	Mississippi, Alabama	3	948
Jul	Massachusetts	1	——
Jul	South Carolina	1	980
Aug	Texas	3	948
Oct	Alabama, Florida	2	972
Nov	Florida	1	——
1917 Sep	Florida	3	958
1918 Aug	Louisiana	3	955
1919 Sep	Florida, Texas	4	927
1920 Sep	Louisiana	2	975
Sep	North Carolina	1	——
1921 Jun	Texas	2	979
Oct	Florida	3	952

Table 29—Continued

		States affected	*Highest damage potential*	*Minimum pressure (mb)*
1923	Oct	Louisiana	1	985
1924	Sep	Florida	1	985
	Oct	Florida	1	980
1925	Nov	Florida	1	——
1926	Jul	Florida	1	980
	Aug	Louisiana	3	955
	Sep	Florida, Alabama	4	935
1928	Sep	Florida, Georgia, South Carolina	4	929
1929	Jun	Texas	1	982
	Sep	Florida	3	948
1932	Aug	Texas	4	941
	Sep	Alabama	1	979
1933	Jul			
	Aug	Florida, Texas	2	975
	Aug	North Carolina, Virginia	2	971
	Sep	Texas	3	949
	Sep	Florida	3	948
	Sep	North Carolina	3	957
1934	Jun	Louisiana	3	962
	Jul	Texas	2	975
1935	Sep	Florida	5	892
	Nov	Florida	2	973
1936	Jun	Texas	1	987
	Jul	Florida	3	964
	Sep	North Carolina	2	——
1938	Aug	Louisiana	1	985
	Sep	New York, Connecticut, Rhode Island, Massachusetts	3*	946
1939	Aug	Florida	1	985
1940	Aug	Texas, Louisiana	2	972
	Aug	Georgia, South Carolina	2	970
1941	Sep	Texas	3	958
	Oct	Florida	2	975
1942	Aug	Texas	1	992

Table 29—Continued

		States affected	Highest damage potential	Minimum pressure (mb)
	Aug	Texas	3	950
1943	Jul	Texas	2	969
1944	Aug	North Carolina	1	990
	Sep	North Carolina, Virginia, New York, Connecticut, Rhode Island, Massachusetts	3*	947
	Oct	Florida	3	962
1945	Jun	Florida	1	985
	Aug	Texas	2	967
	Sep	Florida	3	951
1946	Oct	Florida	1	980
1947	Aug	Texas	1	992
	Sep	Florida, Louisiana	4	940
	Oct	Florida, Georgia, South Carolina	2	974
1948	Sep	Louisiana	1	987
	Sep	Florida	3	963
	Oct	Florida	2	975
1949	Aug	North Carolina	1	980
	Aug	Florida	3	954
	Oct	Texas	2	972
1950	Aug	Alabama	1	980
	Sep	Florida	3	958
	Oct	Florida	3	955
1952	Aug	South Carolina	1	985
1953	Aug	North Carolina	1	987
	Sep	Maine	1*	——
	Sep	Florida	1	985
1954	Aug	North Carolina, New York, Connecticut, Rhode Island	3*	960
	Sep	Massachusetts, Maine	3*	954
	Oct	South Carolina, North Carolina, Maryland	4*	938
1955	Aug	North Carolina, Virginia	3	962
	Aug	North Carolina	1	987
	Sep	North Carolina	3	960

Table 29—Continued

		States affected	*Highest damage potential*	*Minimum pressure (mb)*
1956	Sep	Louisiana, Florida	2	975
1957	Jun	Texas, Louisiana	4	945
1959	Jul	Texas	1	984
	Jul	South Carolina	1	993
	Sep	South Carolina	3	950
1960	Sep	Mississippi	1	981
	Sep	Florida, North Carolina, New York, Connecticut, Rhode Island, Massachusetts, New Hampshire, Maine	4	930
1961	Sep	Texas	4	931
1963	Sep	Texas	1	996
1964	Aug	Florida	2	968
	Sep	Florida	2	966
	Oct	Louisiana	3	950
	Oct	Florida	2	974
1965	Sep	Florida, Louisiana	3	948
1966	Jun	Florida	2	982
	Oct	Florida	1	983
1967	Sep	Texas	3	950
1968	Oct	Florida	2	977
1969	Aug	Louisiana, Mississippi	5	909
	Sep	Maine	1	980
1970	Aug	Texas	3	945
1971	Sep	Louisiana	2	978
	Sep	Texas	1	979
	Sep	North Carolina	1	993
1972	Jun	Florida, New York, Connecticut	1	980
1974	Sep	Louisiana	3	952

* Hurricane moving in excess of 30 mph.

SOURCE: P. J. Hebert and G. Taylor, *Hurricane Experience Levels of Coastal County Populations—Texas to Maine*, National Weather Service Southern Regional Technical Report (Silver Spring, Md.: Department of Commerce—NOAA, 1975), 12.

NOTE: Damage-potential figures refer to the Saffir/Simpson scale (Tab. 28); pressures are given in millibars.

APPENDIX D

Hurricane Climatology for the United States Coastline

The recurrence of damaging tropical cyclones is summarized here in two classes: A, those hurricanes that have reached the coast with maximum winds of 74 mph (33 mps) or greater, and B, the great hurricanes, or those that have reached the coast with maximum winds greater than 125 mph (56 mps).

In Figure 146, the return period in years for Classes A and B has been listed for coastal segments—strike areas—80 km in length. Because hurricane-force winds normally extend 80 km or more to the right of the track, the summarization here counts a hurricane whose eye passes across one numbered strike area as a strike not only in the box in which the eye passes inland but also in the one immediately to the right. That is, a hurricane whose eye passed inland in Box 20 would count as a strike on both Boxes 20 and 21.

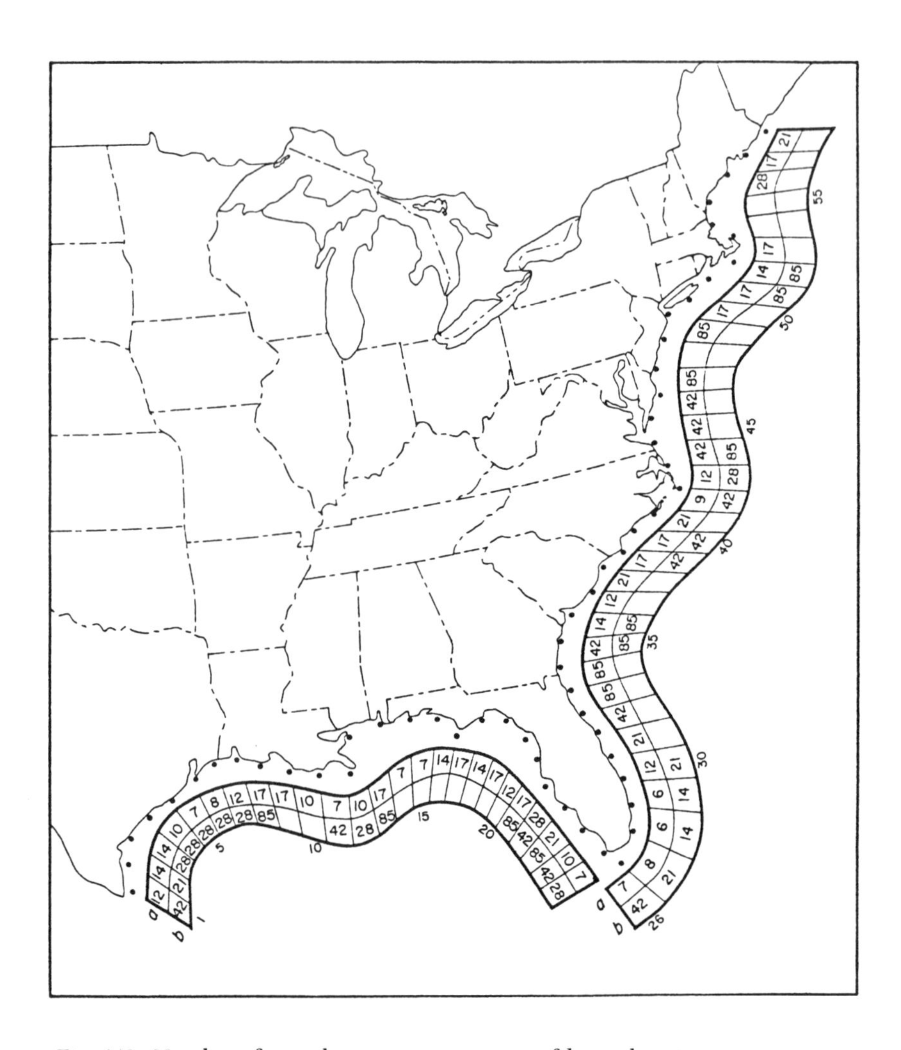

Fig. 146. Number of years between occurrences of *box a*, hurricanes—maximum winds greater than 120 kph— and *box b*, severe hurricanes—with maximum winds greater than 200 kph—for fifty-eight coastal segments (each 80 km long) along the Gulf and Atlantic coastlines of the United States (see Tables 30 and 31). Adapted from R. H. Simpson and M. B. Lawrence, *Atlantic Hurricane Frequencies*, Technical Memo NWS SR-58 (N.p.: Department of Commerce-NOAA, 1971), 5.

Table 30. Number of Hurricanes Reaching United States Mainland During the Period 1886–1970 for Each of the Fifty-Eight Coastal Segments in Figure 146

Sector	*All hurricanes*	*Great hurricanes*	*Sector*	*All hurricanes*	*Great hurricanes*
1	7	2	30	7	4
2	6	4	31	4	—
3	6	3	32	2	—
4	8	3	33	1	—
5	12	3	34	1	—
6	10	3	35	2	1
7	7	3	36	6	1
8	5	1	37	7	—
9	5	—	38	4	—
10	8	—	39	5	2
11	11	2	40	5	2
12	8	3	41	4	—
13	5	1	42	9	2
14	11	—	43	7	3
15	12	—	44	2	1
16	6	—	45	2	—
17	5	—	46	2	—
18	6	—	47	1	—
19	5	—	48	—	—
20	7	—	49	1	—
21	5	1	50	5	—
22	3	2	51	5	1
23	4	1	52	6	1
24	8	2	53	5	—
25	11	3	54	—	—
26	11	2	55	—	—
27	10	4	56	3	—
28	14	6	57	5	—
29	13	6	58	4	—

NOTE: *Hurricane* ≡ winds of 33 mps or higher; *great hurricane* ≡ winds of 56 mps or higher.

SOURCE: Adapted from R. H. Simpson and M. B. Lawrence, *Atlantic Hurricane Frequencies*, Technical Memo NWS SR-58 (N.p.: Department of Commerce—NOAA, 1971), 10.

Table 31. Probabilities for a Hurricane Strike in Any One Year for Each of the Fifty-Eight Coastal Segments Shown in Figure 146

Sector	*All hurricanes (%)*	*Great hurricanes (%)*	*Sector*	*All hurricanes (%)*	*Great hurricanes (%)*
1	8	2	30	8	5
2	7	5	31	5	—
3	7	4	32	2	—
4	9	4	33	1	—
5	14	4	34	1	—
6	12	4	35	2	1
7	8	4	36	7	1
8	6	1	37	8	—
9	6	—	38	5	—
10	9	—	39	6	2
11	13	2	40	6	2
12	9	4	41	5	—
13	6	1	42	11	2
14	13	—	43	8	4
15	14	—	44	2	1
16	7	—	45	2	—
17	6	—	46	2	—
18	7	—	47	1	—
19	6	—	48	—	—
20	8	—	49	1	—
21	6	1	50	6	—
22	4	2	51	6	1
23	5	1	52	7	1
24	9	2	53	6	—
25	13	4	54	—	—
26	13	2	55	—	—
27	12	5	56	4	—
28	16	7	57	6	—
29	15	7	58	5	—

NOTE: *Hurricane* ≡ winds of 33 mps or higher; *great hurricane* ≡ winds of 56 mps or higher.

SOURCE: Adapted from R. H. Simpson and M. B. Lawrence, *Atlantic Hurricane Frequencies*, Technical Memo NWS SR-58 (N.p.: Department of Commerce-NOAA, 1971), 8–9.

APPENDIX E

Global Sources of Tropical Cyclones: Occurrences Observed During the Period 1958 to 1978

Chapters 1, 4, 5, and 12 contain a number of illustrations (Figs. 1, 43, 46, 52, 53, and 134, and Tab. 6) of hurricane climatology and climatic trends, which, for the most part are based upon occurrences during the first half of the twentieth century. Prior to 1943, the detection of tropical cyclone formation was mainly dependent upon observations from ships at sea, and uncertainties existed as to the actual number of tropical cyclones. This was perhaps best illustrated by the virtual doubling of the number of cyclones observed in the eastern North Pacific Ocean when satellite surveillance first became available in the early 1960s.

Figure 147 and Tables 32, 33, and 34 contain what is probably a more homogeneous record of occurrences for a 20-year period ending in 1978. These are based upon analyses conducted at Colorado State University using all available observations from reconnaissance aircraft and weather satellites as well as from the more conventional sources.

It should be borne in mind that these data identify the generating source region and numbers of tropical cyclones that formed in each. Not all cyclones reached full hurricane strength but all had sustained winds of at least gale force. The global sources identified in Figure 147 are based upon summaries of the numbers of cyclones generated in each 5° square of latitude and longitude.

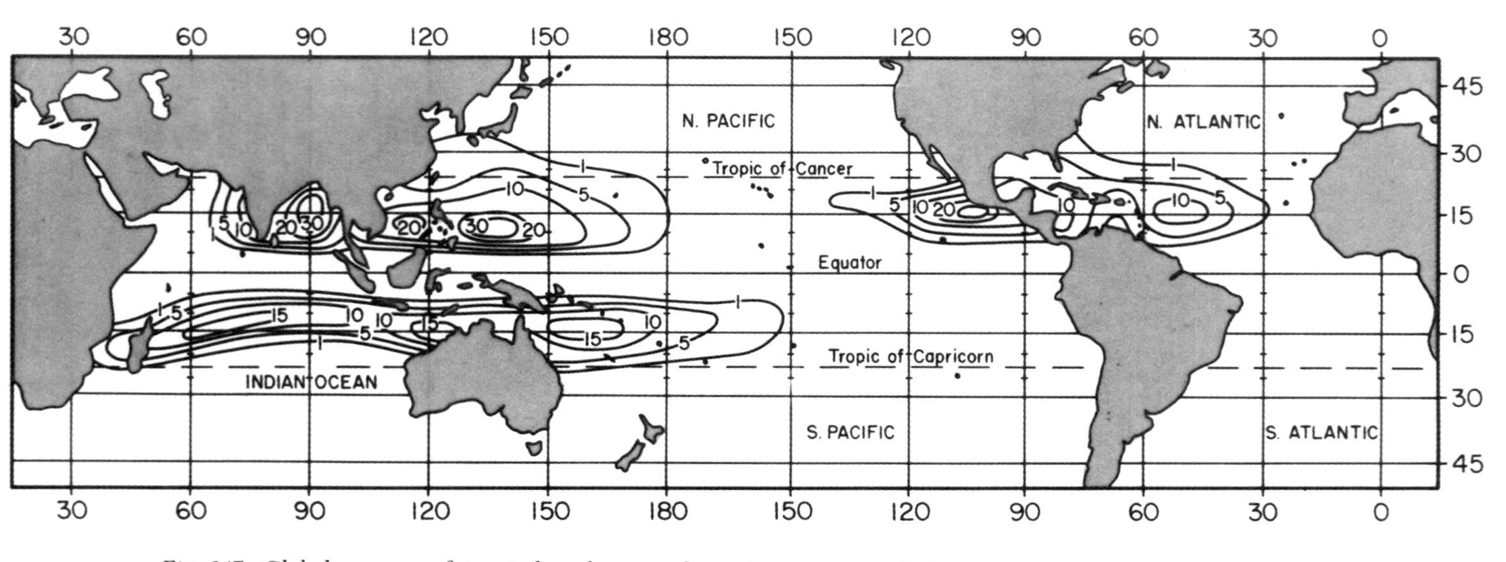

Fig. 147. Global sources of tropical cyclones and numbers generated during twenty tropical cyclone seasons (1958–1978).

Table 32. Total Tropical Cyclones of the Northern Hemisphere, 1958–1977

Year	*Jan*	*Feb*	*Mar*	*Apr*	*May*	*Jun*	*Jul*	*Aug*	*Sep*	*Oct*	*Nov*	*Dec*	*Total*	*Average*
1958	1	0	0	0	2	5	10	9	11	9	4	1	52	4.3
1959	0	0	0	1	2	6	7	11	9	8	2	2	48	4.0
1960	0	0	0	1	3	7	6	14	6	8	2	1	48	4.0
1961	1	1	1	1	4	5	10	5	14	9	6	1	58	4.8
1962	0	1	0	1	3	2	7	11	11	7	4	3	50	4.2
1963	0	0	0	1	3	5	6	5	14	11	0	4	49	4.1
1964	0	0	0	0	3	4	11	15	12	8	10	2	65	5.4
1965	2	2	1	1	4	8	8	8	12	3	4	3	56	4.7
1966	0	0	0	2	2	3	9	13	20	5	7	3	64	5.3
1967	2	1	1	1	2	4	10	12	12	13	3	2	63	5.2
1968	0	0	0	1	2	5	7	17	11	11	6	1	61	5.1
1969	1	0	1	1	1	0	7	13	10	9	4	2	49	4.1
1970	0	1	0	0	4	6	11	10	9	8	7	0	56	4.7
1971	1	0	1	3	7	3	15	11	15	9	4	1	70	5.8
1972	1	0	0	1	4	2	9	12	12	6	4	3	54	4.5
1973	0	0	0	0	0	5	12	8	8	7	5	1	46	3.8
1974	1	0	0	2	4	6	5	14	13	6	4	0	55	4.6
1975	2	0	0	0	2	3	6	11	9	8	6	0	47	3.9
1976	1	1	0	3	2	7	9	15	10	4	0	3	55	4.6
1977	0	0	1	0	3	4	8	4	13	9	4	1	47	3.9
Total	13	7	6	20	57	90	173	218	231	158	86	34	1,093	
Average	0.7	0.3	0.3	1.0	2.9	4.5	8.6	10.9	11.5	7.9	4.3	1.7	54.6	

NOTE: Total for one tropical cyclone season is attributed to the year in which the season began.
SOURCE: From data supplied by William Gray.

Table 33. Total Tropical Cyclones of the Southern Hemisphere, 1958–1977

Year	*Oct*	*Nov*	*Dec*	*Jan*	*Feb*	*Mar*	*Apr*	*May*	*Total*	*Average*
1958	1	1	3	5	7	6	2	0	25	3.1
1959	0	1	4	3	2	7	4	0	21	2.6
1960	0	1	1	9	7	4	0	0	22	2.7
1961	0	1	4	6	8	2	2	0	23	2.9
1962	1	0	4	6	9	5	2	3	30	3.7
1963	0	1	3	7	3	7	1	1	23	2.9
1964	0	2	5	4	5	3	0	0	19	2.4
1965	0	0	3	7	6	6	0	0	22	2.7
1966	0	1	3	5	1	3	2	1	16	2.0
1967	0	2	4	8	7	3	4	0	28	3.5
1968	1	1	3	7	8	2	1	0	23	2.9
1969	0	1	0	5	6	7	3	1	23	2.9
1970	1	3	6	4	7	4	1	0	26	3.2
1971	0	1	6	3	10	3	2	2	27	3.4
1972	1	3	4	10	6	7	3	1	35	4.4
1973	1	3	5	7	4	6	2	0	28	3.5
1974	0	0	2	6	2	5	4	0	19	2.4
1975	0	4	4	8	5	4	3	1	29	3.6
1976	1	0	4	8	9	5	3	0	30	3.7
1977	0	3	4	3	4	4	2	0	20	2.5
Total	7	29	72	121	117	93	41	10	489	
Average	0.4	1.5	3.6	6.1	5.9	4.7	2.1	0.5	24.5	

NOTE: Total for one tropical season is attributed to the year in which the season began.
SOURCE: From data supplied by William Gray.

Table 34. Year-to-Year Occurrence of Tropical Cyclones Forming in Various Source Regions, 1958–1978

Year	*NW Atl.*	*NE Pac.*	*NW Pac.*	*N Indian*	*S Indian*	*Aust.*	*S Pac.*	*Total*
1958	12	13	22	5	11	11	7	81
1959	11	13	18	6	6	13	2	69
1960	6	10	28	4	6	8	8	70
1961	11	12	29	6	12	7	4	81
1962	6	9	30	5	8	17	3	78
1963	9	9	25	6	9	7	7	72
1964	13	6	39	7	6	9	4	84
1965	5	11	34	6	12	7	4	79
1966	11	13	31	9	5	5	6	80
1967	8	14	35	6	11	9	8	91
1968	7	20	27	7	8	7	8	84
1969	14	10	19	6	10	7	6	72
1970	8	18	23	7	11	12	3	82
1971	14	16	34	6	7	14	6	97
1972	4	14	28	6	13	12	10	88
1973	7	12	21	6	4	16	8	74
1974	8	17	23	7	6	10	3	74
1975	8	16	17	6	8	16	5	76
1976	8	18	24	5	9	12	9	85
1977	6	17	19	5	6	7	7	67
Total	176	268	526	121	168	206	118	1583
Average	8.8	13.4	26.3	6.4	8.4	10.3	5.9	79.1

NOTE: Total for one tropical season is attributed to the year in which the season began.
SOURCE: From data supplied by William Gray.

APPENDIX F

Computations of Wave Heights and Tidal Extremes

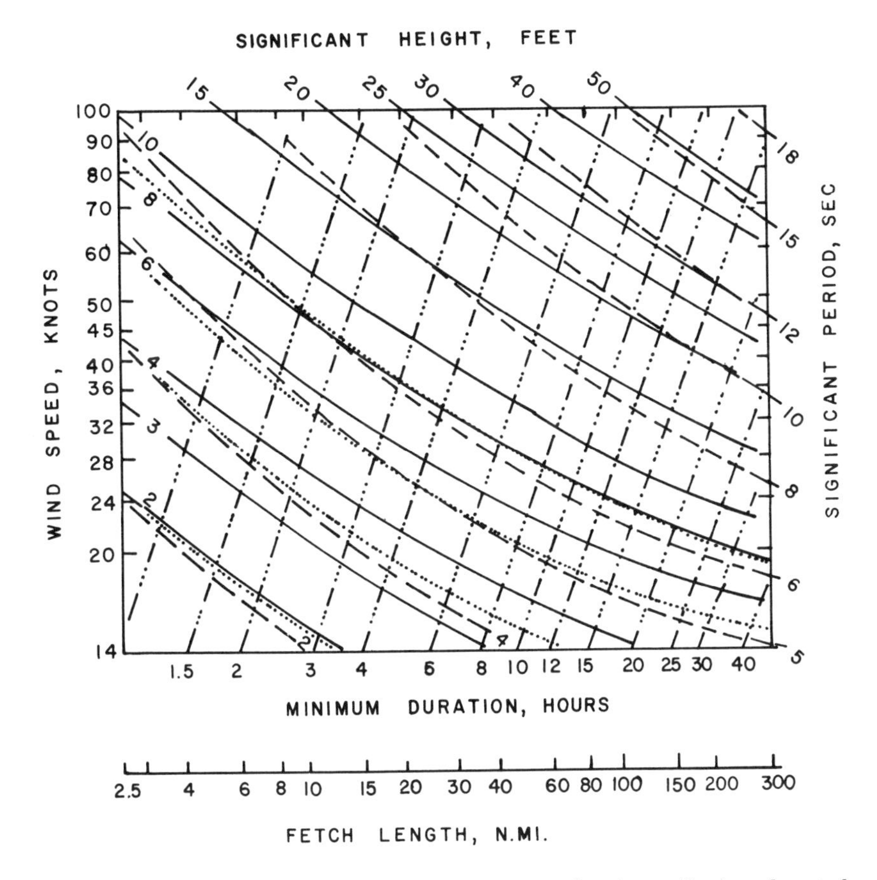

Fig. 148. A nomogram for obtaining significant wave height, H (feet), and period T (seconds) at the end of a fetch as a function of fetch length, F (nm), (or duration of wind, t) and wind speed, V (kt). Entering the nomogram at the mean wind speed (at left), move (to the right) horizontally to the position of fetch length (nm) [or to the intersection with the minimum wind duration (hours) if encountered at a shorter fetch length]. From this point read the value of H (*solid diagonal lines*) and T (*broken diagonal lines*). Legend: t —··· — ··· —; H ———, T — — — —; H^2T^2 .... Adapted from Department of the Army, Corps of Engineers, Coastal Engineering Research Center, *Shore Protection Manual* (Washington, D.C.: Government Printing Office, 1973), Vol. I, p. 3.36.

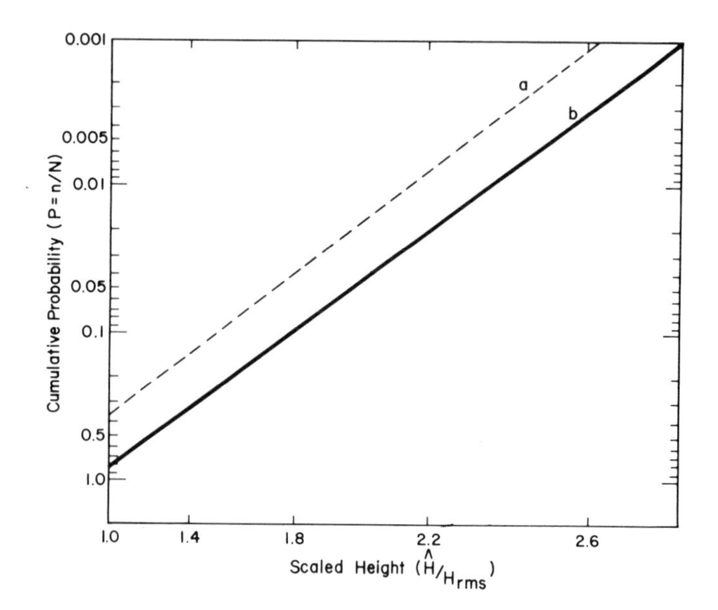

Fig. 149. Theoretical probability of extreme wave heights relative to the root mean square height (where significant wave height, $H_s = 1.416\ H_{rms}$). Adapted from Department of the Army, *Shore Protection Manual*, p. 3.9.

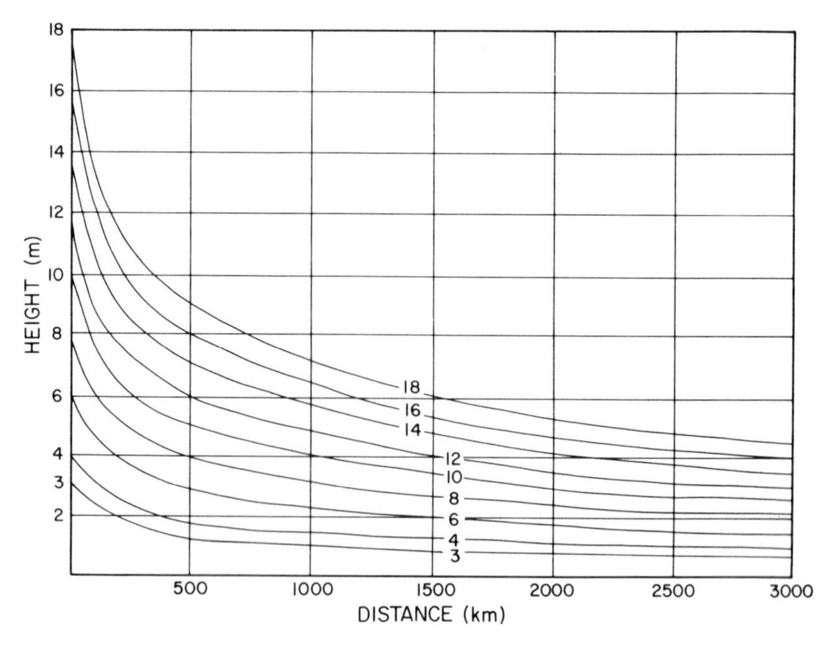

Fig. 150. Decay in wave height as the waves move away from their generating source and become swells. Swell height is expressed as a function of distance from the generating source (or fetch terminus).

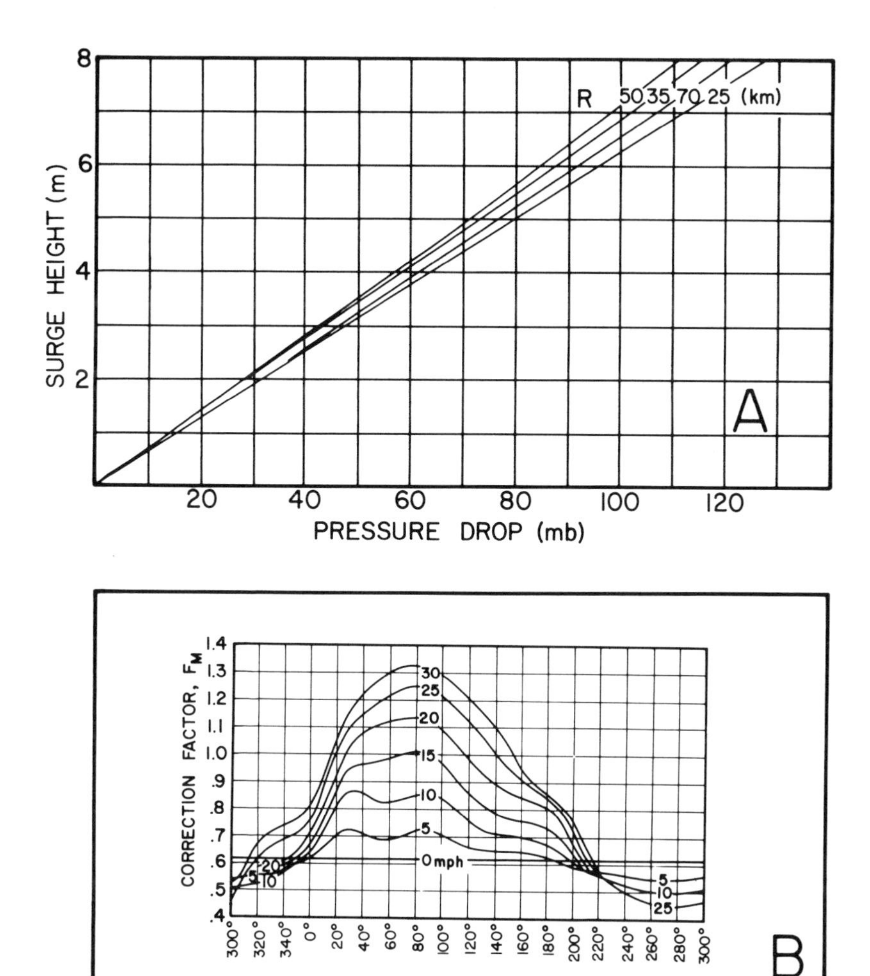

Fig. 151. *Panel A*, the storm surge as a function of pressure drop and radius of maximum winds for a hurricane approaching landfall across a basin with standard bathymetry, perpendicular to a straight coastline at a speed of 13 kt (7 mps). After an application of the shoaling-factor correction (Fig. 152), *Panel B* provides the final coefficient, $F_{M'}$, for coastal crossing angle and the speed of approach. The crossing angle is by definition 0° for a hurricane moving down the coast and 90° when crossing perpendicular to the coast. Adapted from C. P. Jelesnianski, *SPLASH (Special Program to List Amplitudes of Surges from Hurricanes): I. Landfall Storms*, Technical Memorandum II NWS-TDL-46 (Springfield, Va.: Department of Commerce-NOAA, April, 1972), 6, 11.

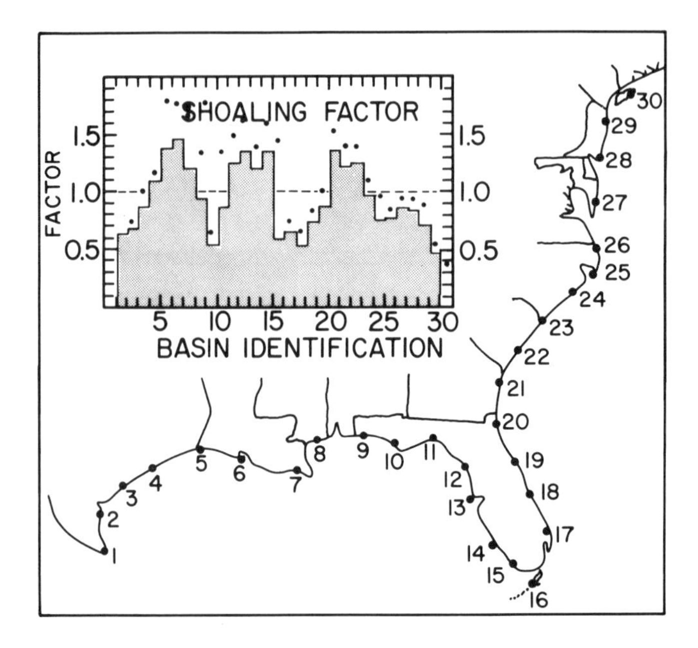

Fig. 152. Shoaling factor for successive open-coast basins from south Texas to New England. Values for individual basins between the numbered positions are average shoaling factors relative to a standard basin located near Jacksonville, Florida. Extreme shoaling factors in each basin are indicated by a *dot*. Adapted from C. P. Jelesnianski, *SPLASH, (Special Program to List Amplitudes of Surges from Hurricanes): I. Landfall Storms*, Technical Memorandum II NWS-TDL-46 (Springfield, Va.: Department of Commerce-NOAA, April, 1972), 8–9, 14.

Table 35. Expected Daily Ranges of Astronomical Tides During Hurricane Season

Location	*Monthly mean (M) and extreme (X) values (meters)*											
	Jun		*Jul*		*Aug*		*Sep*		*Oct*		*Nov*	
Location	*M*	*X*	*M*	*X*	*M*	*X*	*M*	*X*	*M*	*X*	*M*	*X*
Tampico (Mexico)	0.4	0.8	0.3	0.8	0.3	0.7	0.3	0.6	0.3	0.6	0.4	0.8
Galveston (TX)	0.3	0.6	0.2	0.7	0.3	0.7	0.3	0.6	0.3	0.6	0.3	0.5
Mobile (AL)	0.5	0.8	0.4	0.8	0.4	0.7	0.3	0.6	0.3	0.7	0.4	0.8
St. Marks (FL)	0.7	1.5	0.7	1.5	0.7	1.4	0.7	1.3	0.7	1.4	0.7	1.5
St. Petersburg	0.5	1.0	0.4	1.0	0.4	0.9	0.4	0.8	0.4	0.9	0.4	1.0
Key West	0.4	0.9	0.4	0.9	0.4	0.8	0.4	0.7	0.4	0.8	0.4	0.8
Miami	0.8	1.1	0.8	1.1	0.8	1.1	0.8	1.0	0.8	1.1	0.8	1.1
Mayport	1.5	2.0	1.4	1.9	1.4	1.9	1.4	1.8	1.4	1.8	1.4	2.0
Savannah (GA)	2.4	3.1	2.3	3.1	2.3	3.0	2.3	2.9	2.3	2.8	2.3	2.9
Charleston (SC)	1.6	2.2	1.5	2.2	1.5	2.1	1.6	2.0	1.5	2.0	1.6	2.2
Wilmington (NC)	1.1	1.4	1.1	1.4	1.1	1.4	1.1	1.4	1.1	1.3	1.1	1.3

Table 35—Continued

Location	Monthly mean (M) and extreme (X) values (meters)											
	Jun		*Jul*		*Aug*		*Sep*		*Oct*		*Nov*	
	M	*X*	*M*	*X*	*M*	*X*	*M*	*X*	*M*	*X*	*M*	*X*
Hampton Rds (VA)	0.7	1.2	0.7	1.2	0.7	1.1	0.8	1.0	0.8	1.1	0.8	1.1
Reedy Pt (DE)	1.7	2.3	1.7	2.3	1.6	2.1	1.6	2.0	1.6	2.0	1.6	2.0
Sandy Hook (NJ)	1.5	2.2	1.4	2.1	1.4	2.1	1.4	2.0	1.4	2.0	1.5	2.1
New York (Battery)	1.4	2.1	1.4	2.1	1.4	2.0	1.4	1.9	1.4	1.9	1.4	2.0
Bridgeport (CN)	2.0	3.0	2.0	2.9	2.1	2.7	2.1	2.6	2.0	2.8	2.0	2.9
Newport (RI)	1.1	1.7	1.1	1.7	1.1	1.6	1.1	1.5	1.1	1.6	1.1	1.7
Boston (MA)	2.9	4.3	3.0	4.1	3.9	4.0	2.9	3.7	2.9	3.9	2.9	4.2
Eastport (ME)	5.3	7.5	5.4	7.3	5.4	7.1	5.4	6.8	5.4	7.2	5.4	7.5

Table 36. Characteristics of Hurricanes with Major Open-Coast Storm Surges in U.S.

Storm	*Date*	*Position of peak tide*	*Central pressure (mb)*	*RMW (km)*	*Speed of storm (mps)*	*High tide (m)*
1	2 Oct 1893	Mobile, AL	956	32	5	3
2	27 Sep 1894	Charleston, SC	986	48	5	2
3	9 Sep 1900	Galveston, TX	931	26	6	4
4	14 Aug 1901	Mobile, AL	972	61	4	2
5	27 Sep 1906	Mobile, AL	958	121	—	—
6	21 Jul 1909	Galveston, TX	958	35	4	3
7	20 Sep 1909	Timbalier I., LA	931	164	—	—
8	18 Oct 1910	Everglades, FL	955	88	—	—
9	13 Sep 1912	Mobile, AL	993	24	6	1
10	16 Aug 1915	High I., TX	945	60	7	4
11	29 Sep 1915	Grand Isle, LA	931	55	5	3
12	5 Jul 1916	Ft. Morgan, AL	948	92	—	—
13	18 Oct 1916	Pensacola, FL	972	80	7	1
14	28 Sep 1917	Ft. Barrancas, FL	958	59	5	2
15	9 Sep 1919	Key West, FL	927	27	4	2
16	25 Oct 1921	Punta Rassa, FL	952	34	5	3
17	26 Aug 1926	Timbalier I., LA	955	50	4	3
18	18 Sep 1926	Miami Beach, FL	935	45	6	3
19	20 Sep 1926	Pensacola, FL	955	39	4	3
20	16 Sep 1928	West Palm Beach, FL	929	51	7	3
21	28 Sep 1929	Key Largo, FL	948	51	5	3
22	23 Aug 1933	Hampton Roads, VA	971	100	—	—

Table 36—Continued

Storm	Date	Position of peak tide	Central pressure (mb)	RMW (km)	Speed of storm (mps)	High tide (m)
23	7 Sep 1933	Brownsville, TX	949	56	5	4
24	25 Jul 1934	Galveston, TX	975	24	6	2
25	4 Nov 1935	Miami Beach, FL	973	48	8	3
26	31 Jul 1936	Panama City, FL	964	35	5	2
27	21 Sep 1938	Moriches, NY	946	93	>13	—
28	7 Aug 1940	Calcasieu Pass, LA	972	21	2	1
29	11 Aug 1940	Beaufort, SC	970	48	4	3
30	23 Sep 1941	Sargent, TX	958	39	6	3
31	7 Oct 1941	St. Marks, FL	975	34	6	2
32	30 Aug 1942	Matagorda, TX	950	34	7	5
33	27 Jul 1943	Galveston, TX	969	26	2	1
34	14 Sep 1944	Newport, RI	947	42	>13	—
35	19 Oct 1944	Naples, FL	962	63	7	2
36	27 Aug 1945	Matagorda, TX	967	34	3	2
37	24 Aug 1947	Sabine Pass, TX	992	24	2	1
38	17 Sep 1947	Palm Beach, FL	940	35	5	3
39	19 Sep 1947	Biloxi, MS	968	51	10	2
40	15 Oct 1947	Quarantine Sta., GA	974	24	4	2
41	4 Sep 1948	Grand Isle, LA	987	24	6	1
42	26 Aug 1949	New Jupiter In., FL	954	40	6	1
43	4 Oct 1949	Freeport, TX	972	51	9	3
44	30 Aug 1950	Pensacola, FL	979	39	8	2

45	5 Sep 1950	St. Petersburg, FL	958	24	4	2
46	31 Aug 1954	Sakonnet Pt., RI	960	48	>13	—
47	15 Oct 1954	South Port, NC	938	36	>13	4
48	17 Aug 1955	Holden Beach, NC	987	48	6	2
49	19 Sep 1955	Morehead City, NC	960	93	—	—
50	24 Sep 1956	Laguna Beach, FL	975	48	6	2
51	27 Jun 1957	Calcasieu Pass, LA	945	35	7	4
52	10 Sep 1960	East Cape, FL	930	24	>13	4
53	11 Sep 1961	Matagorda I., TX	931	40	4	5
54	8 Sep 1965	Key Largo, FL	948	32	5	4
55	20 Sep 1967	Port Isabel, TX	950	26	5	4
56	17 Aug 1969	New Orleans, LA	909	19	6	8
57	3 Aug 1970	Aransas, TX	945	26	6	3
58	23 Sep 1975	Santa Rosa Beach, FL	955	24	8	5
59	12 Sep 1979	Mobile, AL	943	30	6	4

NOTE: Approximately seventeen storms for which there are no creditable storm surge survey data available have been omitted.

Index